D1192645

WITHDRAWN

St. Scholastica Library
Duluth, Minnesota 55811

SCIENCE AND SOCIETY

Science
and
Society

Past, Present, and Future

Edited by **NICHOLAS H. STENECK**

Ann Arbor
The University of Michigan Press

Q
175.4
.S32
1975

Copyright © by The University of Michigan 1975
All rights reserved
ISBN 0–472–08800–9
Library of Congress Catalog Card No. 73–90887
Published in the United States of America by
The University of Michigan Press and simultaneously
in Don Mills, Canada, by Longman Canada Limited
Manufactured in the United States of America

Preface

THE MANNER in which the symposium "Science and Society: Past, Present, and Future" initially took shape can quite easily be explained. Well over a year before the symposium was to take place, three members of the University of Michigan Kopernik Five Hundredth Anniversary Celebration Committee who had been tentatively designated as the "symposium subcommittee" assembled in the office of one of its members, Albert Hiltner, chairman of the Department of Astronomy, for the purpose of planning a scholarly event that would appropriately mark the upcoming quinquecentennial. After some preliminary discussion, the second member of the subcommittee, Andrew Ehrenkreutz of the Department of Near Eastern Languages and Literatures, suggested, as near as I can recall: "Couldn't we consider exploring a topic that is quite broad in scope and of present interest, such as science and society? Copernicus's humanistic approach to science might provide an opportunity to investigate the past history of this topic and also serve as a guide to its present and even future history." Thus, the symposium was born, or at least its head appeared. Thereafter, as the third member of the subcommittee, I proceeded to the task of delivering or assembling the body that was eventually, in the form of the symposium itself and of the published proceedings contained in the following volume, to become attached to the head.

The task of assembling the program for the symposium led at once to two separate concerns. It was my intent from the

LIBRARY
College of St. Scholastica
Duluth, Minnesota 55811

beginning to give the symposium a focus that would include as many fields of interest as possible, to make the symposium inter-disciplinary, much in the spirit of the Renaissance science of Copernicus. Accordingly, when seeking papers for the symposium, every effort was made to ensure representation from as many disciplines and points of view as possible. At certain crucial points in the planning of the program, I specifically sought out speakers from particular disciplines in order to include points of view that might not otherwise have been represented. It was also my intent, again from the beginning, to impose the least number of restrictions possible on the persons participating in the symposium. The initial request for papers was deliberately general. Speakers were asked simply to suggest " . . . papers that will center around the overall topic, science and society." Beyond this minimal restriction, the speakers were left to explore the topic in the ways that best met their own interests and fields of specialization.

The outcome of these two concerns as applied to the planning of this particular symposium proved in the end to be both rewarding and a source of some difficulty. It was rewarding because the three days of discussion that resulted led to a better understanding of why the topic, science and society, is a problem of ongoing concern to so many scholars. The give and take of three days of discussion helped isolate those areas in which there is likely in the future to be "give" and those areas in which vested interests will make the going rough. It helped to explain the origins of our present ecological dilemma and to point the way to the variety of solutions that we can expect to hear about in the very near future. It helped those involved to realize that what is for them a pressing concern, may be of less priority to someone else. Many of these more obvious conclusions were evident to all at the time of the symposium. However, the celerity of such events also leaves many things unsaid, many loose ends—in the form of parallel arguments, common concerns, diametrically opposed suggestions, and so forth—to be tied. Accordingly, I have added an introductory chapter in which an attempt is made to cull from the proceedings some of the more obvious generalizations. To these generalizations, no doubt, many more can be added.

The difficulty that is associated with this symposium has to do with its diversity, a diversity that stems in large part from its interdisciplinary orientation. (While theoretically desirable, inter-disciplinary endeavors are difficult to realize effectively in practice.) Although tied to a common theme, each paper in this volume represents a particular and individual statement in that each scholar, each representative of a discipline, attacks the relationship of science to society from a different starting point. There is no common definition for science or for society that applies throughout, and it is necessary to wipe the slate clean of any preconceived notions at the beginning of each paper in order to fully appreciate the assumptions and goal of each author. The topic science and society does not conjure up a common image with which all scholars can immediately identify. Then again, this volume has no consistency in style. In response to my rather open-ended invitation—which did set out approximate time limits—I received both brief statements that were designed to provoke response and lengthy articles that were meant to document, sometimes in great detail, a particular thesis or idea. Moreover, each scholar responded in a style that was suited to his own discipline and his own appraisal of the nature of the program. Such discrepancies could, of course, have been eliminated by a strict editorial hand; however, I have resisted this as it would have made the final product in print something that it was not in spoken word. The symposium "Science and Society: Past, Present, and Future" was, for better or for worse, interdisciplinary. It provided a forum for an extremely diverse group of scholars to give air to their ideas on science and society in whatever way they saw fit. And if in the end the results do not seem to stand as a unified whole, this is due to the fact that the topic under consideration has relevance to so many disciplines in so many different ways.

Help in planning the symposium was drawn from many quarters. I could not have assembled such a varied program without the suggestions that were happily forthcoming from many colleagues. Henryk Skolimowski of the Department of Humanities at the University of Michigan took an active role in planning the symposium, especially in drawing together the scholars for the panel on the future of science. Oliver Holmes of the Department of History at Michigan suggested a number of the names that ultimately appeared on the program and also expressed a constant interest in the progress of our endeavor. Our uncertain beginning could not have been successful without the support of the University of Michigan, particularly from the Office of the President and President Robben W. Fleming, and then later without the support of two granting agencies who shared our initial vision, the National Science Foundation and the American Council of Learned Societies. Finally, a special note of thanks is due the members of the University of Michigan Kopernik Five Hundredth Anniversary Celebration Committee for permitting me to turn what was for all of them a very important anniversary into an event that transcends the narrow limits of a birthday celebration in the traditional sense.

NICHOLAS H. STENECK

Contents

I   Introduction

NICHOLAS H. STENECK

PERHAPS the best way to begin an introduction to a volume that ranges as broadly as the one that follows is with a statement, however general, that will help identify the common ground to be covered. Such a statement will provide a common thread that can subsequently be unraveled in our search for whatever lessons are to be learned and conclusions are to be drawn from the proceedings of the symposium on "Science and Society: Past, Present, and Future" that comprise the substance of this volume. Having come this far, however, it must immediately be noted that the common ground to be found in this volume has little to do with any implicit understanding of what is meant by the terms *science* and *society*. Among the papers herein presented, there is little consistency in this regard, and it would prove futile to begin by exploring the manner in which various writers use these terms in the course of their comments. Nonetheless, there is something that seems to hold the comments of the participants at the symposium together; namely, their general acceptance of a premise that summarizes in a particular way the past four hundred years of Western history—this volume assumes as its bias, Western science and society—and the bearing that history has on the future of mankind. Simply stated (the statement covering the common ground), this premise holds that over the past four to five hundred years man has evolved a way of dealing with nature (science) that (1) has altered his relationship to and understanding of nature, thereby, (2) producing a crisis that (3) must be resolved if mankind (society) is to have any future.

Such a premise, if considered in very general terms, might seem evident and in need of little explanation. The environmental and related problems about which we daily hear are obviously to a great extent the result of applied scientific discovery and, equally as obvious, a threat to the future of society. But if one probes more deeply and seeks to understand just what this premise implies, it is soon discovered that it implies different things to different people. Each participant has very specific feelings about man's development over past centuries and his course for the future; each has specific ideas regarding the most crucial elements of our general premise. Sometimes these feelings and ideas come into direct conflict with one another, but more often than not they simply represent different facets of the many-sided topic, science and society. Accordingly, this introduction attempts to draw these many facets into some form of unified whole that will give the reader a brief indication of the general direction our symposium took. In drawing the many facets together, I have been forced to pass over many of the specific points of disagreement. These I leave to the readers of this volume as fertile ground for subsequent debate.

Man after Science

"Before there was Copernicus, there was a context." With these words, Paul W. Knoll of the University of Southern California opened the discussion on science and society: past, present, and future. His intent, as expressed in these opening words and as elaborated in greater detail in his paper, was to discuss, in the most tangible way possible for a case where biographical details are few, the impact of a surrounding society on a future man of science, the impact of "context," in both its social and intellectual forms, on the development of science. This theme is a present one but not a constant one in the papers that follow. In his commentary on Knoll's paper, David C. Lindberg attempts to pin down some of the more elusive elements of context that Knoll consciously avoids, looking particularly at comparative anthropological data (p. 48) and the nature of the "externalist" approach to the history of science (p. 49). In a later commentary, Orren Mohler suggests that the product of the printing industry, the book, is both a "necessary and sufficient" element in the assimilation of science into our "ways of think-ing and living" (p. 82). Necessary, because without it the free exchange of ideas and the crystallization of systems of thought into fixed and easily studied form are impossible; suffi-cient, because it has yet to be surpassed and is not likely to be surpassed as a mode of communicating scientific ideas (p. 85). Also in line with this general theme, L. Pearce Williams (p. 211) makes a case for the influence on science of the relative concerns of French and German educational systems and their respective scientific establishments with an eye toward some ex-planation of the fall of French science and the rise of German science over the course of the nineteenth century. And it would seem as though the era of society's influence on science has not yet passed. John M. Fowler and Alasdair MacIntyre express the fear that science is in very real danger of succumbing to the pressures of context—astrology and the pseudosciences ac-cording to MacIntyre (p. 357), both right-wing (mystic) and left-wing (political) elements according to Fowler (pp. 200–201). How-ever, before turning to the problem of the future of science, which it seems by general accord will be written in harmony with context, the direction of causality must be turned around and the question raised: "How has man been changed by science?"

The most obvious answer to this latter question lies in the realm of knowledge; there can be little doubt that man after science, that is after the scientific revolution, is in many ways more learned about the universe that surrounds him than man before science. With Copernicus, Robert S. Westman argues (p. 71), we have not merely "astronomical innovation" but also the "fulfillment of the old Platonic-Pythagorean hope: the dis-covery of a rational, systematic order in the universe." But the fulfillment of this dream has not been effected without corre-sponding changes in man's other thoughts. To know more about our universe is to know, in theory at least, more about our place in the universe, and, as Edward Rosen points out, this affects man's image of himself. Moreover, the corresponding change is not necessarily one of diminution, as has so often been thought. Even though it is true that after Copernicus man is forced to expand his understanding of the size of the universe and therefore to decrease his understanding of his own relative size, this does not mean that his place in the universe becomes less important. In fact, according to Rosen just the opposite is possible. Through the optimism and faith in science of men such as Kepler, man's place in the universe comes to be understood as that of a unique observer, traveling from place to place to make his observations (p. 65). Man is so much the better for having science, not so much the worse.

The optimistic way in which Kepler and others interpret their new understanding of the universe reflects a growing faith in and enthusiasm for science that extends far beyond the realm of knowledge. Faith in science and in its ability to uncover the secrets of the universe soon becomes a way of life, a spirit that will infect future generations of scientists. This spirit is the one that Wilhelmina Iwanowska portrays so graphically in her description of the rigors of the scientific life (p. 78), the same spirit that underlies her suggestion that "the right to research and to knowledge . . . is a natural human right" (p. 79). It is the same spirit that makes the pursuit of science, in Stephen E. Toulmin's opinion, similar in many ways to the monastic life of prior centuries (p. 114). And it is the same spirit that extends today even to some elements of ghetto life, as Thomas J. Cottle so vividly captures it in the words of the young boy, Keith Downey:

> You see how much scientists do for people. That food laboratory we saw there has to be an important place. When they get done with their work there won't be a single person in this country going to starve any more. . . . They're the ones who'll do the work so that pretty soon, like that one man said, a person can swallow a couple of pills and have all the food he needs that day. Or maybe that week too. That's the day, man, I want to see (pp. 218–19).

Faith in science produces an optimism that leads many of the writers in this volume to assume that our only hope for the future lies with some form of science and that the bond between science and society, once set, cannot be broken. Science has become a way of life to which, for better or for worse, we have pinned our hopes for the future.

The consequences of science becoming a way of life for much of Western civilization are far-reaching. Jerzy A. Wojciechowski explores just one of these in his discussion of "The Ecology of

Knowledge." According to Wojciechowski, science brings to man's understanding of knowledge a new dimension, the dimension of progress. Knowledge now becomes a tool that enables " . . . man to master nature and put its wealth at man's disposal" (p. 264). And this, coupled with the fact that "the more man progresses, the more he is forced to progress" (p. 282), leads to the dizzying rate of change and progress to which we are tied today.

The idea of progress closely parallels a second consequence of the scientific way of life, one that I think it is safe to say few would ignore, the almost unquestioning acceptance of the role technology can and should play in our lives. Technology soon becomes the most obvious evidence in support of the idea of progress, the proof of the pudding, so to speak, of the scientific way of life. After science, man becomes intellectually geared to what he can produce, obvious and substantial progress.

The process of becoming committed to a scientific way of life has its negative side too. Doing science, according to Henryk Skolimowski, is at once a liberating activity and an activity that makes us "captives of new kinds of prejudices" (p. 126). The scientific method liberates man's thought from a universal philosophy that was not entirely committed to the understanding of the physical world but at the same time ties it to a way of thinking that takes as its basic tenet the "atomization of phenomena" (p. 128). To be a scientist, in the classic or pre-1970s sense, is to be a specialist, one who is committed to only a limited number of goals and who does not and almost cannot view problems in a universal perspective. The fruits of his labor—objective knowledge, which comes from a commitment to a Newtonian morality (Toulmin, pp. 117–19), and technological progress, which finds room in our culture due to the bourgeois-capitalist nature of the industrial revolution (Hooker, pp. 314–15)—become his justification, and all else takes on secondary importance.

It is at this point, the point where the scientific method, the idea of progress, and technology start to fuse, that the circle begins to close. What began as an intellectual endeavor, as pure science, takes on through the idea of progress and the reality of technology, a material form, and that material form in turn becomes the justification for further intellectual activity. A particular form of thinking, scientific thinking, assumes goals that are not only attainable, but over the course of the centuries since the scientific revolution, attained, and this in turn gives further license, in the form of financial support, institutional programs, social respectability, and so forth, to that form of thinking. As the circle begins to close, science becomes a self-perpetuating force, one that multiplies with little outside help and even less control. It becomes, in the twentieth century, *the* way of life. It is also at this point that the entire structure, like the wonderful, one-horse shay, starts to fall apart at the seams.

The Period of Crisis

That our present age is one of crisis seems to be a generalization that hardly needs justification. Likewise, there was little doubt among the participants of this symposium concerning the reason why we are in a state of crisis. We are where we are today because the scientific revolution was, to quote Dennis C. Pirages, an "unbalanced revolution." It tipped the scales that measure the balance of society too far in one direction, the direction of science. One path, the scientific path, has been trodden by too many people for too long a period of time. As a result, it is wearing thin—in the form of ecological imbalance and spiritual impoverishment—while other paths have become unfamiliar through years of neglect and second-class citizenship. Such single-mindedness leads to crisis because it fails to take into account the interlocking nature of the parts of the universe. To quote a premise advanced by C. A. Hooker and assumed in many of the papers that follow: "Modern science . . . forces on us the conclusion that life is an integrated (coherent) single system; each part is intimately linked to every other part" (p. 310). We are in a state of crisis because we have ignored this premise.

The disregard for the oneness of life that has led to our present state of unbalance is illustrated in several ways in the papers that follow. There is, of course, the unchecked development of technology, a development that has for the most part taken place without adequate regard for the long-range needs, both physical and spiritual, of a technological society. One awesome consequence of the development of technology is, in Iwanowska's opinion, the development of modern armaments (p. 81). To this can be added other and equally as obvious consequences: the energy crisis, pollution, and so forth. That our environment is at present unbalanced and that the development of science in the form of technology has in part been responsible for this unbalance are premises that seem to inform most present-day discussions of our present state of crisis.

The unbalanced nature of the scientific revolution extends beyond the obvious problems produced by unchecked technology. In fact, several writers regard the latter as merely a symptom of more deeply rooted infections. Dennis C. Pirages places much of the blame for the present ecological crisis on a laissez-faire attitude toward science that stems from the relative neglect of the social sciences in the scientific revolution. The conditions of the scientific revolution, he argues, "directed development of inquiry outward into the physical world much to the neglect of analysis of social problems and human behavior" (p. 231). Scientists, whose primary concerns lie with theory and technological development, have in many instances been left with the task of making—or neglecting—decisions that apply more properly to the realm of the political or social scientist. As a result, some very crucial decisions, i.e., the acceptance of the automobile, have been made without adequate consideration being given to all of the consequences involved (p. 238).

The effects of the unbalanced development of science are also felt in the realm of education. Cottle vividly describes the wide gulf that presently exists in our society between a member of the younger generation who expects science to be a panacea for all future ills and a member of the older generation who knows better. The faith of the former is based largely on his training in school and exposure to a neighboring institute for scientific research; the bitterness of the latter is based on experience. Fowler goes on to tie the present failure of science to win converts in part to the single-mindedness of present-day science education, a failure that is evident in "the falling curves of research support and the accompanying unemployment, reduced graduate student enrollment, and so forth" (p. 202). Such single-mindedness, he argues, is evident in the assumed goal of science education, producing more scientists (p. 203), and in its methods of instruction (p. 204). It is on these areas that Fowler focuses his attention for science education in the future.

Michael W. Ovenden and Skolimowski probe even more deeply to the epistemological realm to discover the unbalancing effects of the scientific revolution. "Our philosophy . . . is the filter" through which all information about the world passes and through which we seek to make sense of the world (Ovenden, p. 370). It ". . . acts like a sieve and thus controls the flow of phenomena which will become the subject of our understanding and upon which our understanding and ultimately action will be based" (Skolimowski, p. 127). And as our philosophy becomes narrowly scientific, so too the manner in which we view the world becomes restricted and unbalanced. At the very least, such restriction and unbalancing means that we ignore other forms of knowledge, forms that would use the "slagheap" of information that presently goes untapped (Ovenden, p. 371). More seriously, science provides " . . . a new set of preconceptions which is a new set of prejudices through which we are compelled to view the world for fear of being ostracized from a society that is governed by scientific reason" (Skolimowski, p. 127). In brief, for these writers mankind is unbalanced to his epistemological core, which means that the course for the future, to turn to the next section, implies some drastic changes.

The Future of Science

The place that science should or will occupy in the society of the future leads to a perplexing series of questions that seem to find answers in part in the past, in part in our present state of crisis, and in part in ideas and suggestions that are as yet untried. It raises issues that were constantly under discussion at the symposium, even during the first and second days. Just as speakers on the future found it impossible to discuss the future without dipping into the past, so too speakers on the past and present felt compelled to look at times into the future. In the process numerous suggestions were made regarding the future of science, and it is these that are the subject of this section. I begin with the views of those who see few changes in store for science in the future.

Iwanowska believes that science must be allowed to develop in the future in much the same way that it has in the past. "We cannot stop research in nuclear physics [even considering the threat of the arms race]; it would be absurd! Science must grow" (p. 81). But this does not obviate the need for control:

> ... technological and industrial applications of science must be placed under efficient controls. It is incumbent upon the political, economic, and social sciences to keep pace with the progress of science and technology (p. 81).

As a practicing scientist, Iwanowska would relegate the control of the application of science to other disciplines, thereby leaving science to continue its development as principally a rational and intellectual activity. The latter form of science should not be limited.

MacIntyre carries this line of reasoning one step farther, not only placing social considerations outside the realm of science, but in fact arguing that the inclusion of such considerations in the realm of science could prove detrimental in the long run. He resists the notion that science ought to be changed, largely out of fear of the consequences of such changes. If science were changed, it could be toward an image " ... of science as primarily an instrument for producing natural and social effects rather than as a central form of rational inquiry" (p. 357), and this MacIntyre thinks could prove harmful to science. Such concerns lead him to posit a twofold priority for the future: first, the conscious support of those activities (e.g., the study of the history and philosophy of science) that help discern "the relationship of the rationality of the sciences to other forces of rationality" (p. 362), and second, the assurance that superstition does not "flourish by default" (p. 362). MacIntyre, too, would choose to keep science a discrete intellectual activity in the future, leaving social problems to other and more practical realms. (It should be noted that by relegating social problems to other realms, MacIntyre does not attempt to ignore them. Quite the contrary is in fact true, as became obvious during the discussion of his paper. Such concerns are omitted from his brief paper due to the limits of the topic.)

MacIntyre's concern about the consequences of including social considerations under the domain of science is repeated in another paper, but this time from an entirely different point of view. Pirages takes as his starting point for changing science the assumption that social considerations are already in the hands of the scientific community. This leads him to warn that if there is no change, "science and technology will continue to legislate in this manner, acting as the master rather than the servant ..." (p. 239). In order to avoid such an eventuality, change is needed, change that will divest the scientific community of its power to control problems of social consequence. In his suggestions for "Righting the Balance" (p. 243), Pirages deals with the delicate problem of how control and the decision-making process can effectively be dislodged from their present locus within the scientific establishment and portioned out to "new institutions" that will "bring science back under social control" (p. 245). This will involve the utilization of "participatory technology," that is, "citizen participation in the public development, use, and regulation of technology" (p. 245).

Such suggestions find a great deal that is compatible with the changes suggested by another speaker on an entirely different topic, science and education. Speaking from the standpoint of a scientist interested in improving science education, Fowler argues that scientists should feel an obligation to educate the citizenry that Pirages would place in the controlling position. Fowler looks at an informed public as a friend of science, not a foe, one that will be more understanding toward science if it appreciates the fact that there is a "humaneness" to the "scientific way of life" (pp. 208, 210).

> ... the citizenry of the future needs to feel more at home with science, to know how a scientist works, to recognize a scientific argument, even learn some of the basic strategies of science, such as estimation, the handling of uncertainty and error, and to acquire habits of observation and questioning (p. 206).

If, as Pirages suggests, the future of science is in part to be determined in public debate, then Fowler wants to ensure that the public is capable of making intelligent decisions.

What we have seen advocated to this point in the papers of Iwanowska, MacIntyre, and Pirages is a course for the future that stresses the role forces outside science will play in overcoming our present state of crisis. In contrast to this view, several of the papers that follow take an entirely different approach to the future, stressing the role that science and scientists must assume in order to be more responsive to the needs and conditions of future society. These speakers firmly believe that the domain of science must be broadened in the future, not limited, and that scientists must be capable of assuming the responsibility for a multitude of problems that extend well beyond the confines of objective knowledge.

Wojciechowski takes up the latter point from the standpoint of science as knowledge. Since knowledge is not isolated but rather is deeply tied to all forms of existence, it cannot be pursued without considering and making room for its consequences.

> Knowledge always has a consequence which goes beyond the act of knowledge (p. 273).

> The more man knows the more mindful he must be about nature (p. 278).

> The greater the ecological predicament, the greater is the need to plan; and the more future oriented becomes man's attitude (p. 286).

The need to plan and to be future oriented leads Wojciechowski to "the ecology of knowledge," a notion that "forces man to consider himself as a part of a bigger whole" (p. 287). The ecology of knowledge will lead to "the evaluation of long-range effects of changes in the intellectual environment" (p. 293), to the study of "the whole field of rational activity with the view of elucidating the nature of the various branches of knowledge" (p. 294), and, perhaps most importantly, "to the idea of planning" (p. 296), planning that will keep the delicate balance between man and nature and yet allow the intellectual freedom and creativity that is our "most important safeguard against the leveling effect of the common denominator, unavoidable in all planning . . ." (p. 301). As a pursuer of knowledge, the scientist cannot avoid such considerations; science cannot remain in the future an isolated intellectual activity.

Wojciechowski regards planning for the future as both essential and a potential hazard, a "mixed blessing." To plan is to reduce options and hence to limit freedom. This is precisely the point that Hooker takes up as he suggests means via which we can plan for the future without ushering in "a society with the most powerful establishment ever created—knowledge *is* power in 1984" (p. 327). Planning can be effective only if science takes under its domain all of the contingencies that have bearing on our future. Science and scientists must be bound "inextricably into other human processes, cultural, political, legal, etc." (p. 331). Science must increase, not defer, its "social-moral" responsibilities (p. 333). Scientists must act so as to ensure a nonauthoritarian methodology and educational system, thereby promoting a "foundationless" and "thoroughly naturalized" epistemology (p. 333). And in the end, the scientist has an obligation to ensure that "science increases our freedoms—intellectual and social—rather than decreases them" (p. 331). Hooker develops these ideas within the context of science. "I am not antiscience," he writes at one point (p. 328). "I believe that society will be relatively better cared for if the development of societal management systems is dominated by men of science [rather] than [by] others with more obvious vested interests." Our hope for the future still lies with the scientist, but with an entirely different kind of scientist. Change is essential.

As was mentioned above, several writers pointed to the scientific method as partially responsible for our present state of crisis. Here then is one final area of possible concern for the future, one more area for potential change. Ovenden addresses himself to this issue in a very limited way, Skolimowski in a much more sweeping way. Ovenden believes that the present "analytical method" of science, "the belief that the way to understand things is to break them down into the smallest possible pieces . . . " (p. 364), "has usurped the functions of a philosophy" (p. 370), and in so doing, has put intellectual constraints on the development of science. He therefore suggests that we need to turn to "new sources of awareness," sources that can be tapped by turning from analysis to "intuition" (p. 371). Ovenden regards intuition as a spark that will boost the scientific intellect; a source of new ideas that will from time to time jolt the normal analytic processes into new paths of inquiry. It will keep science from becoming set in its ways, from becoming stale; it will not, however, replace the normal processes of science. It is at this point that he and Skolimowski part company.

Skolimowski goes one step beyond the other speakers previously discussed and changes present-day science, the science of the scientific revolution, into something that may in fact no longer be science. I suggest this interpretation because in his desire to bring man and nature into some kind of accord, he is willing to sacrifice something that all other writers keep, namely, scientific objectivity. "We must make science produce knowledge that is humanly and thus morally enhancing. We must adjust science to human morality and not conversely" (p. 134). And this, Skolimowski argues, can only be done by changing "not only an idiom *within* science," but "the whole idiom *of* science" (p. 126). What idiom of science will be evolved? Skolimowski believes that we must reject our objective and abstracted universe for one that is Copernican in scope, for a cosmos that entails " . . . the aesthetic and the moral orders which [are] all fused together in the visible movements of the celestial spheres" (p. 131). Science under such an idiom will be "subject to norms that are intersubjective . . . [but not] regulated by the canons of scientific objectivity . . . " (p. 134). To attain such goals "will require enormous changes in the nature and method of science" (p. 134), but it is only by such changes that we shall "be able to eliminate the spurious and pernicious dichotomy between the human morality and the alleged morality of science" (p. 135).

The Future of Man

To this point in the discussion of the future we have been dealing with only one of the variables contained in the topic, science and society. There remains yet another variable to be discussed and that is society. Are there changes in store for society in the future? Will society in the form of individuals, in the form of man, have to adjust to new ways of living in order to lessen the problems of the present age? Curiously, this topic occupied far less time during the symposium. Whereas it was readily assumed that drastic changes could be advocated and implemented in the realm of science, very few suggestions were made regarding the makeup of society. Even A. Hunter Dupree's and Gerald M. Platt's speculations regarding the possibility of drawing together Parsonian social theory and molecular biology and the natural consequences of such speculations in terms of a "Darwinian trajectory" (Dupree, p.

148) and the relationship between culture and society (Platt, p. 186) led to little discussion regarding the pliability of human nature. The latter seems almost to comprise a given, an element of the problem that is understood and not subject to great change. More than once it was suggested that in our readjustment to overcome present problems, we must act to accommodate science to man and not the reverse. There are, however, at least two papers that depart from this position.

Unlike most speakers at the symposium, Nolan Pliny Jacobson sees the changes that are needed to eliminate our present crisis as already sown and sprouting in society in the form of what he calls the "new life-style of self-correction" (p. 255). This "new life-style" arises from the complexity of present life: the loss of control that "pervades all levels and areas of life," the broadening of our "angles of perception," the "multinational, multiracial, multicultural" nature of modern science, etc. (pp. 251–52). It is a life-style that produces a new kind of man.

> Something has happened to man's capacity to believe. More and more, the balance shifts in favor of *methods* of inquiry, *ways* of believing, a *logic* of discovery, rather than the concepts to which such investigation gives rise (p. 253).

Jacobson thus completely reverses the process that was discussed in the previous section. He sees man changing vis-à-vis science, not science changing to accommodate the ideals of man. The "process of interaction and communication is transforming man by transforming the conditions that nurture and control his growth." And the result of this transformation is a "creature who of necessity now can survive only by correcting the way he has been reared" (p. 254), a creature who is "self-corrective."

Jacobson assumes that the changes that affect human nature are and will be largely unplanned and unpredictable. He agrees with the common assumption that "the course of history is largely outside our control" (p. 251). But is this assumption a necessary one? It is precisely on this point that Carl Cohen focuses as he lays out in very broad terms the outlines of yet another "Copernican revolution," a revolution that will concern "neither the nature of the universe, nor the nature of knowledge, but rather the nature of ourselves, what we are and can be"

(p. 99). We are at a point in time, Cohen argues, where we can seriously question the nature of the given, human nature, and work in very definite ways to change that given. We possess, or will possess shortly, the tools necessary to mold man to the image that we desire. But what image should we mold? How are we to deal with this newly found power to shape the future of mankind? "The reflexivity of the situation," Cohen suggests, "is intellectually amusing; the possibilities for good and evil are so immense as to boggle the mind" (p. 100).

Cohen is perhaps correct in assuming that we shall soon be faced with the awesome reality of deciding how our children and children's children will be bred, of deciding what type of race we want to leave to posterity. Possibly this eventuality is still distant enough to permit the present generation a brief reprieve from deciding the issue. Be this as it may, we must today face squarely the prospect of changing man, the other variable, as we search for solutions to our present dilemma. If we take lessons from the past seriously, then we should have learned that science did not develop in vacuo and that as a result many things, including man, are different today because of science. Even if our present mode of living is unbalanced, that mode of living represents several centuries of scientific and human evolution going hand in hand to their present level of development. We have for several centuries been pursuing a course that has given free reign to science to develop and that has forced man to develop to accommodate.

Now we want to reverse all of this, to turn science around, to add to it dimensions that have been lost, to bring to it moralities that it has forgotten. But can we do this without taking into consideration our other variable, man? I am not speaking here about changing man the practicing scientist, or even man the intellectual, but of that larger percentage of society that has only recently come to enjoy and expect the fruits of a technological society. Can we change science without paying due regard to this other variable? I am of the opinion that we cannot. We cannot expect scientists to stop producing better and better products at a greater and greater cost in energy, when society itself, which has been reared on a "progress-is-our-most-important-product" ethic, has come to expect such products. We cannot expect the automobile industry to stop turning out cars in lieu of mass transportation, when the American public wants cars and refuses to use buses. True, some changes can be effected in this direction. We will give up phosphates and nonreturnable bottles, put two people in a car to conserve gas, recycle papers and glass, but all of this is done within the same mental framework, the framework that constantly expects more that is better, faster. It is also true that in time the mores of a new science will filter down through society and once again change man to conform to the image of science that we create. But this takes time, something we have very little of today. It seems reasonable to suggest, therefore, that we start burning the candle at both ends; working at both ends of the nexus, science and society, in order to expedite a smooth transition into tomorrow. How should we go about this? Here I suggest is a topic for yet another symposium on science and society, this time with the emphasis squarely on the latter and its development in the future.

SCIENCE AND SOCIETY: PAST

II The World of the Young Copernicus: Society, Science, and the University

PAUL W. KNOLL

BEFORE there was Copernicus, there was a context; and it is this background upon which I should like to focus today as we mark the five hundredth anniversary of the birth of Nicholas Copernicus.[1] I do not wish by this to imply that the details of Copernicus's biography are unimportant,[2] or that his *De revolutionibus orbium coelestium* is not rightly regarded as an epoch-making achievement by historians of science.[3] But since these two elements have traditionally received considerable attention, I thought it appropriate to begin this interdisciplinary session on the interrelationship of science and society with some consideration of the factors which help account for the Copernican achievement.

Even though this study shall in passing refer to details of Copernicus's life and system, our chief attention will be upon three elements of fifteenth-century Poland which touch closely the development of the individual whom we honor. They are, first, the changing character of the society which underlay the public activity of Copernicus; second, the practice of philosophy and science, particularly astronomy, at the Polish university where he enrolled; and third, the growing effect of humanism upon the curricula and the outlook of that university.

I

Little by little during the fifteenth century, the society which had developed in medieval Poland was beginning to change. That older society was one characterized by a relatively free peasantry, by a limited urban stratum which was partly the reflection of native elements and traditions and partly the result of interminglings from abroad, and by a relatively small aristocracy of great nobles who had originally been the faithful supporters of the monarch.[4]

Over the years, however, the peasantry had become increasingly bound to the land and forced into a semifree status which corresponded closely to earlier serfdom in the medieval west. It was partly the development of great estates in Poland for the effective exploitation of the steadily growing international grain trade which had brought the peasantry to, in M. M. Postan's apt phrase, "the very threshold of enslavement and impoverishment."[5]

Very nearly the same magnitude of change may be observed in the cities. In addition to the aristocracy and clergy who had earlier dominated Polish urban life, at least three groups of burghers were becoming increasingly important in towns such as the royal capital Cracow, the great port city of Gdańsk, or the commercial and administrative center of Toruń. First were the merchant patricians who by 1500 had effectively assumed control over the political and economic fortunes of the cities. Ethnically mixed, both as to geographical extraction and as to language, this was the social class from which Copernicus came. The second group were the numerous artisans and craftsmen who were required to service the needs of the population and who constituted a potential threat to the interests of the patricians. Finally, there were the poor and the unskilled who in this period also, as in all times and all urban places, lived on the margins of society.[6]

Elsewhere in society, during the last three-quarters of the fourteenth and throughout the fifteenth centuries, the older nobility was gradually being replaced by a newer landed gentry whose economic position was based upon the aforementioned grain trade and whose growing political dominance was partly the result of two dynastic extinctions in the space of less than a decade and one half. Although the early monarchs of the Jagiellonian dynasty had attempted to check this ascendancy, they had ultimately been unsuccessful. By the birth of Copernicus in 1473, royal power was increasingly circumscribed; and the middle gentry (as distinct from the great magnates who were to dominate the seventeenth and eighteenth centuries) were the effective arbiters of Polish political and social development.[7]

Distinctly then, the picture which emerges from this very brief survey of Polish society in the time of Copernicus is one of flux and change. Inevitably, the world view which had characterized medieval Poland would have changed, for the foundations upon which it rested were being profoundly altered. Had there never been a Copernicus, therefore, there might still have been some kind of scientific revolution. That there was this altered society, however, accounts in part for the way Copernicus was able to view the universe in a different manner and for the way, as Professor Barbara Bieńkowska of Warsaw has shown, in which the Copernican system was eventually accepted.[8]

Nevertheless, it was not socioeconomic change alone which underlay the Copernican achievement. It was also the progress of philosophy and science, particularly astronomy, at the University of Cracow which was a contributing factor.

II

At the University of Cracow, which had been founded in 1364 by King Casimir the Great,[9] the early emphasis had been upon canon and civil law. But the beginnings of the institution had been difficult, and after the death of the royal patron in 1370, the school fell into a decrepitude from which it was not rescued until the last decade of the century. Urged on by Magister Bartholomew of Jasło, Queen Jadwiga and her husband Władysław Jagiełło took steps to revive the university.[10]   In 1400 the *studium* was refounded;[11] and in the following decades, the Jagiellonian University (taking the name from its second founder and restorer) rapidly emerged as one of the leading institutions of central Europe. Particularly after the University of Prague was wracked by the troubles of the Hussite period,[12] Cracow and the University of Vienna stood preeminent in the region.

Now there was a difference from the preceding century, however, for in the Jagiellonian *studium,* theological study had been added to the earlier law and medical curricula. In the lectures within this new faculty and within the arts faculty, philosophy and the exact sciences (particularly mathematics and astronomy) developed during the fifteenth century to become the specialty of the university.[13] Let us trace some of that process so that we may more fully appreciate its possible impact upon Copernicus.

Central to the sciences in this period was arithmetic and geometry. Throughout much of the early part of the fifteenth century, lectures in the former were based chiefly upon John of Murs's *Arithmetica communis* (contained today in five manuscripts in the Biblioteka Jagiellońska),[14] upon John of Hollywood's *De algorithmo* (contained today in two Cracow manuscripts),[15] and upon the *Algorismus minutiarum* of John of Linières (five manuscripts of which remain in Cracow).[16] These works were gradually replaced later in the century by treatises of Polish masters, such as the *Algorismus proportionum* of Sędziwój of Czechło and the *Algorismus minutiarum* of Marcin Król of Żurawica.[17] Also in geometry, older medieval commentaries upon Euclid were gradually replaced by Polish works, the most important of which was Marcin Król's *Opus de Geometria* (1450).[18] In general, therefore, it is apparent that a specific Cracovian approach was being developed in the field of mathematics.

Yielding in the same fashion to increasing native expertise, astronomy at Cracow was not simply a derivative curriculum, but was more and more based upon solid Polish achievement. In the first place, a strong Polish tradition in astronomy was already two hundred years old. Beginning from the early thirteenth century, Witelo, Franco de Polonia, Nicholas of Poland, John of Grotków, and John of Poland had all made their contributions.[19]

Beyond this, however, the study of astronomy at the refounded university was given considerable impetus by the foundation of a special chair in astronomy and mathematics. A private citizen in Cracow, Nicholas Stobner,[20] established this endowment in 1405, and from at least 1415 onward a consistent pattern of lectures in geometry, optics, astronomy, and mathematics was delivered annually. That these lectures were not simply recapitulations of previous works is reflected in the revision, from 1428 onward, of the so-called Alphonsine tables. The result was a new set of observational data, the *Tabulae resolutae de mediis et veris motibus planetarum super meridianum Cracoviensem*.[21] These tables continued in use with revisions until the early sixteenth century, and Copernicus undoubtedly made use of them.

Among the incumbents of the Stobner chair was probably

the aforementioned Sędziwój of Czechło, who was active at the university from 1429 to 1431.[22] His commentaries on optics and planetary motion proved so popular that they eventually replaced John Peckham's *Perspectiva communis* and Gerard Sabbionetta's *Theorica planetarum* as standard texts in Cracow.[23] Another name closely connected with the Stobnerian tradition is Lawrence of Racibórz, who also was associated with the arts faculty of Cracow.[24] He gave annual lectures from 1420 to 1427 upon Sabbionetta's *Theorica planetarum*. Unfortunately, almost all of his work has perished.[25]

Elsewhere in the *studium* the growth of astronomical expertise was even more marked. The foundation in 1459 of a second chair in this field, though one devoted more to astrology, initiated four decades of the study of theoretical and applied astronomy during which the University of Cracow emerged, as Aleksander Birkenmajer has written, as the "international center of astronomical education at the end of the middle ages."[26] Four individuals in particular are associated with this preeminence: Marcin Król of Żurawica, Marcin Bylica of Olkusz, Jan of Głogów, and Wojciech (Adalbertus) of Brudzewo.

Marcin Król, promoted to *magister artium* at Cracow in 1445, later studied at Prague, Leipzig, and Bologna. Although he was eventually appointed to a chair of medicine at Cracow, his real interests lay in arithmetic, geometry, astronomy, and astrology. (These were apparently derived in part from his arts study, in part from his year in Bologna where he attended the astronomy lectures in 1448/49.) His most important work was a harsh critique of the Alphonsine Tables entitled *Summa super Tabulas Alphonsi* (alternatively *Correctiones Tabularum Alphonsi*).[27] In this work he argued that the actual positions of the planets did not correspond to the data given in the Alphonsine tables. In trying to correct these, however, he set forth several interpretations which anticipated the later, and better known, work by the Viennese astronomer and mathematician Georg Peuerbach, the *Theoricae novae planetarum*.[28] One of Marcin's last recorded acts was to found the aforementioned astrological chair.

Marcin Bylica, who was born about 1433, began his study at the University of Cracow in 1452. As a student of Marcin Król, he became in 1459 the first incumbent of the newly founded chair. Shortly thereafter, he traveled to Italy, where he studied at Padua and Bologna, and became the close friend of Johannes Müller of Königsberg, known as Regiomontanus. Much of the rest of his life was spent outside of Poland, particularly in Hungary;[29] but he retained close contacts with the University of Cracow. During his lifetime he sent the *studium* many books, including a heavily annotated copy of Regiomontanus's *Tabulae Directionum* and a copy of Peuerbach's *Theoricae novae planetarum,* which were widely consulted at the university.[30] Sometime before his death in 1494, he bequeathed to the *Alma mater Jagiellonica* his large library of mathematical, astronomical, and astrological works, as well as his extensive collection of astronomical instruments.[31] His influence at Cracow, therefore, was less immediate than that of his predecessors, but he continued the tradition which culminated in the following two figures.

Jan of Głogów finished the arts course at Cracow in 1468, immediately began to lecture on astronomy and astrology, and remained as a professor there for forty years until his death in 1507. His extensive writings include numerous horoscopes, many astrological handbooks, as well as several pure astronomical treatises.[32] In these latter, he combined increasingly accurate observational data with interpretations of celestial phenomena which implicitly challenged accepted views. For example, in three works, his *Introductorium in tractatum sphaerae Joannis de Sacrobusco* (written before 1506), his commentary upon the *Theorica planetarum* of Gerard Sabbionetta (before 1483), and his *Interpretatio Tabularum resolutarum ad meridianum Cracoviensem* (ca. 1488), he raised older Averroist challenges to the Ptolemaic system of cycles and epicycles and even suggested that the sun, as the most dignified of the planets, may be thought to control the motion of the other planets. (In this context the earth was not considered to be a planet.)[33]

Wojciech of Brudzewo was a student of Jan of Głogów. He became a master at Cracow in 1474, where he taught until his death in 1495, lecturing on mathematics, optics, and astronomy. In this last field, he did not comment upon the work of Gerard of Sabbionetta which had been largely superseded as a text at Cracow. Instead, he based his lectures on Peuerbach's *Theoricae novae planetarum* and in 1482 composed his own treatise upon this theme, *Commentariolum super Theoricas novas planetarum.* This commentary soon became the accepted text in theoretical astronomy at Cracow.[34] In this, and in other lesser works, he went even further than Jan of Głogów in challenging accepted astronomical theories. He attacked the Ptolemaic explanations for the motions of the moon, suggested that some recent Aristotelian commentators had provided a basis for accepting the earth's motion around its own axis, and explored the possibility that an understanding that motion could be relative might overcome sense evidence of solar motion.[35] In all these arguments, his reasoning was partly based upon observational data and partly derived from the implications of natural philosophy at Cracow in the preceding generations.

With Wojciech of Brudzewo we approach the age of Copernicus. At the same time, we are able to see the extent to which astronomical study at Cracow had gradually progressed. As both the fifteenth-century German chronicler Hartmann Schedel and his contemporary, the Hungarian historian A. Bonfinius, recognized, the Jagiellonian University was indeed an important center for astronomical study.[36] But this was not the only influence which worked upon Copernicus. Philosophy also played an important role at Cracow.

III

It has long been a commonplace to say that scholastic thought in the fourteenth and fifteenth centuries was undergoing substantial modification from what it had been in the Age of Aquinas. But if it is no longer possible to say with Étienne Gilson that philosophy after Thomas had come to the end of its journey, one may nevertheless recognize that sharp challenges to the Thomistic approach were developing. One of these came from the Oxford school of William of Ockham; the other, from Paris, and bore the imprint of the influence of John Buridan.

Buridan, who was born sometime before 1300, spent nearly his whole career of forty years in the arts faculty at the University of Paris. Before his death in 1358, he had composed commentaries upon nearly every work of Aristotle, with particular attention to questions of the philosophy of nature. Although he consistently dealt with Aristotelian problems, he—with his followers such as Nicole Oresme, Lawrence of Lindores, Marsilius of Inghen, Albert Rickmersdorf of Saxony, and others—frequently developed approaches which were distinctly un-Aristotelian.[37]

For example, Buridan based the reasoning which underlay his writing upon the assumption that there were three criteria for truth: faith (*fides*), natural reason (*ratio naturalis*), and authority (*auctoritas*), particularly that of Aristotle. The twofold implication of this division was to loosen the ties between theology and philosophy (though in a less drastic manner than in Ockham) and to de-emphasize the opinions of the philosophers (again, particularly Aristotle) in relation to what natural reason might on its own derive from prime principles, self-evident propositions, and the demonstrable principles of the philosophy of nature.

Not only were the approaches un-Aristotelian, however; in some instances even the answers began to take on a new character. Chiefly because Buridan was himself greatly interested in natural philosophy, his work and that of his successors increasingly focused upon physics and the laws of motion.[38] The most critical approach to Aristotle came on the question of the theory of natural and enforced motion. Buridan developed in embryo that which Oresme elaborated into the impetus theory that was to influence philosophers and scientists throughout Europe, but particularly in Cracow.[39]

The influences we have described above were mediated to Cracow from the University of Prague.[40] Late in the fourteenth century, manuscripts of the works of Buridan and his followers began to appear in the Polish capital; and between about 1415 and 1450 several important philosophical works were written by Polish professors, all of whom betray strong Buridanian influences in their thought. Some of these treatises are attributable to specific individuals, while others are anonymous. The most important characteristics of these works in our present context deserve mention, for they reveal the kind of teaching on the laws of motion which Copernicus might have encountered.[41]

Andrew Wężyk's *Questiones disputatae super octo libros Physicorum Aristotelis* argued that projectiles continued to move after leaving the hand of their original mover because they had received impetus from the causative agent. In his *Puncta super octo libros physicorum Aristotelis,* Andrew of Kokorzyn modified

Buridan's theory slightly to make it more Aristotelian; he suggested that the motion of a projectile in flight is produced both by its impetus and by the continued action of the air upon it. A purer impetus theory is, however, contained in Benedict Hesse's *Questiones super octo libros Physicorum Aristotelis,* a work by the same title falsely attributed to Benedict Hesse, and an anonymous *Exercitium super octo libros Physicorum Aristotelis.* Later in the fifteenth century in other works, there was a clear attempt to bring Buridan's ideas into harmony with Aristotle on the issue of motion, thus compromising somewhat the achievement of the earlier decades. Several of these works do, however, apply the theory of impetus to the heavenly spheres, thus implicitly making celestial and terrestrial motion identical. Finally, in connection with this problem of motion, it should be noted that in an anonymous 1433 work, *Questiones super quatuor libros De coelo et mundo Aristotelis* and in the Cracovian glosses (ca. 1490) upon Pierre d'Ailly's *Tractatus super quatuor libros Meteorum Aristotelis,* six explicit arguments in favor of the earth's motion are presented.

It is apparent from the foregoing, therefore, that in addition to the high level of astronomical theory and practice at Cracow, there was also a philosophical challenge to older ways of thinking which had gained ground within the university. The third element of the age which was to influence Copernicus, though perhaps only indirectly, was the spirit of Italian renaissance humanism.

IV

The first Polish contacts with Italian humanism came early in the fifteenth century during the conciliar period. Although the activities of the Polish delegation to Pisa in 1409 have not been thoroughly investigated,[42] we do know considerably more about Polish participation in Constance. There Poles were active in both the affairs of the council and in its related activities.[43] The leading members of the delegation were Paulus Vladimiri of Brudzeń, rector of the University of Cracow, and Andrew Laskarz, Bishop-elect of Poznań. The former spoke eloquently against the ambitions of the Order of the Teutonic Knights, expressing an understanding of international law which upheld the right of pagans to resist forcible conversion to Christianity.[44] Both of these men had personal contacts with the leading Italian humanistic figures at the council, particularly Francesco Cardinal Zabarella and the famed educator Pier Paolo Vergerio. Bishop Laskarz, who was a fervant conciliarist, was influenced by these meetings and by the widely heralded news of Poggio Bracciolini's sensational discoveries to take copies of antique literature and *trecento* humanistic works back to Poland with him.[45]

The council of Basle from 1431 to 1449 proved to be more important than Constance in the growth of Humanism as an element in the early Renaissance in Poland.[46] The members of the Polish delegation had extended contacts with Italian humanists, and many humanistic works were obtained at this time which later found their way into Polish collections, particularly the library of the University of Cracow. These included Petrarch's *Rerum familiarum (libri XII—XXIV)* and *De remediis utriusque fortunae,* and L. A. Seneca's *Epistolae ad Lucilium.*[47]

In our present context, the most important member of the Polish delegation was Mikołaj Lasocki (ca. 1380–1450).[48] He was educated in Cracow and in 1423 became a secretary in the royal chancery. From that time on he acted as a diplomat throughout Europe. Between 1425 and 1434 he was sent to Rome three times. Next he spent four years at Basle, during which he traveled to Emperor Frederick III, and to Hungary. His humanistic interests are reflected in his orations, which were constructed according to those ideals of *eloquentia* current in fifteenth-century Italy, and in his friendships in Italy and at Basle. He was associated with John Aurispa and Aeneas Sylvius Piccolomini, met and made friends with Poggio Bracciolini, and established close contacts with Guarino Guarini of Verona, with whom he corresponded and to whom he sent several young Poles for study.[49]

The beginning of humanistic interests in fifteenth-century Poland may also be traced to the influence of Italian humanists in the north;[50] while a third medium through which Italian influences penetrated Poland in this period was the extensive political and cultural contacts between Poland and Hungary. (This important topic unfortunately lies outside the scope of this present study.)[51] It was this triplex influence of native Poles in contact with humanism abroad, the direct penetration of Italian influences into the country, and the indirect mediation of these same cultural currents through Hungary that created conditions by the mid-fifteenth century in which the first flowering of the Polish renaissance took place. Four individuals represent this: Zbigniew Cardinal Oleśnicki, Johannes Długosz, Gregory of Sanok, and Jan Ostroróg.

The man who became Poland's first cardinal enrolled in 1406 in the arts faculty at Cracow, where history and the Ciceronian style seemed to interest him most.[52] His rise as diplomat and royal secretary was rapid, and in 1423 he was named Bishop of Cracow. For thirty-two years this post was the basis of his position as the preeminent influence on Polish political and cultural life. He was an unremitting enemy of the Teutonic Order in Prussia, the advocate of a personal union between Poland and Hungary and closer ties between Poland and Lithuania, a firm defender of the principle of Christendom against the Turks, and a strong conciliarist who, like Aeneas Sylvius Piccolomini, later made a successful and graceful return to papal allegiance. His cultural influence was equally significant. He patronized learning wherever it was found in Poland and supported promising students. Thus he vigorously defended the University of Cracow and founded the Jerusalem hostel for students there. In addition, he corresponded extensively with Aeneas Sylvius, and supported humanistic forms and ideals in his letters and writings.

Johannes Długosz (1415–80) was the greatest medieval Polish historian and remained unsurpassed until Adam Naruszewicz in the eighteenth century.[53] Although he studied in the arts faculty at Cracow, mastering particularly rhetoric and logic, financial difficulties forced him to end his studies, and he never gained an academic degree. He eventually came to the attention of Cardinal Oleśnicki who employed him in his chancery. From 1436 to the end of his life he was a canon in the cathedral chapter in Cracow. In many instances he served as a royal diplomat and throughout his career he was a privileged if unofficial participant in the corridors of power. His access to royal archives and to collections of historical material which were prepared in some instances especially for him gave his *Annales seu cronicae inclyti regni Poloniae* in twelve books both authority and reliability. In addition, he prepared long historical studies on the insignia of the

Polish kingdom, lives of all Polish archbishops and bishops, and a detailed catalog of the benefices belonging to the diocese of Cracow.[54] It is not traditional to speak of Długosz within a humanistic context; usually he is seen as the culmination of Polish and European medieval historiographical characteristics and traditions. Recent attention to the nature and context of the early Renaissance in Poland has begun to suggest, however, that this judgment needs modification.[55] Długosz quite consciously modeled his *History* upon Livy's *Ab urbe condita.* Not only in the style and wording, where his debt to Livy can be shown in many instances, but also in his close sympathy to the patriotic inspiration of Livy does Długosz reflect an increasing awareness and understanding of the antique in Poland.[56]

The same phenomenon is particularly clear in the writings and career of Gregory of Sanok.[57] Born in 1407 in a recently settled colonial area of the San River, the three most important intellectual influences upon him as a youth were (1) his traditionally oriented elementary studies in Cracow; (2) his five years as a wandering scholar in Germany, where, as he later remarked, the effects of Bracciolini's discoveries during Constance still reverberated;[58] and (3) his studies in the liberal arts faculty at the University of Cracow from 1428 to 1433. Already attracted to the revival of classical antiquity and rising in the ecclesiastical hierarchy of his day, he then spent some years in Italy at the papal court of Eugenius IV and elsewhere. Here he purchased a handsome copy of Boccaccio's *Genealogia deorum gentilium* which had once been the property of the French humanist Gontier Col (now MS Cracow, B.J. 413).[59] His humanistic interests were deepened by this Italian visit, and upon his return in 1439 to Cracow he gave a series of lectures at the university commenting upon Vergil's *Buccolics.* During the next decade he had a great many contacts with the Hungarian courts of John Hunyadi and his son, the eventual King Matthew Corvinus, and with the famed humanists there, Pier Paolo Vergerio and Bishop John Vitez.[60] For the last two and one-half decades of his life, Gregory served as Archbishop of Lwów. There he attended to affairs of his see while pursuing his own literary work and presiding over a humanistic circle.

Some of his works were historical, such as his two-book description of King Władysław Warneńczyk's expedition against the Turks; others were polemical, political, and rhetorical. Unfortunately, his commentaries on classical authors, particularly Vergil, have perished. From other evidence, however,[61] we know that the humanism which Gregory represented was uncompromisingly antique in its orientation. It was secular, derived from the pre-Christian past, and it celebrated the eloquence and achievement of that time untempered by Christian tradition. Gregory's physical isolation from the cultural center of Poland, Cracow, meant that he and his humanism had less direct effect upon the early Polish renaissance than did others. Nevertheless, his secular approach was not an isolated phenomenon in this period. It is revealed also in the thought and work of Jan Ostroróg.

This political writer, orator, and royal administrator studied for two years (1453–55) at the University of Erfurt, then traveled to Bologna where he earned a doctor's degree in both canon and civil law in 1459 and lectured on the decretals for the next two years.[62] He also learned Greek and gained a reputation as both a *humanista* and a *legista*.[63] In 1461 he returned to Poland and for four decades was active in all areas of political life. In the last years of his career his chief role was a *starosta* (almost a territorial vice-regent) for Greater Poland. The theme which ran through all he did as royal representative was the defense of secular and lay interests of the state as over against ecclesiastical interests and prerogatives.

Nowhere is this so clearly shown as in his *Monumentum pro Rei publicae ordinatione*, a treatise which he prepared at least by 1475 and perhaps as early as 1460.[64] It draws heavily upon historical arguments and the examples of strong, effective governance in antiquity. In this impassioned appeal for reform and restructuring of the state, Ostroróg defends the position and prerogatives of the monarchy against the ambitions of the oligarchy of the nobles and the higher clergy. His conception of the state is surprisingly national and secular, and his program of administrative and social reforms assumes an efficient executive, one law for all classes, common taxation, universal military service, and the Polonization of all aspects of Polish life.

From what has been said thus far in this section, it is clear that after some slow beginnings, humanistic activity had clearly developed in fifteenth-century Poland. It was not widespread (medieval scholastic traditions were far more important and still dominant in Polish culture), but in certain circles it was beginning to make itself felt. One of these was the University of Cracow which served in the late fifteenth century as the nexus of this new cultural force.

We have previously noted with almost every figure who is associated with early Polish humanism some connection with the University of Cracow. It is apparent that the medieval tradition of the *artes liberales* was very strong and exercised an influence similar to that enjoyed by the *artes* in the twelfth-century *studia* of western Europe. Gradually, however, a specific concern for antiquity on its own terms and for the *studia humaniora* in an Italian sense may be seen. For example, throughout the 1420s and 1430s in the academic addresses given by professors recommending students in the arts for degrees and in the maiden speeches given by the new licentiate or bachelor, one finds a growing emphasis upon the humane arts versus more practical and mechanical concerns. Frequently, as in the use of Ovid by Łukasz of Wielki Koźmiń in 1412,[65] or in the appeals to Lucian in 1420 and to Quintilian in 1429 by Mikołaj Kozłowski,[66] the use of classical material is sharply reminiscent of the *ars oratoria* in fifteenth-century Italy.

But the clearest appeals to humanistic concerns within the university were to come from Rector Jan of Ludzisko in the 1440s. He was born of peasant background about 1400 and studied at the University of Cracow.[67] He received his bachelor's degree in 1419 and became a master in 1422. For a year after that he studied astronomy, then went to Italy where he pursued medicine at Padua, receiving his doctor's degree in 1433. He remained in Italy for the next seven years, spending some time as a student of Guarino's in Ferrara, and some time also in Rome. By the time he returned to Cracow he was deeply committed to the cult of antiquity. In 1440 he lectured on eloquence—not rhetoric—in the liberal arts faculty. The same year he was designated as the university's official orator and for the next seven years, as orator and sometime rector of the faculty, he advocated curricular reform for the *studium* which would include more antique literature, both poetry and history.

In his orations, eight of which are extant,[68] he touched upon all of the various concerns of the active academic in the fifteenth century. His address in June, 1440, was a panegyric upon that *eloquentia* which aimed—as did the best of Italian oratory in this period—at the whole man; spirit as well as intellect, his divinity as well as his humanity.[69] Later that same year, Jan revealed himself as a firm conciliarist when in a second oration he welcomed an embassy from Basle. His most famous oration came in 1447, when in the presence of King Kazimierz Jagielloñczyk, he opened the academic year with a vivid description of the plight of the peasantry and a spirited appeal to both ruler and faculty to ameliorate their condition by social reform.[70] His other addresses are of lesser merit, but all reveal a passionate embrace of quattrocento humanistic ideals.

The role of Jan of Ludzisko as reformer of the curriculum was characterized by a singular lack of success during his association with the university. But two years after his disappearance from historical record (except for our knowledge of his death about 1460), his efforts found support from within and without the university, for in 1449 the *studia humaniora* were established as a new arts curriculum, to be taught in a parallel faculty, the *Collegium Minus,* so called to distinguish it from the *Collegium Maius,* a term applied to the studium as a whole.[71]

This new foundation included, among other positions, a chair of poetry, in which lectures were to be given on Vergil, Ovid, Horace, Terence, and Stacius, and a chair of rhetoric in which were to be read Quintilian's *De Institutione oratoria,* the *Rhetorica* of Cicero, and the *Chronicon Polonorum* of Vincent Kadłubek, the twelfth-century Polish historian.[72] In the following two decades, particularly under the leadership of Jan Dąbrówka (d. 1472), the *Collegium Minus* enrolled about 40 percent of the total student body,[73] and all who were associated with the university at any stage of their studies or careers came into some kind of contact with humanism.

The figure of Jan Dąbrówka deserves special mention in the present context.[74] Born about 1405, he became Master in Arts at Cracow in 1427 and lectured for six years thereafter on rhetoric. At that point he began the formal study of Aristotelian philosophy, scholastic theology, and canon law. By 1458 he was both *doctor decretorum* and *theologiae doctor*. In his numerous writings, on the *trivium* and *quadrivium*, law, theology, the sciences, and the historical writings of Vincent Kadłubek, he reveals a deep knowledge and appreciation of classical literature and the works of Petrarch, Pier Paolo Vergerio (whose educational treatises he adapted to the Cracow *studium*)[75] and other Italian humanists. On numerous occasions he served as rector and vice-chancellor of the university. Under his leadership there was firmly established in the university the humanistic tradition of criticism — both in the narrow area of pure scholarship and in the broader area of the philosophy of man. His death in 1472 marked the end of the beginning for the early Renaissance in Poland and was as important in its own way as the birth of Copernicus in the following year.

V

In 1491 there enrolled in the arts faculty of the University of Cracow "Nicholas the son of Nicholas of Torun." For four years he was a student there, leaving without a degree in 1495 to continue his study in Italy. We know almost nothing directly about Copernicus's experiences in those years, and most conclusions are in one degree or another inferential. We know the word, for example, of Jan Brożek that Copernicus was a student of the distinguished astronomer Wojciech of Brudzewo,[76] but this can not be verified. Thus, what has been discussed in some detail previously in this study can not be neatly or confidently related to Copernicus. Much is uncomfortably tentative. What may reasonably be said by way of conclusion, however, is that in a time of change when the social foundations of one world view were being profoundly altered; that at a university where the technical level of science and the subtlety of philosophical debate was abreast of that of any *studium* in Europe; that at an institution where the new critical attitude of humanism had deeply penetrated the curriculum: that in this time and place, the still-unformed youth whose *De revolutionibus* was to be a revolution spent years which were as fundamental to his development as were the later experiences in Italy. This was the context from which he came.

NOTES

1. It is inappropriate in a serious scholarly gathering such as this to descend to the polemical debate as to whether Copernicus was a German or a Pole. There has been intense heat and little light shed on this subject in the past. But, in this context, see Aleksander Birkenmajer, "Dookoła Kopernika. List Otwarty do pana doc. dr Hansa Schmauch w Malborgu," in French translation in Birkenmajer, *Etudes d'histoire des sciences en Pologne* (Wrocław, Warsaw, Cracow, and Gdańsk, 1972 [*Studia Copernicana*, IV]), pp. 579–82. I use the form Nicholas Copernicus on the pragmatic ground that he is best known by this name in the United States.

2. The most reliable short biographical treatment of Copernicus is now contained in *Polski Słownik Biograficzny* (Cracow [later also Wrocław and Warsaw], 1935), XIV (1968), 3–16. With the publication of Marian Biskup, ed., *Regesta Copernica* (Wrocław, Warsaw, Cracow, and Gdańsk, 1973 [*Studia Copernicana*, VII]), scholars now have a reliable, complete calendar of the data relating to Copernicus's life.

3. Among the numerous studies in English of this type, I have found A. R. Hall, *The Scientific Revolution* (London, 1954); T. S. Kuhn, *The Structure of Scientific Revolutions* (Chicago, 1962); and Kuhn, *The Copernican Revolution* (Cambridge, Mass., 1957), particularly useful.

4. This is a vastly oversimplified description of Polish society up through the end of the thirteenth and the early part of the fourteenth centuries. More detailed analyses may be found in the appropriate sections of *Historia Polski* (Warsaw, 1954–), published by the History Institute of the Polish Academy of Sciences, vol. I, i (1955). In English, the chapters on the twelfth and thirteenth centuries by Alexander Bruce Boswell, in *The Cambridge History of Poland*, edited by W. F. Reddaway et al., 2 vols. (Cambridge, 1941–50), I, 43–59, 85–107, are still useful.

5. M. M. Postan, "Economic Relations Between Eastern and Western Europe," in Geoffrey Barraclough, ed., *Eastern and Western Europe in the Middle Ages* (London, 1970), p. 169. For the problem of the peasantry during this period in general, the most important recent Polish work is Kazimierz Tymieniecki, *Historia chłopów Polskich*, 2 vols. (Warsaw, 1965–66). II: *Schyłek średniowieczna*.

6. Polish towns in this period have been the subject of a good deal of investigation. Among numerous works, including individual town

histories, see especially Stanisław Piekarczyk, *Studia z dziejów miast polskich w XIII-XIV* w. (Warsaw, 1955); and Stanisław Herbst, "Miasta i mieszczaństwo Renesansu polskiego," in *Odrodzenie w Polsce,* 5 vols. (Warsaw, 1955–60), I: *Historia,* 336–61.

7. The crucial developments of the fourteenth and early fifteenth centuries are best traced in Stanisław Gawęda, *Możnowładztwo małopolskie w XIV i pierwszej połowie WV wieku. Studium z dziejów rozwoju wielkiej własności ziemskiej* (Cracow, 1966); while the older work by Henryk Lowmiański, *Uwagi w sprawie podłoża społecznego i gospodarczego unii jagiellońskiej* (Wilno, 1935), is still reliable for the later period.

8. See particularly her brilliant book *Kopernik i heliocentryzm w polskiej kulturze umysłowej do końca XVIII wieku* (Wrocław, Warsaw, Cracow, and Gdańsk, 1971 [*Studia Copernicana,* III]). This same thesis, that old opponents of a system are not converted but are replaced by new adherents who are the product of a changed society and intellectual orientation, is clearly reflected in the several articles of *Colloquia Copernicana I,* ed. Jerzy Dobrzycki (Wrocław, Warsaw, Cracow, and Gdańsk, 1972 [*Studia Copernicana,* V]), particularly Hans Blumenberg, "Die Kopernikansche Konsequenz für den Zeitbegriff," pp. 57–78; Juan Vernet, "Copernicus in Spain," pp. 271–92; and Jolan Zemplen, "The Reception of Copernicanism in Hungary," pp. 311–56.

9. See my article "Casimir the Great and the University of Cracow," *Jahrbücher für Geschichte Osteuropas,* XVI (1968): 251–68. The documents relating to this first founding of the university printed by Stanisław Krzyżanowski, "Poselstwo Kazimierza Wielkiego do Awinionu i pierwsze uniwersyteckie przywileje," *Rocznik Krakowski* IV (1900): 1–111, should now be supplemented by Leon Koczy, ed., *University of Cracow: Documents Concerning Its Origins* (Glasgow, 1966).

10. Maria Kowalczyk, "Odnowienie Uniwersytetu Krakowskiego w świetle mów Bartłomieja z Jasła," *Małopolskie Studia Historyczne,* VI, iii/iv (1964): 23–42.

11. *Codex diplomaticus universitatis studii generalis Cracoviensis,* 5 vols. (Cracow, 1870–1900), I, 118. See also Zofia Kozłowska-Budkowa, "Odnowienie Jagiellonskie Uniwersytetu Krakowskiego (1390–1414)," in Kazimierz Lepszy, ed., *Dzieje Uniwersytetu Jagiellońskiego w latach 1364–1764* (Cracow, 1964), pp. 37–56. The older account by Casimir Morawski, *Histoire de l'Université de Cracovie, moyen age et renaissance,* 3 vols. (Paris and Cracow, 1900–1905), I, 55–81, is still useful.

12. See František Šmahel, "Husitská universita," in *Stručne dějiny University Karlovy* (Prague, 1964), pp. 44–76.

13. The intellectual history of the University of Cracow in the fifteenth century remains yet to be studied. Numerous scholars, particularly in Poland (many to be cited below) are at work on specific aspects of this task, but an extended synthetic treatment has not yet been attempted.

14. See W. Wisłocki, *Katalog rękopisów Biblioteki Jagiellońskiego*, 2 vols. (Cracow, 1877–81), MSS 546, 568, 1865, 1927, 1970.

15. *Ibid.*, MSS 1851, 1860.

16. *Ibid.*, MSS 551, 552, 573, 602, 662.

17. *Ibid.*, MSS 602, 1844, 1859, 1927. See the general account by J. Dianni, *Studium matematyki na Uniwersytecie Jagiellońskim do połowy XIX wieku* (Cracow, 1963), pp. 15–23, and his more specific study "Pierwszy znany traktat rękopiśmienny w literaturze matematycznej w Polsce, Algorismus minutiarum Martini Regis de Peremislia," *Kwartalnik Historii Nauki i Techniki* XII (1967): 272–83.

18. This work (MSS Cracow B.J. 1865 and 1968) was early edited by Ludwik Birkenmajer, *Marcina Króla z Przemyśla Geometria praktyczna* (Warsaw, 1895).

19. See Mieczysław Markowski, "Astronomia i astrologia w Polsce od X do XIV wieku," in *Historia Astronomii w Polsce* (in press). The fundamental works on Witelo by Aleksander Birkenmajer are now available in French translation in his *Etudes d'histoire des sciences en Pologne,* pp. 97–434.

20. And not Jan Stoebner, a Prague bachelor from 1379, as argued by F. Karlinski in the mid-nineteenth century. See A. Birkenmajer, "Les astronomes et les astrologues silesiens au moyen age," in *Etudes d'histoire des sciences en Pologne,* p. 456, n. 62.

21. See Markowski, "Astronomia i astrologia," and the list of manuscripts cited in his *Burydanizm w Polsce w okresie przed kopernikańskim* (Wrocław, Warsaw, Cracow, and Gdańsk, 1971 [*Studia Copernicana*, II]), p. 263, n. 96.

22. See Jerzy Wiesiołowski, "Sędziwój z Czechla (1410–76), Studium z dziejów kultury umysłowej Wielkopolski," *Studium Źródłoznawcze* IX (1964): 78f.

23. Both treatises are contained in MS Cracow B.J. 1929.

24. He and his writings are discussed by J. Rebeta, "Miejsce Wawrzyńca z Raciborza w najdawniejszym okresie krakowskiej astronomii XV wieku," *Kwartalnik Historii Nauki i Techniki* XIII (1968): 553–64. See also the eulogy upon him given by Matthew of Labiszyń in

1448 in Maria Kowalczyk, *Krakowskie mowy uniwersyteckie z pierwszej połowy XV wieku* (Wrocław, Warsaw, Cracow, 1970), p. 135, n. 33, p. 186 (MS Cracow B.J. 2231).

25. See Jerzy Zathey, "Colligite fragmenta ne pereant. Contribution aux recherches sur l'histoire de l'enseignement à l'Université de Cracovie au XVe s.," *Medievalia Philosophica Polonorum* X (1961): 99. Other minor figures associated with the Stobner chair are discussed by Markowski, *Burydanizm w Polsce,* p. 229.

26. A. Birkenmajer, "L'Université de Cracovie, centre international d'enseignement astronomique à la fin du moyen age," in *Etudes d'histoire des sciences en Pologne,* pp. 483–95.

27. The work, MS Cracow B.J. 1927, ff. 501–637, remained unedited.

28. See the recent work by Zdzisław Kuksewicz, "Marcin Król z Żurawicy," *Materiały i Studia Zakładu Historii Filozofii Starożytnej i Średniowiecznej* I (1961): 118–40.

29. See the very useful article of Leslie S. Domonkos, "The Polish Astronomer Martinus Bylica de Ilkusz in Hungary," *The Polish Review* XIII, iii (1968): 71–79. The history of Polish-Hungarian cultural relations during the Renaissance is of fundamental importance in understanding the dissemination of Italian humanism in east central Europe. For a general overview, see the cooperative Polish-Hungarian symposium, *Renaissance und Reformation in Polen und Ungarn* (Budapest, 1964). More specifically, see Zathey, "Martin Bylica z Olkusza, profesor Akademie Istropolitany," *Humanizmus a Renesancia na Slovensku v 15-16 storoči* (Bratislava, 1967), pp. 40–54.

30. MSS Cracow, B.J., 597 and 599, respectively.

31. See the discussion of this library and these instruments in Zofia Ameisenowa, *Globus Marcina Bylicy z Olkusza i mapy nieba na Wschodzie i Zachodzie* (Wrocław, Warsaw, and Cracow, 1959 [also in English translation: *The Globe of Martin Bylica of Olkusz and Celestial Maps in the East and in the West* (1959)]). The most useful short treatment of Bylica is by A. Birkenmajer in *Pol. Słow. Biog.,* III, 166–68 (also in *Etudes d'histoire des sciences en Pologne,* pp. 533–36, in French).

32. For a short biography, see *Pol. Słow. Biog.,* X, 450–52. The best introduction to his works is now Władysław Seńko, "Wstęp do studium nad Janem z Głogowa," *Materiały i Studia Zakładu Historii Filozofii Starożytnej i Średniowiecznej* I (1961): 9–59 and III (1964): 30–38.

33. See the discussion of these points in L. Birkenmajer, *Stromata Copernicana, Studia, poszukiwania i materiały biograficzne* (Cra-

cow, 1924), pp. 125ff.; A. Birkenmajer, "Les astronomes et les astrologues," p. 464.

34. The biography and works of Wojciech are now best treated in Ryszard Palacz, "Wojciech Blar z Brudzewa," *Materiały i Studia Zakładu Historii Filozofii Starożytnej i Średniowiecznej* I (1961): 172–98.

35. See particularly the introduction to L. A. Birkenmajer, ed., *Commentariolum super Theoricas novas planetarum* . . . (Cracow, 1900). Some of the most crucial points are further discussed by the same author in his *Stromata Copernicana*, pp. 83–103.

36. "Cracoviae est celebre gymnasium multis clarissimis doctissimisque viris pollens, ubi plurimae ingenuas artes recitantur. Astronomiae tamen studium maxime viget. Nec in tota Germania, ut ex multorum relatione satis mihi cognitum est, ille clarius reperitus." Hartmann Schedel, *Liber chronicarum* (Nürnberg, 1493), p. 267. "A coniectoribus et astrologis, quibus referta Cracovia est." Mentioned by S. Katona, *Historia critica regum Hungariae*, XVII (Buda, 1793), p. 258. Each cited in A. Birkenmajer, *Etudes d'histoire des sciences en Pologne*, pp. 491 and 464, n. 86. The implications in both quotations for Polish and German cultural history have been hotly disputed. See Markowski, *Burydanizm w Polsce*, pp. 242–44.

37. For general treatments on Buridan, see Pierre Duhem, *Le systèm du monde. Histoire des doctrines cosmologiques de Platon à Copernic*, 10 vols. (Paris, 1913–59), IV, 124–42, VI, 697–729; and E. Faral, *Jean Buridan, maître ès arts de l'Université de Paris* (Paris, 1950).

38. See, for the general problem of motion in the late middle ages, Annaliese Maier, *Studien zur Naturphilosophie der Spätscholastik*, 5 vols. (Rome, 1949–58), II: *Zwei Grundprobleme der scholastischen Naturphilosophie* (1958); and III: *An der Grenze von Scholastik und Naturwissenschaft* (1952); and, very briefly, but cogently, Edward Grant, *Physical Science in the Middle Ages* (New York, 1971), pp. 50–54.

39. The first Polish scholar to call attention to Buridan's influence in Poland was Kazimierz Michalski, "Jan Buridanus i jego wypływ na filozofię scholastyczną w Polsce," *Sprawozdania z Czynności i Posiedzeń AU w Krakowie* XXI, x (1916): 25–34. See also his "Zachodnie prądy filozoficzne w XIV wieku i stopniowy ich wypływ w środkowej i wschodniej Europej," *Przegląd Filozoficzny* XXXI (1928): 15–21.

40. See Markowski, *Burydanizm w Polsce*, p. 200.

41. An enormous amount of Polish scholarship in the past three decades has been devoted to the philosophical and scientific tradition at the

University of Cracow in the fifteenth century. For the works to be discussed below, see Markowski, "Poglądy filozoficzne Andrzeja z Kokorzyna," *Studia Mediewistyczne* VI (1964): 55–136; idem, "Krakowskie Komentarze do "Fizyki" Arystotelesa zachowane w średniowiecznych rękopisach Biblioteki Jagiellońskiej," *Studia Mediewistyczne* VII (1966): 107–24; S. Swieżawski, "Filozofia w średniowiecznym Uniwersytecie Krakowskim" in *Historia kultury średniowiecznej w Polsce,* 2 vols. (Warsaw, 1963), I, 129–59; Paweł Czartoryski, *Wczesna recepcja "Polityki" Arystotelesa na Uniwersytecie Krakowskim (Wrocław, Warsaw, Cracow, 1963);* and especially Markowski, *Burydanizm w Polsce,* pp. 120–222.

42. But see the article of Władysław Abraham, "Udział Polski w soborze pizańskim," *Rozprawy Polska Akademii Umiejętności, wydział hist.-filozoficzny* XLVII (1905), which to my knowledge has not been superseded.

43. I am in fundamental disagreement with Antoni Karbowiak, *Dzieje wychowania i szkół w Polsce,* 3 vols. (St. Petersburg and Lwów, 1898–1923), II, 454, whose judgment was that "at the Council of Constance, Italian humanism did not exercise any influence upon the Poles." For example, Bracciolini's discovery of Quintilian's *De institutione oratoria* was known and appreciated by the Poles. See the letter of Guarino to Poggio in November, 1417, in Eugenio Garin, ed., *Il pensiero pedagogico dello umanesimo,* in *I classici della pedagogia italiana,* vol. II (Florence, 1958), p. 322.

44. See particularly Paulus Vladimiri, *Tractatus de potestate papae et imperatoris respectu infidelium* in M. Bobrzyński, ed., *Starodawne Prawa Polskiego Pomniki,* V (Cracow, 1878). This rector of the University and chief antagonist of the Order is justly noted for his contribution to political theory, but his motives at Constance were more political in nature. The recent study by Stanislaus Belch, *Paulus Vladimiri and His Doctrine Concerning International Law and Politics,* 2 vols. (London, The Hague, and Paris, 1965), does not adequately reflect this. See the penetrating criticism of Howard Kaminsky in *Speculum* XLII (1967): 718–20.

45. On Bishop Laskarz in general, see *Pol. Słow. Biog.,* I, 103–6. His library is discussed in the context of learned libraries in Poland in the fifteenth century by W. Szelińska, *Biblioteki profesorów Uniwersytetu Krakowskiego w XV i początkach XVI wieku* (Wrocław, 1966).

46. Two works have been particularly important on this topic: T. Zegarski, *Polen und das Basler Konzil* (Poznan, 1910); and Ignacy Zarębski, "Zur Bedeutung des Aufenthaltes von Krakauer Univer-

sitätsprofessoren auf dem Basler Konzil für die Geistesgeschichte Polens," *Vierteljahresschrift für Geschichte der Wissenschaft und Technik,* V (1960), *Sonderheft,* pp. 7–25.

47. See Zathey, Anna Lewicka-Kamińska, and Leszek Hajdukiewicz, *Historia Biblioteki Jagiellońskiej,* vol. I: *1364–1775* (Cracow, 1966), pp. 35f.

48. The following is based upon the biographical information found in *Bibliografia Literatury Polskiej (Nowy Korbut),* II (Warsaw, 1964), 437–38. These biographical data are derived from Długosz, *Vitae episcoporum Poloniae: Catalogus episcoporum Vladislaviensium,* in *Opera Omnia,* I (Cracow, 1887), 539–40.

49. See, for details, J. Brüstigerowa, "Guarino a Polska," *Kwartalnik Historyczny,* XXXIX (1925), 70–80.

50. See, for example, Józef Garbacik, "Paolo Veneto, filozof-dyplomata i jego pobyt w Polsce w r. 1412," *Prace Historyczne* IV (1960): 17–31, and the work on Aeneas Sylvius Piccolomini by Zarebski, "Stosunki Eneasza Sylwiusza z Polską i Polakami," *Roz. Akad. Umiejęt.,* Series II, XLV (1939), 281–437; and his "Z dziejów recepcji humanizmu w Polsce," in H. Barycz and J. Julewicz, eds., *Studia z dziejów kultury polskiej* (Warsaw, 1949), 147–71. In this present context, I am also interested in Jacopus Publicius, a Florentine orator who taught at the University of Cracow in 1469. Passing reference to his activity in Poland is made by Barycz in *Pol. Słow. Biog.,* IX (in an article on Jan z Oświęcimia).

51. For an example of these contacts, see the forthcoming study on the humanist Bishop John Vítez of Varad by L. S. Domonkos in *The New Hungarian Quarterly.*

52. Surprisingly, the only extended biographical treatment of Cardinal Oleśnicki remains M. Dzieduszycki, *Zbigniew Oleśnicki,* 2 vols. (Cracow, 1853–54); it is, however, badly outdated. More up-to-date information, together with a short discussion of the cardinal's works, is to be found in *Bibliografia Literatury Polskiej,* III, 32–35.

53. There is an immense literature on Długosz. For an up-to-date guide, see my comments in *The Rise of the Polish Monarchy, Piast Poland in East Central Europe 1320–1370* (Chicago, 1972), pp. 5ff. I have found Jan Dąbrowski, *Dawne Dziejopisarstwo Polskie (do roku 1480)* (Wrocław, Warsaw, and Cracow, 1964), pp. 189–240, particularly helpful.

54. There is a new critical edition of Długosz's *History* in progress, and it is the intention of the Institute of History of the Polish Academy of Sciences eventually to replace the nineteenth-century edition of his *Opera Omnia,* ed. A. Przezdziecki, 14 vols. (Cracow, 1863–87).

See Jadwiga Karwasińska, "Sur l'élargissement de la base documentaire de l'histoire du Moyen Age en Pologne," *Acta Poloniae Historica* XXVI (1972): 189.

55. Wanda Semkowica-Zarembina, "Elementy humanistyczne redakcji 'Annalium' Jana Długosza," in *Medievalia–w 50 rocznicę pracy naukowej J. Dąbrowskiego* (Warsaw, 1960); Władysław Madyda, ' Johannes Longinus Długosz als Vorläufer des Humanismus in Polen," *Renaissance und Humanismus in Mittel- und Osteuropa, eine Sammlung von Materialien,* ed. J. Irmscher, vol. II (Berlin, 1962), 185–91.

56. "Dlugossius multis operis sui locis Titi Livii, rerum Romanarum scriptoris, scribendi modum, pugnarum descriptiones et similia secutus est," in *Annales seu Cronicae . . . Dlugossii* (Warsaw, 1964–), I, 329. See also Madyda, "Johannes Długosz," pp. 187ff.

57. The basic biographical source on Gregory of Sanok is Phillipus Callimachus, *De vita et moribus Gregorii Sanocensis, archiepiscopi leopoliensis,* ed. I. Lichonska (Wrocław, 1963). See also the popular biography by Andrzej Nowicki, *Grzegorz z Sanoka 1406–1477* (Warsaw, 1958), which must, however, be used with great caution; see the comments of Zarębski in *Wiadomości Historyczne* I (1958): 355–61.

58. Callimachus, *De vita . . . Gregorii,* p. 41.

59. Zarębski, "MS B.J. 413," *Biul. Bibl. Jag.,* XV, i–ii (1963).

60. Tadeusz Sińko, "De Gregorii Sanocei studiis humanioribus," *Eos* VI (1900): 241–70; and Domonkos, "Bishop John Vitez of Varad," *The New Hungarian Quarterly* (in press).

61. See *Pol. Słow. Biog.,* IX, 86–89.

62. Biographical data are in *Pol. Słow. Biog.,* XI; and W. Voisé, *Jan Ostroróg–wybitny pisarz polityczny polskiego Odrodzenia* (Warsaw, 1954).

63. *De vita . . . Gregorii,* p. 64.

64. Anna Strzelecka, "Uwagi w sprawie daty powstania oraz genezy 'Monumentum' Ostroroga," in *Prace z dziejów Polski feudalnej ofiarowane Romanowi Grodeckiemu* (Warsaw, 1960), pp. 251–64.

65. MS Cracow B.J., 2215, ff. 251–55.

66. MS Cracow, B.J., 2216, ff. 105–112, 173–79. These speeches have, since I used the manuscripts in 1966 and 1970, now been discussed by Kowalczyk, *Krakowskie mowy uniwersyteckie,* with whom, however, I differ on the extent of tradition versus innovation reflected in the speeches.

67. See *Pol. Słow. Biog.,* X, 461–62, for biographical data. The study

by B. Nadolski, "Rola Jana z Ludziska w polskim Odrodzeniu," *Pamiętnik Literacki* XXVI (1929): 198–211, is excellent.

68. MS Cracow, B.J., 126. Since I have examined this manuscript, the extant works of Jan have received a modern critical edition, *Johannis de Ludzisko Orationes*, ed. J. S. Bojarski (Wrocław, Warsaw, and Cracow, 1971).

69. I have in mind the recent, brilliant work of Charles Trinkhaus, *"In Our Image and Likeness": Humanity and Divinity in Italian Humanist Thought*, 2 vols. (Chicago, 1970). Professor Trinkhaus has here convincingly elaborated several themes of eloquence and wisdom which have been central to the historical writing on the Italian renaissance for the past two decades.

70. See Franciszek Bujak, "Mowa Jana z Ludziska do Króla Kazimierza Jagiellończyka z roku 1447," in *Księga pamiątkowa ku czci Władysława Abrahama*, vol. II (Lwów, 1931), pp. 217–33. This speech has several times been translated into Polish, most notably by B. Nadolski, ed., *Wybór mów staropolskich* (Wrocław, 1961).

71. The details of this process for curricular reform are best traced in Morawski, *Histoire de l'Université*, II, 220–36; and Zarębski, "Okres wczesnego humanizmu," in Lepszy, ed., *Dzieje Uniwersytetu*, pp. 172–75.

72. Details on the structure of the Collegium Minus are found in a document, *Conclusiones Maioris Collegii*, written by Jan Dąbrówka (see below) in 1466: MS Cracow, Archiwum U.J. 63, printed in *Codex diplomaticus universitatis studii generalis Cracoviensis*, 5 vols. (Cracow, 1870–1900), III, 47. Its importance was largely unrecognized by Karbowiak, *Dzieje wychowania a szkół*, III, 309ff.

73. Karkowiak, "Studia statystyczne z dziejów Uniwersytetu Jagiellońskiego 1433/4–1509/10," *Archiwum do dziejów literatury i oświaty w Polsce* XII (1910): 77–81; Zarębski, "Okres wczesnego humanizmu," pp. 176–79.

74. Among the older treatments of Jan Dąbrówka, see Morawski, *Histoire de l'Université*, II, 223–28. More recently, however, the Institute of Philosophy and Sociology of the Polish Academy of Sciences has, in its historical section, been concentrating upon the career and writings of Jan Dąbrówka; and in such serial publications as *Studia Mediewistyczne, Medievalia Philosophica Polonorum*, and *Materiały i Studia*, scarcely an issue appears without something devoted to him.

75. W. Szelińska, "Dwa testamenty Jana Dąbrówki," *Studia i materiały z dziejów nauki polski* Series A: V (1962): 18–21, describes the

text of an inventory of Jan's library and notes Vergerio's works. For the more explicit statement in the text, see Zarębski, "Okres wczesnego humanizmu," p. 171.

76. See A. Birkenmajer, "Copernic philosophe," in *Etudes d'histoire des sciences en Pologne,* p. 623.

COMMENTARY

DAVID C. LINDBERG

PAUL KNOLL has skillfully summarized several of the social and intellectual changes that were transforming Poland in the fifteenth century: the alteration of social structure and economic patterns, the growth of mathematical science and natural philosophy at the University of Cracow, and the incursion of Italian humanism into Polish intellectual life and its subsequent impact upon the curriculum at Cracow. Clearly Professor Knoll deserves only gratitude for bringing to our attention conclusions to which those of us who do not read Polish would otherwise have limited access.[1]

But Professor Knoll has not merely set forth these conclusions for their own intrinsic importance. Rather, he has used them to tantalize us; he has dangled before us the enticing suggestion that there exists some connection between changes in the social structure or developments in the curriculum at Cracow and the astronomical achievements of Copernicus—without spelling out the nature of the connection nor furnishing evidence for its existence. To describe this connection he has employed such words as "context," "background," and "factors which help account for the Copernican achievement"; but these are "weasel words," apparently selected because they skirt the question of causation. I think we would do better to face the issue directly: can it be established (or at least sensibly argued) that the social and intellectual developments described by Knoll are causes of the Copernican achievement? If so, we must add substance to the bare claim, showing how these developments might have produced their effect; if not, then they do not help to account for anything. This is treacherous ground, and perhaps Professor Knoll should be applauded for his good sense in not venturing onto it, but (being in a reckless mood) I cannot resist this opportunity to make a preliminary survey.

Let us begin with the "factor" described by Knoll that is least open to dispute: astronomical, mathematical, and other scientific studies at the University of Cracow. Although it is impossible to identify the particular course of studies pursued by Nicholas Copernicus at Cracow, it is beyond reasonable doubt that here he received his initial training in astronomy and ancillary disciplines. Ludwik Birkenmajer has reconstructed the following list of lectures on astronomy and mathematics offered at Cracow during the years 1491—95:

> 1491 Sacrobosco, *De sphera,* lectured upon by Wojciech of Pniewy.
> 1492 Euclidean Geometry, lectured upon by Bartłomiej of Lipnica.
> 1493 Albertus (Wojciech) of Brudzewo, *Commentariolum super theoricas novas planetarum,* lectured upon by Szymon of Sierpc.
> Tables of Eclipses, lectured upon by Bernard of Biskupie.
> *Tabulae resolutae,* lectured upon by Michał of Wrocław.
> Regiomontanus, *Calendarium,* lectured upon by Marcin of Olkusz.
> 1494 Astrology, lectured upon by Wojciech of Szamotuły.
> 1494—95 Ptolemy, *Tetrabiblos,* lectured upon by Wojciech of Szamotuły.[2]

There is no need to suppose that Copernicus attended all of these lectures in order to acknowledge that his education at Cracow surely furnished him with the basic tools of the astronomical trade. It seems clear, moreover, that the student of astronomy at Cracow acquired not only a solid understanding of Ptolemaic astronomy, but also some familiarity with the criticisms that had been leveled against it in the intervening centuries. In particular, it is clear that there was considerable discussion of the Averroistic complaint that Ptolemaic astronomy was not physically possible.

As for the developments in natural philosophy discussed by Knoll, it is pointless to speculate about what influence John Buridan's theory of impetus may have had upon Copernicus, since the latter does not employ this theory in any of his works. More to the point are other anti-Aristotelian conclusions (I do not see that the *approaches* were un- or anti-Aristotelian, as Knoll suggests), such as the possible diurnal rotation of the earth, discussed by both Buridan and Nicole Oresme. It might be fruitful to consider parallels between these discussions and Copernicus's arguments for the earth's mobility, but this is an intricate matter, which goes beyond present limits of time and space.[3] In any case, it is clear that the role of Copernicus's scientific training in the formulation of the new astronomical and cosmological system was one of preparation. At Cracow Copernicus obtained the skills required for the practice of astronomy and encountered the problems (or certain of them) to which the heliocentric system would ultimately be the solution.

How was this heliocentric solution discovered? It is commonly acknowledged that Copernicus discovered it through the reading of ancient authors whose works had been recovered by fifteenth-century humanism—and here we come to the second element in what Knoll refers to as the "context" of the Copernican achievement. It is Copernicus himself who tells us that he came upon the new idea in the course of reading the ancient philosophers:

> I pondered long upon this uncertainty of mathematical tradition in establishing the motions of the system of the spheres. At last I began to chafe that philosophers could by no means agree on any one certain theory of the mechanism of the Universe. . . . I therefore took pains to read again the works of all the philosophers on whom I could lay hand to seek out whether any of them had ever supposed that the motions of the spheres were other than those demanded by the mathematical schools. I found first in Cicero that Hicetas had realized that the Earth moved. Afterwards I found in Plutarch that certain others had held the like opinion. . . . Taking advantage of this I too began to think of the mobility of the Earth. . . .[4]

Other possible literary sources of the heliocentric hypothesis would have been the *Sand-Reckoner* of Archimedes and George Valla's *De expetendis et fugiendis rebus* (Venice, 1501).[5]

On the surface, then, it would seem that Polish humanism (or Italian humanism imported into Poland) was a major cause of the Copernican revolution. However, two qualifications of the utmost importance are required. First, it is by no means certain (or even likely) that Copernicus first encountered the heliocentric hypothesis while at the University of Cracow or through Cracovian humanism. About 1495 Copernicus left the university without a degree, and in 1496 he departed for Italy, where he spent most of the next seven years. He went first to Bologna, where he assisted the celebrated astronomer, Domenico Maria Novarra, later to Padua and Ferrara.[6] There is evidence to suggest that Copernicus's humanistic training (evidenced, for example, by his translation into Latin of some *Letters* of a Byzantine Greek) occurred principally at Bologna, where Greek was taught (as it was apparently not at Cracow) and where Copernicus purchased a Greek-Latin dictionary published in 1499.[7] It has often been asserted that Copernicus came upon the heliocentric idea before his departure from the University of Cracow, but Edward Rosen has argued convincingly that this is based upon a misinterpretation of the pertinent document.[8] The truth is that we do not know when Copernicus first read of the heliocentric hypothesis; we only know (or surmise with some confidence) that his own heliocentric theory was not formulated until sometime during the period 1508–14.[9] Thus while the humanistic revival of ancient texts seems to have been of crucial importance for Copernicus, the credit may not belong to Polish or Cracovian humanism.

Second, if we are to obtain a balanced picture we must recognize that defense of a heliocentric universe was only a small part of Copernicus's achievement. If the heliocentric idea came to Copernicus through his reading of ancient sources, it remained for Copernicus to develop this vague ancient hypothesis into a complete system of mathematical astronomy. Therefore, while we must recognize the importance of Copernicus the humanist in recovering the heliocentric idea, we must also acknowledge that heliocentrism became a viable option for mathematical astronomers only through the creative technical proficiency of Copernicus the mathematical astronomer. Heliocentrism had been rejected before, and it could have been rejected again had Copernicus not transformed it into an indispensable mathematical calculating instrument.[10]

We come finally to the last of Professor Knoll's three

"factors"—the alteration of social structure and economic life in fifteenth-century Poland. Is it possible that this was causally related to the Copernican revolution? The issue at stake is a historiographic one of the utmost significance, for it divides historians of science into their principal warring camps. Are new ideas to be explained exclusively in terms of individual genius, the internal logic of the discipline in question, and the convergence of previously separate bodies of ideas, or is it possible also to identify social and institutional causes? And if the latter, to what level of specificity is it possible to descend in providing external explanations of conceptual results? It is scarcely possible, in the present state of the art, to give definitive answers to these questions, but perhaps a few tentative remarks will be in order.

I believe that a strong case can be made for the influence of social structure on certain broad categories of conceptual activity. Perhaps the best evidence comes from anthropological studies, which appear to have established that some of the more general aspects of world view depend on the scale and organizational complexity of society. For example, Monica Wilson, who has studied the effect of social structure on witch-beliefs, writes:

> I have suggested ... that witch beliefs are general in small-scale societies with inadequate control of their environment and dominated by personal relationships, societies in which people think in personal terms and seek personal causes for their misfortunes.[11]

More recently, Keith Thomas has attempted to connect witchcraft in sixteenth- and seventeenth-century England to specific defects of the social structure:

> Witch-beliefs are ... of interest to the social historian for the light they throw upon the weak points in the social structure of the time. Essentially the witch and her victim were two persons who ought to have been friendly towards each other, but were not. They existed in a state of concealed hostility for which society provided no legitimate outlet. ... The charges of witchcraft were a means of expressing deep-felt animosities in acceptable guise. ... The great bulk of witchcraft accusations thus reflected an unresolved conflict between the neighbourly conduct required by the ethical code of the old village community, and the increasingly individualistic forms of behaviour which accompanied the economic changes of the sixteenth and seventeenth centuries.[12]

Wide diversity of opinion continues to prevail among anthropologists regarding precisely how witch-beliefs are to be explained, but what seems clear is that one component of the proper explanation will necessarily be the influence of social structure.

The larger issue of the origin of nonmagical world views and rational decision-making procedures has recently been discussed by Robin Horton—again on the basis of anthropological data. Horton argues that a nonmagical world view is characteristic of cultures where there is an "awareness of alternatives to the established body of theoretical tenets,"[13] and that this awareness is dependent on the development of literacy (since "the possibility of checking current beliefs against the 'frozen' ideas of an earlier era throws the fact of change into sharp relief")[14] and on cultural heterogeneity (which ensures contact with alternative interpretations of reality). Awareness of alternatives also gives rise to critical decision-making procedures, which may lead to what is sometimes called "Western rationality." The effects of social scale and social structure are once again evident.

There are many difficult issues to be resolved before it can be confidently described precisely how social structure influences ideas, but it does seem clear that in one way or another (or in many ways simultaneously) the *general* content and structure of man's world view is affected by the character of the society in which he lives. On the other hand, I believe that one must be very skeptical of attempts to give externalist explanations of *specific* features of man's world view. Returning to Copernicanism, I see no prospect that we will ever be able to explain in social terms why or how Copernicus substituted a heliocentric for a geocentric cosmology. We can furnish convincing externalist explanations for the existence of the specialized role of astronomer, and of social support for astronomical activity;[15] we may also explain in terms of social structure the development of critical decision-making procedures, such as those applied to intellectual systems in the West since roughly the sixth century B.C.; but I do not believe that we can successfully descend with our externalist explanations (apart from an occasional rare exception) to the details of cosmological or astronomical systems. In the case before us, it is quite clear that in attempting to reform astronomy Copernicus was not departing from the methodological canons of the past astronomical tradition; on the contrary, he was applying ancient techniques and criteria to problems internal to the astronomical tradition. He discovered a new solution (or rediscovered an old one, but in either case not a new *kind* of solution) to a perfectly traditional astronomical problem. This, of course, places us on the internalist's home terrain, and I believe that we are compelled to accept his explanation in terms of the internal logic of the astronomical enterprise and, perhaps above all, the individual genius of Nicholas Copernicus.

NOTES

1. On science at Cracow and the development of Polish humanism, see also Aleksander Birkenmajer, *Études d'histoire des sciences en Pologne* (Wroctaw, 1972), which contains French translations of several of Birkenmajer's articles published originally in Polish; and Eugeniusz Rybka, *Four Hundred Years of the Copernican Heritage* (Cracow: Jagellonian University, 1964), based principally on Polish sources.
2. I have taken this list from Rybka, *Four Hundred Years*, p. 56.
3. On this question, see Marshall Clagett, *The Science of Mechanics in the Middle Ages* (Madison: University of Wisconsin Press, 1959), pp. 583–614.
4. Nicholas Copernicus, *De revolutionibus*, Preface and Book I, trans. John F. Dobson and Selig Brodetsky, *Occasional Notes of the Royal Astronomical Society*, vol. 2, no. 10 (May 1947), pp. 4–5.
5. Thomas W. Africa, "Copernicus' Relation to Aristarchus and Pythagoras," *Isis* 52 (1961): p. 406.
6. For a dependable biography of Copernicus, see Edward Rosen's, appended to his *Three Copernican Treatises*, 3d ed. (New York: Octagon, 1971); for the events discussed here, see pp. 316–30.
7. *Ibid.*, pp. 324–25.
8. *Ibid.*, p. 316. For an instance of the misinterpretation against which Rosen argues, see Rybka, *Four Hundred Years*, pp. 64–65.
9. Rosen, *Three Copernican Treatises*, pp. 338–39, 344–45. For an opposing point of view, see Jerome R. Ravetz, "The Origins of the Copernican Revolution," *Scientific American* 215, no. 4 (October 1966): p. 92.
10. It appears to me that those historians who protest the attempt to explain the Copernican revolution in terms of humanistic concern

for the values and knowledge of Antiquity, urging instead that we find our explanations within the internal history of mathematical astronomy, must have this point in mind. For example, Ravetz's argument that the Copernican revolution was a response to technical problems internal to mathematical astronomy is surely tenable only if applied to the justification and working out of the ancient heliocentric hypothesis, rather than to its discovery; Ravetz develops his position in the article cited above and more fully in J. R. Ravetz, *Astronomy and Cosmology in the Achievement of Nicolaus Copernicus* (Wrocław: Polska Akademia Nauk, 1965).

11. Monica H. Wilson, "Witch Beliefs and Social Structure," *American Journal of Sociology* 56 (1951): p. 313. See also Godfrey Wilson and Monica Wilson, *The Analysis of Social Change based on Observations in Central Africa* (Cambridge: Cambridge University Press, 1945), pp. 88–104; Monica Wilson, *Religion and the Transformation of Society: A Study in Social Change in Africa* (Cambridge: Cambridge University Press, 1971).

12. Keith Thomas, *Religion and the Decline of Magic* (New York: Charles Scribner's Sons, 1971), pp. 560–61.

13. Robin Horton, "African Traditional Thought and Western Science," *Africa* 37 (1967): p. 55.

14. *Ibid.,* pp. 180–81.

15. For an elaboration of this theme, see Joseph Ben-David, *The Scientist's Role in Society: A Comparative Study* (Englewood Cliffs, N.J.: Prentice-Hall, 1971).

III   The Impact of Copernicus on Man's Conception of His Place in the World

EDWARD ROSEN

IF WE wish to understand the impact of Copernicus on man's conception of his place in the world, we would do well to cast a cursory glance backward at the beliefs of his predecessors.

In brief interludes of their otherwise insecure and hazardous life, our earliest ancestors may have wondered how they and their world first came into existence. If they had the leisure and the inclination to ponder this question, their thoughts inevitably perished with their frail frames until they learned how to give their fleeting ideas permanent form by preserving them on papyrus, clay, or some other more or less durable material.

Long before the invention of writing, however, cave dwellers made paintings and created statuettes. These early works of art show keen observation of animal life and a certain interest in the human figure, more particularly of the feminine variety, but no awareness of the heavenly bodies. As long as the struggle for survival from day to day was uppermost in the minds of our forebears, those intrepid and skillful hunters and food-gatherers may have paid little or no attention to their cosmic surroundings. As they became increasingly familiar with the steady succession of night and day as well as the recurring cycle of the seasons, they may have developed a reassuring confidence that the world as they experienced it had always been so and would always remain so. Such a conviction would tend to stifle curiosity about the origin of all things.

In Egypt, however, where the low-lying land is annually

submerged beneath the overflowing Nile, the primordial condition of the world was imagined as a great flood. Thus, an unusually high inundation in the historical period was described as covering "this land to its limits. It stretched to the two borders as in the First Time."[1] In the beginning, then, according to this diluvial conception there was water everywhere. As it receded, the top of a high hill emerged, and then the lower features of the landscape, and finally the familiar visage of the countryside. A Sicilian Greek who visited Egypt around 60 B.C., wrote that

> When the world was first being fashioned, the land of Egypt with its productive soil could best of all have effected the origin of mankind. For even now only there may certain living creatures be seen coming into existence in a strange way, whereas no other country generates any such beings. . . . Indeed, in our times when Egypt is inundated, in the lingering waters living creatures are still clearly seen being developed. For, when the river begins to recede and the sun dries out the surface of the mud, living animals are said to take shape, some fully mature but others only halfformed and conjoined with the very earth.[2]

Beneath the solid earth there was thought to stretch a subterranean body of water, over which the sun traveled in its night boat during the hours of darkness after setting in the west. When the sun, regarded as a deity, rises in the east, "he makes the light of day according to their [men's] desire, and he sails by in order to see them."[3] In this version of the Egyptian cosmology, the human race is central, its wishes are paramount, and a view of it is the motive for the sun's ceaseless migration.

By contrast, in that other great Near Eastern hearth at which our so-called Western civilization first lit the flaming torches of its intellect and imagination, the divine Creator exclaims:

> Blood I will mass and cause bones to be.
> I will establish a savage, "man" shall be his name.
> Verily, savage-man I will create.
> He shall be charged with the service of the gods
> That they might be at ease![4]

For Nicholas Copernicus (1473–1543), who, although he was never ordained a priest of the Roman Catholic church, was nonetheless a canon of the Cathedral Chapter of Frombork

(Frauenburg) in the diocese of Varmia (Ermland), the Egyptian and Mesopotamian ideologies faded into the background while the center of the stage was held by the thinking of the ancient Hebrews. In the impressive story of creation at the beginning of the Hebrew Bible (long miscalled the "Old Testament"), the final item in the lengthy catalog of what was created is the human race: "male and female created he them" (Gen. 1:27). For the purpose of making absolutely clear the significance of mankind's terminal place in the list, the unknown author of this section of Genesis has the divine Creator declare explicitly to the first human couple:

> Be fruitful, and multiply, and replenish the earth, and subdue it: and have dominion over the fish of the sea, and over the fowl of the air, and over every living thing that moveth upon the earth. . . . I have given you every herb bearing seed, which is upon the face of all the earth, and every tree, in the which is the fruit of a tree yielding seed; to you it shall be for meat (Gen. 1:28–29).

In this opening story everything was made for the sake of mankind, the lord of creation. But according to the entirely different and considerably earlier tale of creation that starts at verse 4 in Chapter 2 of Genesis, the divine Creator "took the man and put him into the garden of Eden to dress it and keep it" (Gen. 2:15). Hence, the lord of creation is not to loll at his ease, but to exert himself and labor. Like the Egyptians, the Hebrews regarded man as central in the universe and, like the Mesopotamians, they enjoined him to work.

The second and older creation story in Genesis was tacked on to a later account, from which it differs radically. The physical environment of the older tale is an arid region, the primordial substance being dry soil, for the divine Creator "had not caused it to rain upon the earth, and there was not a man to till the ground" (Gen. 2:5). This catalog of things created includes "the earth and the heavens" (Gen. 2:4). But no mention is made of the sea, whereas in the later account, under Mesopotamian influence, "the gathering together of the waters called he Seas" (Gen. 1:10). In the earlier tale, "out of the ground made the Lord God to grow every tree that is pleasant to the sight, and good for food. . . . And out of the ground the Lord God formed every beast of the field, and every fowl of the air" (Gen. 2:9, 19). But no fish were created in this waterless setting, so unlike the later account where "God said, Let the waters bring forth abundantly the moving creature that hath life," and "God created great whales, and every living creature that moveth, which the waters brought forth abundantly, after their kind," and where God also said "Be fruitful, and multiply, and fill the waters in the seas" (Gen. 1:20–22). Fertile soil being unavailable, according to the earlier account, "the Lord God planted a garden eastward in Eden" (Gen. 2:8). How far eastward is made clear when we are informed that

> a river went out of Eden to water the garden; and from thence it was parted, and became into four heads. . . . The third river . . . goeth toward the east of Assyria. And the fourth river is Euphrates" (Gen. 2:10–14).

The realization that Genesis contains not a single creation story but two such tales, one wet and one dry, is an achievement of modern Biblical scholarship. But in Copernicus's time a famous student of the Hebrew Bible and author of an extensive commentary on it, writing about Genesis 2:3, remarked:

> Now Moses proceeds with a clearer description of man, after first repeating what he had said in the first chapter. Although these statements appear to be unnecessary, nevertheless the repetition is not altogether unnecessary, because he wishes to continue his account in a connected manner.[5]

Martin Luther (1483–1546), professor of theology at Wittenberg University, evidently recognized the repetition in Genesis, Chapters 1–2. But, his critical faculties being lulled by the traditional belief that Moses was the sole author of Genesis as well as of the four next books of the Bible, Luther convinced himself that the two discrepant accounts are linked "in a connected manner." As against an allegorizing previous commentator, Luther insisted that "Moses is writing a history and, what is more, one that deals with matters long since past."[6]

Like Luther, Copernicus accepted the Hebrew teaching that a divine Creator made the world and mankind in the beginning.[7] But Copernicus was undoubtedly aware of the divergence between the views of the two groups of thinkers discussed by our Sicilian visitor to Egypt:

> With regard to the first origin of mankind two opinions have arisen among the most highly respected philosophers of nature and histo-

rians. Some of them, suggesting that the universe is uncreated and imperishable, have declared that the race of men has likewise existed from eternity, without [their process of] begetting ever having had a beginning in time; others, however, believing the universe to have been created and to be perishable, have stated that men likewise had their first origin at a definite time.[8]

Taking part in this perennial debate, which still rumbles on in our day, the foremost Christian theologian of the Middle Ages asserted that "there are many imperishable things in the world, for example, the heavenly bodies. . . . Therefore, the world did not begin to be."[9] Nevertheless, Thomas Aquinas (ca. 1225–74) continued, "its eternal duration is not from inner necessity, and therefore cannot be rigorously proved."[10] On the other hand, although the proposition

> that the world had a beginning cannot be demonstrated or known, the statement is plausible. To bear this in mind, moreover, is useful. For, in a matter of faith someone may perhaps attempt a proof by adducing arguments that are not binding. These may furnish an object of ridicule to non-believers, who think that we accept articles of faith on the basis of such arguments.[11]

Without being shackled by prior commitment to a revealed dogma to which he was obliged to adhere, Aquinas's ancient mentor Aristotle, in a treatise on logic, declined to come down categorically on either side of the antinomy, the universe is or is not eternal:

> Moreover, there are problems concerning which there are opposing arguments (for they raise the question whether the matter is or is not so, since there are persuasive considerations on both sides) and with regard to which, since they are basic, we make no decision, in the belief that it is hard to supply the cause, for instance, whether or not the universe is eternal.[12]

On the other hand, in his astronomical work *On the Heavens* (i. 9–ii.1) Aristotle unhesitatingly pronounced the universe to be eternal.

Closer in spirit to Aristotle's straddling of this question in his *Topics* than to his definite stand in the *Heavens* is Buddha's reply to a monk who asked about the length of an aeon:

> Suppose, brother, there were a city ... filled up with mustard-
> seed. ... Therefrom a man were to take out at the end of every
> hundred years a mustard-seed. That great pile of mustard-seed,
> brother, would in this way be sooner done away with and ended
> than an aeon. So long, brother, is an aeon. And of aeons thus long
> more than one has passed, more than a hundred have passed, more
> than a thousand, more than a hundred thousand. How is this?
> Incalculable is the beginning, brother, of this faring on. The earliest
> point is not revealed of the running on, the faring on. ....[13]

Whereas Buddha declared the time of the First Event to be incalculable, Copernicus resolved to "leave the question whether the universe is finite or infinite to be discussed by the natural philosophers." The form given to this resolution by Copernicus is clearly modeled on an expression in the Vulgate,[14] and equally clear is his determination to present himself to his readers as a professional astronomer rather than as a natural philosopher (*physiologus*). Despite this disclaimer, the argument from the end of which the foregoing resolution is quoted had a profound influence on one of his readers, who published the first partial translation of Copernicus's *Revolutions* into English.

Although it has lately become fashionable in some uninformed or poorly informed quarters to look upon Copernicus as oriented philosophically in a neoplatonist or neopythagorean direction, fundamentally he remained faithful to Aristotle. From the thinking of the Stagirite, Copernicus diverged only under compulsion; so to say, that is, only as a result of his recognition of the true cosmic status of the earth. The basic principles of Aristotle's cosmology, which made the earth the stationary center of the universe, could not without alteration be reconciled with Copernicus's earth, which is our earth, which was outside the center and revolved around the center.

In addition, Copernicus's earth also rotated daily around its own axis. Hence, the observed diurnal rotation of the heavens from east to west was changed by Copernicus to a mere appearance, an optical illusion, due to the real eastward rotation of the earth. As the Fifth Assumption of Copernicus's youthful *Commentariolus* declared:

> Whatever motion appears in the firmament arises not from any
> motion of the firmament, but from the earth's motion. The earth

together with its circumjacent elements performs a complete rotation on its fixed poles in a daily motion, while the firmament and highest heaven abide unchanged.[15]

By way of useful contrast to Copernicus's correct account, we may recall the traditional explanation offered by a highly influential philosopher and commentator on Genesis:

The sun is the cause of day and night, revolving above the earth's hemisphere by day, and under the earth by night.[16]

Since Copernicus's "firmament and highest heaven abide unchanged," they possess one of the properties required of an infinite by Aristotle, who said "The infinite cannot possibly be moved at all."[17] Because Aristotle believed that the observed daily rotation of the heavens was the real thing (to quote from one of our contemporary commercial jingles), his stars swung around at a fast pace and therefore lacked the immobility associated with his infinite. Hence his outermost sphere of the stars had to be finite. To the inescapable question, what was farther out than the outermost, Aristotle replied (as Copernicus compressed the Stagirite's answer): "Beyond the heavens there is said to be no body, no space, no void, absolutely nothing."[18] "In that case," runs Copernicus's rejoinder, "it is really astonishing if something can be held in check by nothing." Now comes the master stroke that hit a responsive chord in the sympathetically attuned brain of his first English translator, Thomas Digges (ca. 1546–95). For, Copernicus continues:

If the heavens are infinite, however, and finite at their inner concavity only, there will perhaps be more reason to believe that beyond the heavens there is nothing. For, every single thing, no matter what size it attains, will be inside them, but the heavens will abide motionless.

The inner concavity of Copernicus's heavens was formed by what was then regarded as the sphere of the fixed stars. Like the overwhelming majority of his predecessors, Copernicus may still have believed that the stars, scattered about the heavens at equal distances from their common center within, were all embedded in an invisible, because transparent, spherical surface. Nevertheless, by suggesting "that the Heaven were indeede infinite vpwarde, and onely fynyte downewarde in respecte of his sphericall concavitye" (to quote Digges),[19] Copernicus may have ignited the spark that exploded so brilliantly in the diagram prepared by his Elizabethan translator, who drew the stars dispersed at unequal distances, and devised the following legend for them:

> THIS ORBE OF STARRES FIXED INFINITELY UP EXTENDETH HIT SELF IN ALTITUDE SPHERICALLYE, AND THEREFORE IMMOVABLE.[20]

By the same token, in his text Digges referred to:

> that fixed Orbe garnished with lightes innumerable and reachinge vp in *Sphaericall altitude* [Digges' italics] without ende. Of whiche lightes Celestiall it is to bee thoughte that we onely behoulde sutch as are in the inferioure partes of the same Orbe, and as they are hygher, so seeme they of lesse and lesser quantity, even tyll our sighte beinge not able farder to reache or conceyve, the greatest part rest by reason of their wonderfull distance invisible vnto us.[21]

Thus was Copernicus's universe expanded limitlessly and populated with countless stars, their brightness diminishing as their distance increased beyond the range of unaided human vision.

Yet Digges could write "In the myddest of all is the Sunne,"[22] without stopping to consider that in an infinite universe there can be no "myddest of all." If we now pause to ask ourselves why Copernicus refrained from proclaiming the universe to be infinite, we may well believe that he wished to sidestep Digges's logical fallacy in attributing a center to the infinite. For if the universe's dimensions were immense, even similar to the infinite, then there could be a center. Such a pivot was absolutely necessary for a theoretical and practical astronomer like Copernicus, who still believed that "the motion of the heavenly bodies is ... circular or compounded of circular motions."[23] Since such finite circles require a center, and since any center is incompatible with an infinite extension, Copernicus was logically precluded from proclaiming the universe to be infinite. But he did the next best thing available to him. He put the thesis that "the universe is spherical, immense, and similar to the infinite" among "those propositions of natural philosophy which seemed indispensable as principles and hypotheses."[24]

The necessity of a center had disappeared from the cosmic vision of a thinker who lived a century earlier than Copernicus. Cardinal Nicholas of Cusa (1401–64), who was neither a theoretical nor a practical astronomer, conceived the universe as an

> eternal circle, infinite, without beginning or end, indivisibly one and most capacious. And because this circle is a maximum, its diameter also is a maximum. And since there cannot be more than one maximum, that circle is all the more one, because its diameter is its circumference. But an infinite diameter has an infinite middle. The middle, however, is the center. Clearly, therefore, the center, diameter, and circumference are identical.[25]

The center vanished not only from Cusa's infinite, but also from all finite circles and spheres within Cusa's infinite:

> Neither the earth nor any [other] sphere has a center. For, the center is a point equidistant from the circumference, and no sphere or circle can possibly be so perfect that one even more perfect could not be given. Evidently no center can be given without a truer and more precise [center] being available. Exact equidistance from various [points] cannot be found outside God, because He alone is infinite equality. He, therefore, blessed God, is the center of the universe. He is the center of the earth and of all the spheres and of everything in the universe. At the same time He is the infinite circumference of everything.[26]

Having displaced the earth from the center of the universe and put God there instead, Cusa felt constrained to deny that the earth could be forever at rest:

> The earth, therefore, which cannot be the center, cannot altogether lack motion. For it is also necessary that the earth should move in such a way that it could move infinitely less. Hence, just as the earth is not the universe's center, neither is the sphere of the fixed stars (nor any other sphere) its circumference. Yet it is also true that when we compare the earth with the heaven, the earth seems nearer to the center and the heaven to the circumference.[27]

The foregoing banishment of absolute centrality from mathematical and physical figures of less than divine rank, and the ascription of motion to the earth, were incorporated by Cusa in his treatise *On Learned Ignorance.* A few years after completing this paradoxical work in 1440, Cusa drew the appropriate cosmological conclusion from the second of the foregoing theses: "the earth cannot be stationary, but moves like the other heavenly bodies."[28] In thus attributing to the earth these two properties, mobility and the status of a heavenly body, Cusa anticipated Copernicus by nearly a hundred years. Some scholars have claimed that Copernicus inherited these two basic propositions of his astronomy and cosmology directly from Cusa. But let us bear in mind that Cusa's remarkable enunciation of these two valuable principles remained unknown not only to Copernicus but also to everybody else for four hundred years, since Cusa's manuscript note containing them was not discovered until 1843, exactly three centuries after the death of Copernicus.

Those who profess to see an immediate link between Cusa and Copernicus may agree to set aside Cusa's unpublished manuscript note, but they may still insist on emphasizing Cusa's assertion in his *Learned Ignorance* that the earth "cannot altogether lack motion." This statement was printed twice in Copernicus's lifetime, in the editions of Cusa's *Works* which were published at Strasbourg in 1488 and at Paris in 1514. Had Copernicus known that motion was ascribed to the earth by so renowned a cardinal, would he not have leaped with joy at the opportunity of introducing Cusa's name honorifically in his *Revolutions?* As a Roman Catholic canon, Copernicus dedicated this work to the reigning pope; he embellished it with an encouraging letter from another cardinal; and he avowed the valuable stimulation of a bishop, who was his close personal friend. Let us never forget that Copernicus went to a great deal of trouble to inform his readers that he laid no claim to originality in classifying the earth as a moving body. Thus, in his Dedication Copernicus candidly declared:

> I began to be annoyed that the movements of the world machine, created for our sake by the best and most systematic Artisan of all, were not understood with greater certainty by the philosophers, who otherwise examined so precisely the most insignificant trifles of this world. For this reason I undertook the task of rereading the works of all the philosophers which I could obtain to learn whether anyone had ever proposed other motions of the universe's spheres than those expounded by the teachers of astronomy in the schools. And in fact first I found in Cicero that Hicetas supposed the earth to move. Later I also discovered in Plutarch that certain others were of this

opinion. I have decided to set his words down here, so that they may be available to everybody:

> "Some think that the earth remains at rest. But Philolaus the Pythagorean believes that, like the sun and moon, it revolves around the fire in an oblique circle. Heraclides of Pontus and Ecphantus the Pythagorean make the earth move, not in a progressive motion, but like a wheel in a rotation from west to east about its own center."

This magnificent Dedication was composed in June, 1542, as Copernicus's last addition to the *Revolutions,* the printing of which had been begun in the preceding month. In attributing the concept of the earth's axial rotation to Heraclides and Ecphantus, the Dedication confirmed by a quotation from pseudo-Plutarch what Copernicus had already said in the text of the *Revolutions* (I, 5) about "Heraclides and Ecphantus, the Pythagoreans, and Hicetas of Syracuse," who "rotated the earth in the middle of the universe, for they ascribed the setting of the stars to the earth's interposition, and their rising to its withdrawal." In that same chapter Copernicus had written:

> That the earth rotates, that it also travels with several motions, and that it is one of the heavenly bodies are said to have been the opinions of Philolaus the Pythagorean.

Moreover, in the deleted passage, cited above in note 24, Copernicus coupled with the name of Philolaus that of Aristarchus of Samos, justly celebrated as the "Copernicus of antiquity."

The name of Aristarchus may well have been deleted because he would have been indicted for impiety had the recommendation of a leading philosopher been followed up by appropriate legal action in a Greek court. In any case, Philolaus, Heraclides, Ecphantus, and Hicetas were pagans, and so were the writers who preserved their opinions, Cicero and Plutarch. Despite Copernicus's humanistic regard for classical antiquity, he would have attached far greater importance to the authority of a Roman Catholic cardinal, had he been acquainted with Book II, Chapter 11, of Cusa's *Learned Ignorance.*

On the other hand, had Copernicus been familiar with Cusa's unpublished manuscript note, as a professional astronomer Copernicus would surely have been startled by the cardinal's reasoning. From the doubly impressive statement that "the earth cannot be stationary, but moves like the other heavenly bodies," Cusa concluded that

> Therefore the earth rotates around the celestial poles, as Pythagoras says, approximately once in a day and night, but the eighth sphere [the sphere of the stars, rotates] twice, and the sun a little less than twice, in a day and night.

Let us charitably ignore Cusa's reference to what "Pythagoras says," remembering Pythagoras's avoidance of public utterance and written statement. Let us instead focus our attention on the difference in the rates of rotation. For the unsophisticated observer, who takes the apparent daily rotation of the heavens at face value, this celestial gyration once in a day and night implies the motionlessness of the earth. On the other hand, for Hicetas, Heraclides, Ecphantus, Philolaus, Aristarchus the Copernicus of antiquity, and Copernicus the modern Aristarchus, the real diurnal rotation of the earth deprived the heavens of that motion. Of course, the same time difference would be preserved by Cusa's starry sphere rotating twice a day as against the earth's once a day. Moreover, the slight, but important, excess in the length of the solar day over the sidereal day was indicated by Cusa's having the sun rotate "a little less than twice in a day and night." Nevertheless, Cusa's bizarre suggestion of a once-daily terrestrial rotation and a twice-daily celestial rotation fell on deaf ears and was never repeated.

On the other hand, Cusa's revival of the infinite universe, proclaimed in antiquity by the Roman philosophical poet Lucretius, evoked a characteristically enthusiastic response from that gifted and tragic seer Giordano Bruno (1548–1600). We have lately been reminded that Bruno was neither a mathematician nor a scientist.[29] But he was the first writer in the sixteenth century to pen a paean in praise of Copernicus, after which he exclaimed:

> It is wonderful, O Copernicus, that from the immense darkness of our age, when the entire light of philosophy lies extinguished, as well as the light of the other subjects which depend on it, you could emerge to enunciate somewhat more boldly what had been said in more subdued tones by Nicholas of Cusa in his book *On Learned Ignorance* in the age immediately preceding [your own]. For you relied on the defence that if your correct belief were not strong

enough in itself to be accepted, at least it would be admitted in the guise of a hypothesis in view of the greater convenience which it permits in astronomical computations. Here I shall quote the words which your divine genius inspired in you.[30]

And then Bruno proceeded to repeat long sections of Copernicus's Dedication and his cosmological discussion in *Revolutions,* I, 11. This massive tribute to Copernicus was included by Bruno in his *De immenso.* His last and greatest work, the *De immenso,* was published shortly before Bruno was incarcerated in the dungeons of the Inquisition for eight years and then, together with his books, burned at the stake, partly for advocating what would nowadays be called ecumenism. When Bruno's *De immenso* was published in 1591 at Frankfurt am Main, not very far away at Tübingen University the Copernican cosmology was being explained to an astronomy class. The most talented pupil present was Johannes Kepler, who later made the first substantial improvements in the Copernican cosmology. Although Kepler did not know Italian and therefore was acquainted only at second hand with the dialogues written in the vernacular by Bruno, that unfortunate man's *De immenso* posed no language barrier to Kepler, who was thoroughly familiar with it.[31] Through Bruno, Kepler learned about Cusa, whose writings he did not study carefully. In particular, Kepler characterized the Cusan-Brunonian doctrine of the infinite as "that dreadful philosophy."[32]

Kepler anticipated Sigmund Freud's evaluation of Copernicus's impact on man's conception of his place in the world. According to the founder of psychoanalysis,

It is to the excessive narcissism of primitive man that we ascribe his belief in the omnipotence of his thoughts and his consequent attempts to influence the course of events in the external world by the technique of magic. After this introduction I propose to describe how the universal narcissism of men, their self-love, has up to the present suffered three severe blows from the researches of science. (*a*) In the early stages of his researches, man believed at first that his dwelling-place, the earth, was the stationary centre of the universe, with the sun, moon and planets circling round it. In this he was naively following the dictates of his sense-perceptions, for he felt no movement of the earth, and wherever he had an unimpeded view he found himself in the centre of a circle that enclosed the external world. The central position of the earth, moreover, was a token to him of the dominating part played by it in the universe and appeared to fit in very well with his inclination to regard himself as lord of the world. The destruction of this narcissistic illusion is associated in our minds with the name and work of Copernicus in the sixteenth century. . . . When this discovery achieved general recognition, the self-love of mankind suffered its first blow, the cosmological one.[33]

We have no time now to ponder Freud's second and third severe blows to human self-esteem (Charles Darwin's *Descent of Man* from animal forebears, and Freud's own claim that "the ego is not master in its own house"). But let us listen to Kepler's answer to those who, like Freud, thought that Copernicus demeaned man by displacing him and his habitat from the center of the universe:

> In the interests of that contemplation for which man was created, and adorned and equipped with eyes, he could not remain at rest in the center. On the contrary, he must make an annual journey on this boat, which is our earth, to perform his observations. Thus it is apparent that it was not proper for man, the inhabitant of this universe and its destined observer, to live in its inwards as though he were in a sealed room. Under those conditions he would never have succeeded in contemplating the heavenly bodies, which are so remote. On the contrary, by the annual revolution of the earth, his homestead, he is whirled about and transported in this most ample edifice, so that he can examine and with utmost accuracy measure the individual members of the house.[34]

In conclusion, we may today, as we gratefully celebrate the five-hundredth anniversary of the birth of that profound and daring thinker Nicholas Copernicus, agree with his great follower Johannes Kepler that our conception of man's place in the world—its unique observer from a noncentral moving platform—was bequeathed to us as an imperishable legacy by Copernicus.

NOTES

1. S. G. F. Brandon, *Creation Legends of the Ancient Near East* (London, 1963), p. 16.

2. Diodorus of Sicily, *Library of History* i. 10, 3–7.

3. John A. Wilson, trans., "The Instruction for King Meri-ka-re," in *Ancient Near Eastern Texts*, ed. James B. Pritchard, 2d ed. (Princeton: Princeton University Press, 1955), p. 417b.

4. E. A. Speiser, trans., "Tablet VI of the Creation Epic" (Enuma elish), in *Ancient Near Eastern Texts*, p. 68a.

5. Jaroslav Pelikan, ed., *Luther's Works*, vol. I Lectures on Genesis, Chapters 1–5 (St. Louis: Concordia, 1958), p. 82.

6. *Ibid.*, p. 90.

7. Copernicus, *Revolutions*, I, 10, last sentence.

8. Diodorus of Sicily, *Library of History* i. 6, 3.

9. Thomas Aquinas, *Summa Theologiae*, ed. T. Gilby, VIII (New York and London, 1967), p. 66 (Ia, 46, 1, 2).

10. *Ibid.*, p. 70.

11. *Ibid.*, p. 80.

12. Aristotle, *Topics* i. 11, 104b:12–16.

13. *The Book of the Kindred Sayings* [Sanyutta-nikaya], Part II, Nidana-Vagga (Chap. 15, 1, 6) Pali Text Society, Translation Series, No. 10 (London, n.d.), p. 122, No. 6.

14. Copernicus, *Revolutions*, I, 8: *Sive ... finitus sit mundus ... disputationi physiologorum dimittamus;* Eccles. iii. 11: *mundum tradidit disputationi eorum.*

15. Edward Rosen, *Three Copernican Treatises* 3d ed. (New York: Octagon, 1971), p. 58.

16. Philo, *Questions and Answers on Genesis* (i. 84), Loeb Classical Library, Philo, Supplement I, trans. Ralph Marcus (Cambridge: Harvard University Press, 1961), p. 53.

17. Aristotle, *Heavens* i. 7, 274b: 29–30.

18. Copernicus, *Revolutions*, I, 8, compressing Aristotle *Heavens* i. 9.

19. Thomas Digges, *A Perfit Description of the Caelestiall Orbes* (London, 1576), reprinted in *Huntington Library Bulletin*, no. 5, 1934, p. 91.

20. *Ibid.*, p. 78.

21. *Ibid.*, pp. 88–89.

22. *Ibid.*, p. 87.

23. Copernicus, *Revolutions*, I, 4.

24. Nicholas Copernicus, *Complete Works*, I (London, Warsaw, and Cracow: Macmillan, 1972), *The Manuscript of Nicholas Copernicus' On the Revolutions, Facsimile,* fol. 13^r, lines 1–3.

25. Nicolaus de Cusa, *Opera omnia,* I, eds. Ernst Hoffmann and Raymond Klibansky (Leipzig, 1932), 43:1–7; *De docta ignorantia,* I, 21.

26. *Ibid.*, I, 101:3–11; *De docta ignorantia,* II, 11.

27. *Ibid.*, I, 100:15–20; *De docta ignorantia,* II, 11.
28. *Sitzungsberichte der Heidelberger Akademie der Wissenschaften,* philosophisch-historische Klasse, XX (1929–30), 3. Abhandlung, p. 44, lines 7–8.
29. Lawrence S. Lerner and Edward A. Gosselin, "Was Giordano Bruno a Scientist?" *American Journal of Physics* XLI (1973): 24–38.
30. Jordani Bruni nolani, *Opera latine conscripta* (reprint ed., Stutt-gart-Bad Cannstatt: Frommann-Holzboog, 1961–62), I, pt. 1, 381–82 (*De immenso,* III, 9). Here Bruno has forgotten his own extremely valuable contribution to the proper understanding of Copernicus. For in an earlier work Bruno had correctly pointed out, and he was the first to do so, that the greater computational convenience of Copernicus's hypothesis was a claim advanced not by Copernicus but by "a certain preliminary Address, stuck in by an ignorant and insolent jackass." See Edward Rosen, "Was Copernicus a Hermetist?" *Minnesota Studies in the Philosophy of Science* V (1970): 169.
31. Edward Rosen, *Kepler's Conversation with Galileo's Sidereal Messenger* (New York: Johnson, 1965), p. 45, n. 402.
32. *Ibid.*, p. 37.
33. Sigmund Freud, *Complete Psychological Works,* Standard Ed., ed. James Strachey, vol. XVII (London: Hogarth, 1955; reprint ed., 1968), pp. 139–40.
34. Rosen, *Kepler's Conversation,* pp. 45, 148.

COMMENTARY

ROBERT S. WESTMAN

PROFESSOR ROSEN has written an interesting but puzzling essay. In the first place, we might reasonably expect "The Impact of Copernicus" to refer primarily to the writings of thinkers who lived *after* the death of the great astronomer. About 75 percent of his paper, however, deals with those who lived considerably before Copernicus. Second, it is also reasonable to anticipate that "Man's Conception of His Place in the World" would lead us into a discussion of such topics as the nature of man and his potentialities; his place in the hierarchy of being, that is, with respect to God and the angels above and the lesser animals and plants below; and Renaissance views of human life and the immortality of the soul. Such, at least, is the sense in which the phrase is normally used by historians of Renaissance thought.[1] Professor Rosen, however, is not so much interested in telling us how *post*-Copernican thinkers coped with the moral and theological issues raised by the new theory as he is with *pre*-Copernican cosmogonies and discussions of the magnitude of thc universe. A title more accurately descriptive of the contents of his paper, therefore, might be: "Copernicus and Ancient and Medieval Views of the Creation and the Infinitude of the Universe With Occasional References to Man's Conception of His Place in the World."

Beyond the matter of entitling the essay, it is never quite clear how its various themes are related to one another. Beginning with "our earliest ancestors" and their anthropocentric musings about the creation, a passing reference is made to Copernicus's place in the Hebraic tradition: "Like Luther, [Why Luther?] Copernicus accepted the Hebrew teaching that a divine Creator made the world and mankind in the beginning." Next, we are taken, leapfrog fashion, into a string of passages compiled from an unlikely and unexplained assortment of writers, including Diodorus of Sicily, Aristotle, Thomas Aquinas, and the Buddha, who engage (each other?) in what Rosen calls "that perennial debate, which still rumbles on in our day" as to whether the creation of the world can be proved by reason alone or whether this is solely a matter of faith, as Aquinas ultimately maintains. Rosen's discussion of *when* the world was created, and if it was created at all, next gives way to a confusion between the length of time since the creation (calculable or incalculable) and the world's extension in space (finite or infinite). This permits Rosen to link Copernicus and Buddha as though they were addressing themselves to the same topic: "Whereas Buddha declared the time of the First Event to be incalculable, Copernicus resolved 'to leave the question whether the universe is finite or infinite to be discussed by the natural philosophers.' " And now, having accomplished this transition, Rosen arrives at the question which seems to concern him most of all, namely, the infinitude of the universe.

Rosen's primary aim is to show that Copernicus did not

wish to relinquish the conception of a closed, finite universe. In this regard, three reasons are advanced. First, Rosen wishes to argue that Copernicus was still an Aristotelian in most of his cosmological assumptions including his presuppositions about the status of the region beyond the heavens. In contemplating Aristotle's claim that the extra-cosmic region has no existence, however, Copernicus is disturbed that something which moves can be bounded by that which possesses no being. He then suggests the idea that "If the heavens are infinite, however, and finite at their inner concavity only, there will perhaps be more reason to believe that beyond the heavens there is nothing." Rosen believes that this passage was the basis for Thomas Digges's assertion "that the Heaven were indeede infinite upwarde, and onely fynyte downwarde in respecte of his sphericall concavitye"[2] —an interesting and possible inference. But Rosen now points to a conceptual difficulty in Digges's translation. For Copernicus's "In medio vero omnium residet Sol,"[3] Digges faithfully translates "In the myddest of all is the Sunne" and Rosen, who is anticipated by Thomas Kuhn in this criticism of Digges, correctly argues that in an infinite universe there can be no single point which alone is equidistant from all points on the periphery.[4] Rosen, then, is suggesting that Copernicus may have been aware of "Digges's logical fallacy" and, therefore, decided to avoid the issue by leaving it to the natural philosophers.

Linked with Digges's fallacy, Rosen finds another difficulty which he believes would have deterred Copernicus from entertaining the notion of an infinite universe. Baldly stated, the argument is that there can be no astronomy of circles in an infinite universe because such a universe has no absolute center. Rosen writes: "Since . . . finite circles require a center, and since any center is incompatible with an infinite extension, Copernicus was logically precluded from proclaiming the universe to be infinite." Now this argument is specious for it assumes that there can be one and only one viable reference frame, namely, a spherical universe with respect to whose center all other motions must be related. This is certainly an understandable assumption in view of Copernicus's pronounced Aristotelian leanings but it overlooks the fact that a heliocentric system, as a local system, can be defined in infinite space although the center of the universe cannot be known. Even Digges's statement is correct if by "myddest of all" he means in the "myddest of the planetes."[5] Copernicus, then, was not prevented on *logical* grounds from considering the possibility of an infinite universe. He simply may not have been aware that his theory was not incompatible with such a conception. More importantly, however, Rosen has given no evidence to show that Copernicus *needed* to worry about this problem. It was not until much later, in the context of discussions of rectilinear inertial motion that the magnitude of the universe became an acute issue. It was sufficient for Copernicus to write that " . . . *by the judgment of the senses* the Earth is to the heavens as a point to a body and as a finite to an infinite an infinite magnitude" (my italics).[6] And, furthermore, there is no reason to believe that Copernicus shared the kinds of epistemological and theological concerns which had provoked the deep interest in infinity found in Nicholas of Cusa and Giordano Bruno.

At the end of his essay, Professor Rosen turns his attention briefly to the impact of Copernicus on Freud's conception of man's place in the world. It is entirely fitting and admirable that he should have chosen to quote Freud although disappointing that he should not have taken the time to interpret the significance of the selected passage. If Copernicus feared to publish his work, perhaps because he was excessively concerned with the smallest details of his theory, it was Freud who looked upon fear as a principal cause of distortions and misery in human relationships. Neurosis was seen as a failure to recognize the differences between present personal relationships and past ones, between inner fantasies and outer realities. While Copernicus had recognized that a geocentric perspective provides only a limited way of understanding the motions of other bodies, Freud realized that egocentric assumptions, which are natural to children and whose residues persist into adulthood, produce a narrow and often distorted view of other persons. He hoped that the "universal narcissism" of human beings, which intrudes into all areas of personal experience, might be replaced by a more "ex-centric" viewpoint in which people might recognize the limitations of their presuppositions about the world and their powers to influence it. Freud's attack on human egocentricity, therefore, did not demean humanity but only the illusions upon which it stands.

This is surely not the place to write the full history of Copernicus's influence on changes in man's self-conception. It is clear, however, that no deeper impact would be felt while the theory was interpreted merely as a calculational tool. But the earliest reactions to the Copernican theory, particularly by Philip Melanchthon and his followers at the University of Wittenberg, tended to emphasize precisely that property. Moreover, Melanchthon, following Luther, had systematically drawn a clear distinction between the world of grace and the world of nature. Displacing the earth from the center of the universe, then, did not necessarily have moral implications because while the habitat of man might be seen from a different viewpoint, his values were not.

It is ironic that the achievement of Copernicus did affect the human self-image, after all, because Copernicus's purpose was really quite limited. Like many revolutionaries, he did not set out to overthrow the old world but merely to reform it; and like many innovators, he did not live to witness the full implications of his creative work. By dying in 1543, he escaped not only criticism but also praise; like a great piece of art or music, his theory was not at once appreciated by many for its profounder significance. The deeper meaning of Copernicus's work for the transformation in man's self-image lay not in the astronomical innovation itself but in its fulfillment of the old Platonic-Pythagorean hope: the discovery of a rational, systematic order in the universe. This was an order that could be understood even though man was not at the center of the world. Kepler, in a characteristic piece of Copernican teleology, makes a virtue of the moving habitat of man: whirling about in his "ample edifice," the earth, man can most easily examine and measure "the individual members of the house." In the end, the impact of Copernicus greatly transcended its original intention by becoming a symbol, a metaphor for transformations in human consciousness.

NOTES

1. See, for example, Paul O. Kristeller, *Renaissance Concepts of Man* (New York, 1972), p. 19.
2. Actually, Digges's translation follows Copernicus's conditional con-

struction quite closely: "Yet yf wee would thus confesse that the Heaven were indeede infinite upwarde. . . ," Thomas Digges, *A Perfit Description of the Caelestiall Orbes* (London, 1576), reprinted in Francis R. Johnson and Sanford V. Larkey, "Thomas Digges, the Copernican System, and the Idea of the Infinity of the Universe in 1576," *The Huntington Library Bulletin,* no. 5 (April 1934), p. 91.

3. Copernicus, *De revolutionibus orbium coelestium,* bk. I, chap. 10.
4. Cf. Thomas S. Kuhn, *The Copernican Revolution* (New York, 1959), pp. 233–35.
5. There is no evidence that Digges possessed full awareness of the implications of his position. It would be interesting to know if he had acquaintance with any of the medieval treatments of the problem of the extra-cosmic void.
6. Copernicus, *De revolutionibus,* bk. I, chap. 6.

IV   The Assimilation of Science into Our Ways of Thinking and Living

WILHELMINA IWANOWSKA

IN what follows, the term science is used to refer to the mathematical and physical sciences, namely, mathematics, astronomy, physics, and chemistry, and their penetration into the earth sciences and biology. The development of modern science started with Nicolaus Copernicus's *De revolutionibus orbium coelestium* (1543), which initiated what is called the scientific revolution. On the basis of the newly founded science, technology began its development toward the present technological revolution. Science enlarged man's knowledge and understanding; developing technology changed the conditions of his individual and social life.

Science before Copernicus

Compared to its present state, science was almost nothing before Copernicus. Aristotelian physics was hardly a science. Mathematics was in a relatively better position; it consisted of arithmetic and Euclidean geometry, which originated in Antiquity and developed through the Middle Ages, mostly in Arabic countries. Astronomy reached a high point in a work of Ptolemy of Alexandria (second century A.D.) entitled, in Arabic translation, *Almagest.* This work presented a kinematic model of the Universe with a motionless earth in the center and all other bodies revolving around it in a daily motion. Besides the sun, the moon and the planets moved around the earth in orbits that were composed of circles: a main one called the deferent and a secondary one called the epicycle. The center of the epicycle lay on the primary circle

73

with the planet, sun, or moon lying on the epicycle. By properly adjusting the diameters of the circles, the periods of revolution, and the velocities of revolution, which were not uniform, Ptolemy succeeded in giving a model that represented the apparent motions of all the then known celestial bodies in a nearly satisfactory way, except that their distances clearly disagreed with observations. The most obvious discrepancy was that of the moon. According to the Ptolemaic model, the moon should be half as far from the earth at the quarters as at the times of the new moon and the full moon. This obviously does not happen, since the angular diameter of the moon appears to be constant. As is known, this discrepancy bothered Copernicus very much. It led him to observe, while at Bologna and with the help of Maria Domenico Novara, the occultation of the star Alpha Tauri by the moon and to state that the distance of the moon does not change appreciably with the phases.

It seems strange today that this and other discrepancies between the Ptolemaic model and the observed facts were disregarded by Ptolemy himself and by his successors for nearly fourteen centuries. This suggests that during these centuries there was no deep interest in the external world, in how it is really built and in how it moves. A formal scheme that could predict the positions of celestial bodies was sufficient. Astronomy did not strive for knowledge of the real world; it was a tool for orienting man in space and for measuring time. Moreover, it was filled with myths and legends and used for making horoscopes.

The Work of Copernicus

The inconsistencies in the Ptolemaic model led Copernicus to look for other systems. He discovered in ancient writings reports of the Pythagoreans's belief that the earth and the planets revolve around the sun. However, none of the Ancients had tried to support this idea with any proof. On the basis of some preliminary considerations regarding the motions of the planets as projected from the observations of his predecessors, Copernicus saw that the heliocentric model of the Universe had promise. Sometime before 1515 he wrote and distributed to his friends a handwritten booklet entitled *"Commentariolus."* In this work he presented, still as a hypothesis, a first sketch of a heliocentric system. It took thirty additional years for Copernicus to test and verify this hypothesis, adding his own observations made with very primitive instruments that he had constructed by himself. Telescopes were not yet known; the astronomical instruments of that time consisted of wooden or brass bars and circles with angular divisions on them. The instruments enabled the observer to measure angular distances between stars and planets and the elevations of these bodies above the horizon. The results of these observations and of his very laborious mathematical reasoning and calculations were incorporated in the *De revolutionibus orbium coelestium (On the Revolutions of the Celestial Bodies),* which was printed in 1543, the year of the death of Copernicus.

The *De revolutionibus* is preceded by a remarkable letter, which dedicates the work to Pope Paul III, and is divided into six chapters or books. The first book presents the assumptions of the new theory: The earth is not the quiescent center of the Universe. The earth has a threefold motion: a daily revolution around its axis, a yearly motion around the sun, and a precessional motion that produces a slow change of the direction of the earth's axis. The sun is the center of the system; the planets, including the earth, revolve around the sun in an order corresponding to the lengths of the periods of the revolutions. The orbits of the planets are circular and only small epicycles are necessary to account for the slight lack of uniformity in the planetary revolutions. The larger epicycles of Ptolemy are no longer necessary since the loops observed in the planetary motions are accounted for by the motion of the observer circling with the earth around the sun. The moon is the only body that orbits around the earth. Copernicus supports these assumptions for two reasons, the greater physical probability of his model and its overall mathematical simplicity.

To the descriptive presentation of his basic assumptions in the first book of the *De revolutionibus,* Copernicus adds some interesting new ideas about gravity as a common property of all celestial bodies that keeps them in their spherical volumes and about the enormous distances of stars in comparison to the dimensions of the planetary system. The latter suggestion leaves the possibility of a finite or infinite universe as an open question.

Whereas the first book of the *De revolutionibus* is easy to read and can be recommended to anyone, the next five books contain laborious mathematical and mostly geometric deductions.

On the basis of his assumption that the earth has a threefold motion, Copernicus meticulously calculates the apparent motions of the sun, the moon, and the planets as they appear to an observer placed on the moving earth. He then compares these theoretical predictions with the observed motions and finds substantial agreement. In the process of pursuing this work, Copernicus also gives a model for scientific method; an idea, a working hypothesis, is proved by means of strict mathematical deduction. The hypothesis leads to consequences, which, in their turn, are confronted with observations or experiments.

As is well known, the reactions to the Copernican theory were very different. Practical applications of the heliocentric theory in the form of tables of planetary positions were accepted with great enthusiasm and appreciation, since these tables conformed to the observations better than any others before. But the theory itself was opposed and rejected by most people, both common and professional, by both churchmen and university scholars, because it seemed strange, complicated, and unacceptable to the senses and to the anthropocentric nature of man's world. This attitude can be traced even to the present day. In many celebrations of the Copernican anniversary he is praised most for his nonastronomical activity, as an administrator, economist, physician, and even poet, being called a humanist and a great man of the Renaissance, though in fact he overthrew the astronomy of the ancients and opposed humanocentrism.

Copernicus's revolutionary work on the structure of the universe, though opposed by the majority of people and placed by the Holy Inquisition in 1616 on the list of prohibited books, where it remained for more than two hundred years, was enthusiastically accepted and promoted by those who were able to understand it properly. A sequence of brilliant names follows Copernicus over the course of the seventeenth century: Galileo Galilei, Johannes Kepler, and Isaac Newton. These men developed and improved, step by step, the knowledge of the real universe. Galileo was the first to use a lens telescope for observations of the sky and to discover support for the Copernican theory in the motions and phases of the planets and their satellites. Kepler was able to discover his kinematic laws of planetary motion from the numerous planetary observations made by Tycho Brahe, only by assuming the heliocentric system. Newton, in his main work, *Philosophiae naturalis principia mathematica* (1687), established the basic principles of dynamics and the law of gravitation, again, from the motions of the planets in a heliocentric system. Thereafter, the laws of Kepler became a direct consequence of Newtonian dynamics.

In this way mechanics, the first branch of physics, was established in the second half of the seventeenth century, building, as it did, on the work of Copernicus published 144 years earlier. In order to apply Newtonian principles of mechanics to the motions of terrestrial and celestial bodies, better mathematical methods were needed. Newton himself applied differential and integral calculus, which were invented independently by Leibnitz, to such problems. Earlier than this Descartes developed analytical geometry. Thus astronomy, physics, and mathematics started their avalanchelike development, which lasts even to the present.

Newtonian mechanics was successfully applied to the precise calculation of the orbits of planets, satellites—natural and now artificial—comets, and asteriods. But when the mutual attractions of these bodies were taken into account, their orbits were not strictly Keplerian ellipses; they were more or less perturbed. Such perturbations in the case of Uranus led to the prediction of two new planets, Neptune and Pluto, even before they were observed. Newtonian mechanics was also successfully applied to the calculation of the orbits of double and multiple stars, the orbits of stars in the Galaxy, and the motions of double and multiple galaxies. It worked well until very big masses, densities, and velocities were met. Then a more sophisticated treatment was necessary and it was Albert Einstein, at the beginning of the twentieth century, who advanced the theories of special and general relativity to deal with these cases, theories which are still being tested and checked in astronomy and at the same time applied to many problems of macro- and microphysics.

The brilliant successes attained in mechanics encouraged physicists to push their investigations and experiments in other directions. As a result, new branches of physics were created: optics; thermodynamics; magnetism; electricity; atomic, molecular, and nuclear physics; quantum mechanics; and chemistry. In each of these endeavors a common method was applied, a scientific method that joined mathematical calculation to experiment and observation. In turn, the new branches of physics were applied to astronomical investigations, again giving rise to new methods, such as are presently used in astrophysics, radio astronomy and space research. Powerful new instruments have been developed: telescopes with spectrographs, radiotelescopes and interferometers, and automatic devices for the detection and measurement of electromagnetic and corpuscular radiation of all wavelengths and all energy ranges. Computers have been applied to the control and processing of measurements and to solving numerical problems. The field of investigation in astronomy has increased enormously; astronomers study now not only the motions of planets, but also their physical states, the structure and evolution of stars and the interstellar medium, the chemical composition and evolution of cosmic matter, the energetic particles in interstellar space and the superdense matter in white dwarfs, neutron stars and the predicted black holes, the processes of nucleosynthesis in stellar interiors, of gravitational collapse and gravitational radiation, and the problem of life in the universe.

This explosion of science has had a profound impact on the human mind by widening the boundaries of the external world and opening new perspectives. For those involved in research this presents an exciting challenge. It demands an exhaustive intellectual effort like that embodied in the old yellow pages of the manuscript of the *De revolutionibus*. Unlike Copernicus, the present-day researcher does not work alone. More commonly he consults and discusses the results of his experiments and observations, of his ideas and reasoning, with his colleagues. Yet, he must also have hours of silence and solitude for his work. Therefore, he likes to work at night, even when he is not an astronomer. If he smokes and drinks coffee at this work, he ruins his health.

The growth of science leads to the need for progressive specialization and creates problems in the storage of data, problems that are now partially solved with the aid of punch cards and magnetic tapes read by a computer. This growth also raises the question: Where is science going? Is it approaching that ideal state where all problems will be solved and everything will be known? If such a state is ever possible, we are still far from reaching it because up to now each discovery and each solution for one problem creates new problems that need to be solved. Or are we approaching a point where science stagnates, a point where no further steps will be possible because the increasing difficulty of problems and the burden of growing information will make them too heavy to be lifted by human brains? It is difficult to predict the future of science. At present we are travelling between these two asymptotes.

The birth and growth of science also enlightens the minds of ordinary people at every level of their education. Beginning in the elementary schools and progressing to the secondary schools and universities, instruction in mathematics and the natural sciences involves more and more of the time and effort of pupils, offering satisfaction to those who are capable and frustration to others. There is a twofold division of human intellects, which are more or less capable of mathematical thinking. Copernicus was aware of this when he dedicated his work to mathematicians and directed some rather harsh words at those who, being unable to understand it and being presumptuous, will condemn him and his work. This duality leads to an early specialization in education and to the separation of mathematical and humanistic types of secondary schools. But even those who are not very strong in mathematics can enjoy the natural sciences when it is presented to them in a descriptive manner. Many examples of this can be seen among amateur astronomers.

To know more about our environment is the natural right and natural need of a human being. The right to such knowledge is as essential as the right to work, the right to rest, and the right to life itself. No limits to this knowledge ought to be set; galaxies and quasars are equally as important as our terrestrial globe, as our country, and as our own body and soul. The right to research and to knowledge cannot be questioned or limited; it is a natural human right. Limitation of knowledge is not the question that faces us now. What we now must question is: how should research and education be organized? What is the most efficient and just way? (Long ago such organization ceased to be an individual matter.) How should massive resources be allocated for scientific research and education within the budgets of the community, of the state, and of the world? How should these resources be divided between different topics and different geographical areas? What projects should get the priority? What systems and programs of education must be followed? These are not easy questions to answer. They outline difficult tasks that will challenge social scientists and economists, governmental and international organizations, especially when asked within the context of a rapidly growing population, the ever-expanding scope of science, and the limited capacity of human intelligence. I don't and would not add at this point, *limited resources.* The amount of resources used for strictly scientific endeavors is still relatively very low, especially when compared to the resources wasted in a destructive way for making and supporting wars, for maintaining armaments, for producing millions of useless things, if not harmful as alcohol and narcotics, for making things of all kinds.

Up to now I have spoken about what is called "pure science." Of no less importance to human life and thought are the applications of science that result in technological progress. Technology is also very young, although we can find imposing technical achievements in the architecture and building of ancient times. Egyptian pyramids; the temples, palaces, bridges and aqueducts of ancient Asian, European and American civilizations; medieval cathedrals and Renaissance buildings; all were produced before Copernicus and independently of modern science. The art of building is very old indeed, older than Newton's statics and mechanics. The ways to plan and erect an edifice were worked out through practice and experimentation and passed in a long and sometimes unbroken tradition from one generation to another. Magnificent buildings were erected, even though a theory based on the laws of physics was missing. Building materials were used in many ways; the tools and machinery were very primitive with labor being performed mainly by slaves taken from conquered countries.

Apart from these achievements, the modern technical revolution has its roots in the development of the modern science that begins with the work of Copernicus. Technology is based on the laws and principles of physics. Modern technological achievements like the vapor engine, which was invented by James Watt at the beginning of the eighteenth century; like the automobile and the airplane, which were invented quite recently; like the electrical devices that were invented by Edison at the end of the nineteenth century; and like today's electronic equipment, which has progressed rapidly from electronic tubes to transistors and sealed circuits, all are direct descendants of science.

It is equally as true that modern technical progress has changed our living conditions in essential ways. In many respects we feel that these changes have been a blessing; our lives are longer, easier, and more comfortable than ever before. The hard human labor of agriculture and industry is being replaced by machines. Yet, there are some aspects of technical progress that are becoming increasingly harmful to human health and happiness. The damage to the natural environment that results from uncontrolled and rapidly expanding industry has become one of the most alarming problems of today. We have to remember, however, that the damage to the natural environment is caused primarily by the growth of the earth's population. Science and technology can only help to find food and resources to maintain and support this growth.

The most serious problem connected with the technical revolution is the application of modern technology to armaments that threaten large areas of our globe with the total destruction of life at the push of a button. We must find a way out of this danger; we must find a way to stop arming, to stop the production of armaments! But we cannot stop research in nuclear physics; it would be absurd!

Science must grow. The technological and industrial applications of science must be placed under efficient controls. It is incumbent upon the political, economic, and social sciences to keep pace with the progress of science and technology. And do not forget this small, essential detail: in order to be happy, a human being needs to have some moral ideal.

COMMENTARY

ORREN MOHLER

THE ESSAY by Dr. Wilhelmina Iwanowska certainly requires only praise as a proper comment. It is very interesting to review again the ideas of Copernicus, as set forth in the great book that was supervised through publication by Osiander, the obscure Lutheran preacher, who, probably, was responsible for an augmented title and a preface which he neglected to sign. The view expressed by Osiander in his unsigned contribution—science is fundamentally an abstract mathematical formulation of hypotheses—is a most modern viewpoint, but it is almost certainly in contradiction with Copernicus's own feeling; although he dedicated his treatise to mathematicians and clearly expressed contempt for nonmathematical minds.

Since the major and general aspects of the assimilation of science into our ways of thinking and living have been extremely well described by Dr. Iwanowska, it may be worthwhile to consider a mechanical aid to the degree of assimilation that has been achieved. Perhaps the existence of handmade books in only very small numbers, and those restricted for the use of few people, may be a part of an explanation for the survival of the Ptolemaic model during the nearly fourteen centuries before its errors began to appear intolerable to Copernicus and other mathematicians. An abundant supply of easily available and widely distributed books is an essential element in the process of assimilation. I would say it is both a necessary and sufficient condition. Books can supply to every individual his own personal records of hypotheses, authors' thoughts about their hypotheses, and their conclusions (if any). All of this is, generally, in a compact form, easily carried and studied wherever and whenever needed, either immediately, or after an interval of many years. One of the most important properties of the book is its permanence—it requires only modest attention and care. Collectively, books constitute a kind of memory, enduring from generation to generation, capable of protecting mankind from an endless, and generally mindless, repetition of the same studies, the same investigations, the same procedures, the same errors. Books, indeed, are foundations for progress. The emergence of modern printing, and processes for producing many copies of a book, preceded the birth of Copernicus by only a few decades.

By the time a publisher was required for *De revolutionibus,* printing establishments were in production within a few hundred miles of Frauenberg. The technology was available and rather highly developed for filling its essential role in the assimilation of Copernicus's ideas into the life and thinking of his community. But the full power of the press was not applied to speeding the adoption of the new idea. In fact, some statistically minded historians have remarked that Copernicus's greatest work is a leading candidate for the title: The World's Worst All Time Seller. It is possible to question the efficacy of a book whose first edition was a mere one thousand copies (of which a considerable number were never sold) and whose subsequent editions appeared at the rate of one new edition per century for the next four centuries. This is far from being unheard of in connection with a book containing ideas of genuine importance. Often the original itself languishes while commentaries proliferate in large numbers. Copernicus hardly can be accused of rushing into print. In fact, his publication record might be considered unsatisfactory in modern academic circles. But he was not academically employed and he, therefore, had a real choice in the matter of publishing. He may have decided that his ideas could wait until their time arrived. It could have been feared theological opposition causing his long delay in granting permission to publish; but it could equally well have been recognition of a certain fragility in his argument—weakness that caused a later astronomer, Tycho Brahe, the greatest of the exclusively optical observers, to deny the validity of the Copernican hypothesis with the scientifically unanswerable argument—disagreement with observation—or, it could have been a long-known abrasiveness, present in some degree, in all scientists.

Then, as now, ideas containing suggestions of obsolescence in existing concepts have rarely been greeted with enthusiasm by the originators of the momentarily acceptable ways of thinking. The idea of the nucleus of the atom becoming a usable source of energy is a recent example. When, forty years ago, physicists began to consider the atomic nucleus seriously, publications setting forth the possibility of obtaining energy from atomic nuclei were treated with both scorn and derision by the men who had developed the ideas fundamental to the success of the proposal. It is possible to listen to recordings of the voice of the man who discovered the atomic nucleus, Lord Rutherford, asserting the utter hopelessness of achieving success in a project dedicated to the release of atomic energy. Robert Andrews Millikan, Nobel prizeman and spokesman for science in the United States during that decade, argued in various publications that man could not realize such preposterous fantasies. They were against nature. In the face of such authoritarian unanimity papers were withheld, or reserved until more suitable times for publication. No one dared to publish a serious book about atomic energy at that time, although the concept had been freely used in science fiction for almost a century, and the general public in 1933 was far more receptive to the idea than were many scientists of high merit and prestige.

Nearly all scientists resist new ideas; it may be part of the code of professional conduct; and, if so, this is a good thing. The heat generated by strenuous resistance distills away lighter elements and encourages survival of only those theories, or parts of theories, capable of withstanding attack by intellectual fire and flood. In the time of Copernicus, scientific debate proceeded mainly through the publication of books detailing the arguments. It is understandable that two and a half centuries were required for his successors—Brahe, Galileo, Kepler, and Newton—to assure the complete assimilation of the heliocentric idea and its consequences into the civilization of Western Europe. It takes a long time to write, publish, read, and to understand significant books. In the following quartermillennium, the printing processes continued to supply the principal machinery for assimilation of scientific ideas, but books became minor arenas for the resolution of conflicts between old and new ideas.

Scientific journals and newspapers, even daily newspapers, are the instruments for modern skirmishers, and they are increasingly the means for transmitting to the public new scientific knowledge to be assimilated. And there are some among scientists who feel, evidently, that announcement of new results should appear first in *Time,* the weekly newsmagazine.

We are now somewhere along the way toward majority assimilation of another important concept and its consequences, as set forth in another important book, Darwin's *Origin of Species.* The entire original edition of 1,250 copies was sold before the close of business on the day of issue, but a storm of controversy breaking over the book on its day of publication has not yet subsided. Recent actions of the California state legislature, and also the program of the Michigan Academy of Sciences, meeting here later this week, are ample evidences of lack of assimilation (or perhaps better, acceptance) of Darwin's proposals by large groups of informed people. The impartial power of the press has never been more clearly demonstrated than it appears as it plays its role in this continuing intense conflict. I have been urging consideration of printing processes as dominant mechanical aids in the presentation of scientific theories and discoveries, and their incorporation (assimilation, if you will) into the day-to-day activities of the people.

Are there other devices, or procedures, in existence, or projected, that might be helpful in shortening the time from the formulation of an important theory to the time when it is effective in our thoughts and lives? One immediately thinks of photography, not yet one hundred years old. It provides a permanent record, suitable for visual display, of almost anything that will disturb, or produce, a radiation field, a stream of atomic, or even subatomic particles. It is often said that one picture is worth a thousand words, and books might be enormously shortened; but pictures without captions are, more often than not, puzzles; and picture writing is notoriously useless except for primitive ideas. The printing process must be called upon if really large numbers of pictures are required for direct visual inspection. The situation is pretty much the same for other methods of transmitting ideas from person to person and place to place. There can be discussion, telephone conversation, electrical storage, instant replays, movies, television, but a serious competitor for the book as a medium for assisting the assimilation of science into ways of thinking and living has not yet appeared.

V   Revolutions and Copernican Revolutions

CARL COHEN

I PROPOSE to ruminate about Copernican revolutions—not *the* Copernican Revolution (although no chapter of intellectual history could be more fascinating) but about the concept of revolution Copernican in character and scope. I aim to do this partly in order to think more clearly about past and present revolutions, and partly in order to speculate about a future revolution. But in good part, also, it is my purpose simply to explore an intellectual and historical category so important that each of its few members, rich in drama and instruction, marks a critical turning for human kind.

Revolution in General

Philosophers and scientists often propose views they honestly believe to be revolutionary; but the proposers are rarely themselves active revolutionaries. "Up to now philosophers have only *interpreted* the world differently," Marx complained in a famous note to himself, and added, "the point, however, is to *change* it."

All of us would like to change the world; very few are they for whom reality does not fall far short of some ideal. Whether we believe (or ought to believe) that what is needed is revolutionary change, or changes only possible through revolution, depends, of course, upon what is meant by "revolution." It is a term bandied about in these days with high irresponsibility.

Most contemporary interest centers on political and social revolutions. We will think more clearly about such revolutions, I submit, if first we examine the concept of revolution more abstractly, its essence and its kinds, subsequently drawing some conclusions about the reality of revolution on the contemporary scene, and the possibility of certain revolutions in the future.

The root meaning of "revolution" is that of turning, turning completely around. So the planets perform revolutions about the sun, and for many engines an important index of the level of activity within is the number of revolutions per minute. In this pure sense, a body moves in one revolution around a complete circular or oval course, returning substantially to the position from which it began. In pursuing such a course, obviously, the revolving body at one point occupies a position diametrically opposed to that from which it started, and this feature of such movement, I presume, led to the use of the term to indicate a total change of circumstances of whatever was in revolution, a drastic alteration of character, or system, or conditions. "Religions, and languages, and forms of government, and usages of private life, and modes of thinking, all," says Lord Macaulay (in Moore's *Byron*) "have undergone a succession of revolutions." It is this feature that has become paramount in our use of the concept; as often happens in language, usage at first metaphorical becomes literal at last.

Nowhere is the use of this dead metaphor, as indicating a drastic turnabout of system or condition, more useful to us than in the sphere of politics; so "revolution" comes to have, more than any other tone, a political one. Ask a common man what a revolution is, and he will tell you, rightly (in so many words), that it is a radical change of government, the overturn of an established political system and its replacement by another. There have been many of these in Western history, but the greatest three of modern centuries, which will be allowed by all, I think, to qualify as paradigms, are (in chronological order): (*a*) the English (Glorious) Revolution of 1688, in which royal despotism is ended, constitutional government established, the supremacy of the Houses of Parliament fixed, and the Bill of Rights declared; (*b*) the French Revolution of 1789, in which absolute monarchy is brought down, the Bastille as its symbol is stormed, and the first French Republic established; (c) the Russian Revolution of 1917, in two chapters, as a result of which the reign of the Tzars is overturned, the Soviet Socialist State established.

But of course, all important revolutions have not been political. We frequently hear reference to the *scientific* revolution, to the *industrial* revolution, to the so-called *managerial* revolution, and others. Allowing without quarrel that these were indeed revolutions, what made them so? Many great changes and discoveries, thrilling in dimension, enormous in impact, are yet not revolutions. The discovery of the New World by Christopher Columbus, though monumental in import, is not the Columbian revolution. The establishment of the science of bacteriology, through the discovery of microorganisms by Louis Pasteur, although among the greatest scientific advances of all time, is not thought of as the Pasteurian revolution. For Christianity the splitting off of the Protestant sects has had consequences too wide and too deep ever to catalogue, yet we do not view those developments as revolutionary; it is the Protestant *Reformation* to which we refer. The real democratization of England came not when King James II was deposed in 1688, but when the great body of its citizenry was enfranchised as the result of legislation in the 1830s and 1860s—the great *reform* bills. Most great advances are rightly viewed as reforms; some, a very few, as revolutions. If we could say somewhat more clearly how this distinction is to be drawn, our political discourse might be much sharpened.

Revolution and Reform

The key to the distinction needed here is to be found, I suggest, in the original idea of a revolution as a turning around. Many changes, some very great, move us crisply from one position to another, from one theory to another, from one state of development to another. A very few changes do that and something else: they overturn an established system, scientific or political, and replace it with another having an essentially different nature, essentially different principles of organization. The splitting of the atom, and the consequent development of nuclear power, brought developments in science and human society, for good or ill, of extraordinary character—yet those advances did not alter fundamentally the already existing theoretical understanding of matter and energy. New scientific instruments—the telescope for Galileo, antibiotic drugs for recent medicine, the bubble chamber for nuclear physics—make possible "breakthroughs" of exhilarating dimensions, but they are not revolutionary, although they are sometimes given that name, through a sort of rhetorical expansion. Similarly, some revolutions, so-called, in many nation-states—Latin America has provided the most well-known examples—have changed some things sharply, and yet in most fundamental respects have altered not very much in the lives of the citizens. *Plus ça change, plus c'est la même chose.*

Revolutions clearly deserving the name are not so common, after all. The industrial revolution really is entitled to that appellation. A complex chain of events, over a fairly extended period, resulted in a complete turnabout in the ways most things were to be made, and in the ways most people were to work and to live their lives. The underlying principles of economic organization were, for most, diametrically altered—from individual to collective production, from rural to urban life. The discovery of non-Euclidean geometry, although perhaps seldom thought of as revolutionary, was truly that, in changing the basic conceptions with which many aspects of geometry, and later physics, were to be dealt with. And the Glorious Revolution, although essentially bloodless, and far from introducing full democracy, was indeed a profound reversal of the principles of government in England. The shift in real authority from the Crown to the Parliament—though its immediate impact was only moderate—was of the most fundamental import. Government, in England, from that time forth, had a very different character, had to be thought about, and acted toward, in very different ways.

Intellectual discourse does frequently distinguish between the *coup d'etat* and the real revolution; the line may be hard to draw sharply, but the distinction is appropriate and important. Of course, those who successfully usurp authority will often claim to be revolutionaries in some honorific sense; long experience instructs us to take such rhetorical flourishes with high skepticism.

Contemporary Revolutions

There is much talk about revolution in the contemporary West. Is all of it no more than rhetoric? Whether the goals or proposals of particular persons or movements do indeed call for a fundamental change in the principles of social organization is a matter to be determined in each case separately, of course; but I submit here some observations concerning two kinds of "revolution" much discussed. The first of these I shall call simply "Marxist," the second, for the sake of a name, "populist," although that term is now not much used. I confine myself to the American scene, but what I say about it can, I think, be much generalized.

Of Marxist revolution not much needs to be said here. I touch not at all on the question of whether such a revolution would be justified or right. What seems clear beyond doubt is that, where events transpire as Marx and his disciples intend (and expect) that they shall, the process *is* genuinely revolutionary. Called for is a thorough overturn of existing economic arrangements, a fundamental change in the principles of possession and management of productive property, and, consequently, the complete elimination of wage labor, and radical changes in the conditions of human work. At least in the economic sphere (and perhaps in all spheres, if Marx was right in holding that economics is the substructure of an edifice of which politics, law, religion and the arts are but superstructure) the basic principles of human organization are to be wholly—dramatically—altered. Whether events in the Soviet Union, or China, or Cuba, have indeed realized, or are now in the process of realizing, this transfiguration, is a matter of much controversy. And whether that is a transfiguration we want, or ought to want, is far from settled. But the honestly professed ideal of a proletarian uprising against the exploitations of an oppressive capitalist economic order is, however well grounded in theory or in fact, unquestionably revolutionary in conception.

The "populist" aspirations of contemporary American and European radicals are not, I think, in the same case. Under the heading "populist" I mean to clump together a variety of organizations and movements which hold, in common, that the abuses and inhumanities of Western governments—the American government most notably—morally demand their overthrow, and their replacement by new regimes genuinely in the service of the people. Such populist revolutionaries share many of the critical views of Marxists—a detestation of private industrial abuses, of gross economic inequalities, and the like. But populists often do not share Marxist convictions about the inevitability of dialectical advance, or the underlying Marxist materialism, and consequent economic determinism. Nor need they accept the Marxist orthodoxies regarding property and its rightful appropriation. Indeed, many American populists are sharply anti-Marxist, and contemptuous of the dogmatism of contemporary Communist parties.

Now again I do not speak here of the rightness of populist principles, or the justifiability of their often bitter criticism of existing governments and leaders. I ask here only whether the spirit of these movements is genuinely revolutionary. The answer, it appears to me, must be no. American government, they contend, is simply not living up to its promises. It purports to be by the people, and for the people, but (say they) turns out in fact to be manipulative, unresponsive, and, for some large minorities, cruelly oppressive. Therefore it is time to make a revolution, in the true spirit of Jeffersonian democracy. The spirit of the new populists—however much the practice of some may fail to accord with it—is best expressed in the original motto of Students for a Democratic Society (SDS) "Let the people decide."

Much in these movements seems to me altogether honorable, admirable, and right. But when the speechmaking is over, and the several concrete proposals and plans of action are put before us, I simply do not find much that is revolutionary in them. Of course representative institutions, when they prove unresponsive, should be made more responsive; of course the exploitation and oppression of minorities, where it exists, needs to be rooted out, and so on. Great changes in American government and society—and elsewhere in the West—are surely called for, as are the efforts to bring them about. But these are not revolutionary changes, they are reforms which, as envisaged, would more fully realize the principles to which virtually all reasonably enlightened citizens honestly subscribe. If (what seems rather unlikely) the present American government, or British government, were overthrown by sincere, radical populists, and new representative assemblies convened, and so forth, it may be that policy changes for the better would come more rapidly than is likely under existing institutions. In view of the procedural insensitivities (and in some cases, the fanaticism) of some populist leaders, that is genuinely doubtful.

But even supposing that such happy consequences would follow an uprising and overthrow, there is little doubt that that overthrow would deserve the name of *coup,* not revolution.

It is worth noting, in passing, that overt efforts to effect such a *coup,* however romantic and unlikely to succeed, tend, in fact, to reduce rather than to increase the responsiveness of existing institutions. By raising general concern about collective self-preservation, such efforts tend to exacerbate, rather than to mitigate, the repressive inclinations of the present order.

Copernican Revolution

Some revolutions are of a very special kind, and that special nature is properly identified by calling them "Copernican." Reflection upon these Copernican revolutions is particularly appropriate just now. If we can isolate that feature or set of features which sets them apart from all other revolutions, social or scientific, we may be the better prepared for a new Copernican revolution which—as it seems to me—is in the offing.

The first self-conscious Copernican revolution was that of Copernicus. He and his successors, though often guarded in statement, knew it for what it was. *De revolutionibus orbium coelestium (On the Revolutions of the heavenly spheres)* is the accepted title of his great work; but there is speculation that he intended to call his book *De revolutionibus* simply—which might, indeed, have been more fitting. Many have pointed to the prefatory note in that work, advising the reader that the "hypothesis" of the book need not be taken as an assertion of fact, but instead as an intellectual tool, only. One would then read *De revolutionibus* not as claiming that the Earth moves, but as the introduction of a mathematical device in which, for purposes of calculation, and for the simplest explanation of the observed behavior of the planets, we do no more than change the origin of the coordinates of our grid. That was not Copernicus's view; the suggestion itself, inserted and retained by others, was not his. A thoughtful reader of *De revolutionibus* cannot miss Copernicus's genuinely revolutionary intent. After noting the speculations of Pythagorean astronomers, and others of earlier times, he writes:

I too began to think of the mobility of the Earth; and though the opinion seemed absurd, yet knowing now that others before me had been granted freedom to imagine such circles as they chose to explain the phenomena of the stars, I considered that I also might easily be allowed to try whether, by assuming some motion of the Earth, sounder explanations than theirs for the revolutions of the celestial spheres might so be discovered. [*De Revolutionibus,* Book I. All quotations here are from a translation by John F. Dobson and Selig Brodetsky, published in 1947 by the Royal Astronomical Society, London, from an almost perfect copy of the first edition of the work.]

This as a preliminary. By the time Copernicus is halfway through the first book of *De revolutionibus* he has shown that by transferring the motion of the Sun to the Earth the risings and settings of the fixed stars will be unaffected, but the observed peculiarities in the behavior of the planets (etymologically: *wanderers*)—their apparent retrogressions and hoverings—will be explained as due "not to their own proper motions, but to that of the Earth, which they reflect." Finally, he says straightforwardly, "we shall place the Sun himself at the center of the Universe." The resultant system is not only elegant and harmonious; it is what we will accept if only, in his words, "we face the facts, as they say, 'with both eyes open.' "

The theory is, indeed, revolutionary. An entirely new set of principles, a new perspective, was to govern our thinking about ourselves, and our universe, thenceforth. With Copernicus's shift in thought men were to change their conception of the structure of the world; not the Earth, but some other body or bodies, far distant, is that about which things in fact revolve. Revolutionary it was beyond doubt, and for the Christian world view of the time it was plain heresy.

It is not, however, the greatness of the shift, or the power of its impact, that marks that revolution as, in the fullest sense, *Copernican.* What was singularly extraordinary, and specifically Copernican about these conceptual changes was this: what had previously been viewed as movement "out there" recognized and measured by fixed observers "down here" now came to be understood as part of a larger system of movements in which neither observed nor observers are stationary. The core passage in *De revolutionibus* is splendidly simple in its presentation of the reversal:

> A seeming change of place may come of movement either of object or of observer, or again of unequal movements of the two (for between equal and parallel motions no movement is perceptible). Now it is Earth from which the rotation of the Heavens is seen. If then some motion of Earth be assumed it will be reproduced in external bodies, which will seem to move in the opposite direction.

Changes in apparent position previously attributed only to the movement of celestial bodies were now to be understood as due in great part to the characteristics of the observer.

"I ascribe movement to the earthly globe," Copernicus wrote to Pope Paul III, and as soon as some people hear that, he observed, "they will cry out that, holding such views, I should at once be hissed from the stage." But no amount of hissing could make the earthly globe stand still, or return us to a Ptolemaic understanding of the movements of the spheres without. The notion that what other things seem to be, seem that way because we are in the condition that we are, is uniquely Copernican. It is a deep and powerful idea, and its impact was bound to reach well beyond the sphere of astronomical science.

Two More Copernican Revolutions

If the first self-conscious Copernican revolution was that of Copernicus, the second was that of Immanuel Kant. Of the great modern philosophers he was the first fully to grasp the force of the Copernican hypothesis generalized. And quite explicitly he introduced his revolutionary account of human knowledge (in the *Preface* to the Second Edition of the *Critique of Pure Reason*) as Copernican in spirit. What we know, he argued, we are able to know partly because the objects of our knowledge have the attributes they do, and partly because we, as knowing subjects, provide the frame for knowing that we do. All phenomena, said he in effect, are the result of a partnership of knower and known, as the apparent movements in the heavens are the result of the movements of both the celestial bodies and that of the Earth. He wrote:

Hitherto it has been assumed that all our knowledge must conform to objects. But all attempts to extend our knowledge of objects by establishing something in regard to them *a priori,* by means of concepts, have, on this assumption, ended in failure. We must therefore make trial whether we may not have more success in the tasks of metaphysics, if we suppose that objects must conform to our knowledge. This would agree better with what is desired, namely, that it should be possible to have knowledge of objects *a priori,* determining something in regard to them prior to their being given. We should then be proceeding precisely on the lines of Copernicus' primary hypothesis. Failing of satisfactory progress in explaining the movements of the heavenly bodies on the supposition that they all revolved around the spectator, he tried whether he might not have better success if he made the spectator to revolve and the stars to remain at rest. A similar experiment can be tried in metaphysics, as regards the *intuition* of objects. If intuition must conform to the constitution of objects, I do not see how we could know anything of the latter *a priori;* but if the object (as object of the senses) must conform to the constitution of our faculty of intuition, I have no difficulty in conceiving such a possibility. Since I cannot rest in these intuitions if they are to become known, but must relate them as representations to something as their object, and determine this latter through them, either I must assume that the *concepts,* by means of which I obtain this determination, conform to the object, or else I assume that the objects, or what is the same thing, that the *experience* in which alone, as given objects, they can be known, conform to the concepts. In the former case, I am again in the same perplexity as to how I can know anything *a priori* in regard to the objects. In the latter case the outlook is more hopeful. For experience is itself a species of knowledge which involves understanding; and understanding has rules which I must presuppose as being in me prior to objects being given to me, and therefore as being *a priori.* They find expression in *a priori* concepts to which all objects of experience necessarily conform, and with which they must agree. [Kemp Smith translation (Macmillan and Co., London, 1950), pp. 22–23.]

This approach, Kant surmises, will provide the touchstone of a new method of thought: "that we can know *a priori* of things only what we ourselves put into them." And this experiment, he concludes, succeeded as well as could be desired, promising to metaphysics, for the first time, "the secure path of a science" [ibid.].

Whether Kant's proofs, in the body of the *Critique of Pure Reason,* establish the success of this revolution in knowledge, as conclusively as did Galileo's proofs establish the success of the earlier revolution in astronomy, remains arguable, of course. But the second, like the first, was indeed conceived as a revolution essentially Copernican in nature.

A third self-conscious Copernican revolution has been proposed by John Dewey, and was specifically called that in the final chapter of his great work of 1929, *The Quest for Certainty.* Dewey argues, in short, that although Kant intended a Copernican revolution, his system does not in fact provide one. He failed, said Dewey, because in developing his so-called revolutionary account of knowledge, he remains in fact within the grip of classical, mistaken conceptions of what knowledge had to be like if it was to be knowledge at all. To know something, for Kant, was to know it certainly, absolutely. The question was not, for him, whether a priori knowledge was in fact possible—he never doubted that—but how in the world we *get* the a priori knowledge we obviously must have. To account for that knowledge he relocated the a priori contribution. It springs, said Kant, not from Divine imprint or external necessity, but from the structure of human intellect. But in his fundamental approach to knowledge, Dewey contends, Kant exhibits in a new guise the age-old error of supposing knowledge to consist of some wonderful correspondence between what was in the human head, and what was out of it—except that that correspondence was by him assured not by modeling the mind after the world, but by modeling the world after the mind. In either case the world comes out looking like a system of necessary truths, locked together by laws having universal and a priori certainty. Kant's work (on this interpretation) is more a revelation than a revolution—it reveals more clearly and more deeply than could that of his predecessors the purely *intellectual* mold into which everything was to be pressed. But all was pressed into that mold still; he simply explained, as no one could before, the alleged necessity of that mold. The knowledge system resulting from this Kantian method, says Dewey, is not really Copernican; it is truly Ptolemaic. As in Ptolemaic astronomy the celestial bodies revolve

about the Earth, in Kantian epistemology the objects of knowledge revolve about the knowing mind. Kant's alleged revolution,
says Dewey, "consisted in making explicit what was implicit in the
classic tradition." That tradition "had asserted that knowledge is
determined by the objective constitution of the universe. But it
did so only after it had first assumed that the universe is itself
constituted after the pattern of reason. Philosophers first constructed a rational system of nature and then borrowed from it the
features by which to characterize their knowledge of it. Kant, in
effect, called attention to the borrowing; he insisted that credit for
the borrowed material be assigned to human reason. . . ." But that
underlying correspondence of reason and nature he never really
questioned. In consequence, there is nothing hypothetical or conditional in Kant's framework of perception and conception. He
took his categories, as Dewey points out, to need no testing, no
experiment; they *must* work, uniformly and triumphantly. They
are as inaccessible to the testings of a public science as are the
hidden commitments of the Ptolemaic astronomers—only differently located.

　　Now, says Dewey, it is past time for a revolution in the
theory of human knowledge that is genuinely Copernican. A great
shift is indeed called for, not simply in the location of the
framework for certainty, but in the nature of the standards to be
applied to judgments themselves. What we know about the world
must receive its warrant not from a priori principles (of any origin)
but from a posteriori findings emerging from the process of
experiment. Of course there will be ideas, hypotheses, directing
experiment; but the function of these conceptual contributions is,
for Dewey, as different from the function of Kantian a priori
principles as is the Copernican from the Ptolemaic astronomy. For
Dewey the directive idea is tentative, uncertain, to be appraised in
view of the outcome of the experiments based upon it; concepts
are not directive in the Kantian sense that they fix with apodictic
certainty the forms within which all things can be known. Knowledge, for Dewey, does indeed flow from a partnership of the
knower and the known; but for Kant that so-called partnership
was in fact a silent subordination, in which the world might play
its cards, but the knowing subject always had the deck stacked.

If these changes, Dewey concludes,

> do not constitute, in the depth and scope of their significance, a reversal comparable to a Copernican revolution, I am at a loss to know where such a change can be found or what it would be like. The old center was mind knowing by means of an equipment of powers complete within itself, and merely exercised upon an antecedent external material equally complete in itself. The new center is indefinite interactions taking place within a course of nature which is not fixed and complete, but which is capable of direction to new and different results through the mediation of intentional operations. Neither self nor world, neither soul nor nature (in the sense of something isolated and finished in its isolation) is the center, any more than either earth or sun is the absolute center of a single universal and necessary frame of reference. There is a moving whole of interacting parts; a center emerges wherever there is effort to change them in a particular direction. [*The Quest for Certainty* (Minton, Balch and Co., N.Y., 1929), pp. 287–91.]

Which of these two revolutions, or alleged revolutions—that of Kant or that of Dewey—is more genuinely Copernican I leave for the reader to judge. Both, I think we will agree, are genuinely revolutionary if their author's accounts of the contexts in which they arise is substantially correct.

A Fourth Copernican Revolution

I conclude on a speculative note. Returning from theories about knowledge to theories about human society, I ask: Does there loom a possible or probable revolution, in the foreseeable future, which would be genuinely Copernican in character and sweep, and with respect to which philosophical reflection (and perhaps even guidance) is now in order? I think there is such a revolution before us, and I make bold here to sketch its outlines. If my speculations hit anywhere close to actual future events, there will be serious tasks for philosophers, professional or other, in the course of that revolution.

The revolution begun by Copernicus himself concerned the system of the world in its astronomical dimensions—what it was like, and what the constituents are of the movement and order that we perceive. Kant and Dewey proposed profound changes in our conceptions of knowledge—what it is like, and what the constituents are of the knowing situation, the correct relations of knower and known, whatever the subject matter. The context to which I now move is that of human community and its planetary environment; it is a context far smaller than the astronomical, far greater than the epistemological. I suggest that we may be coming to a new conception of what that community is or ought to be, flowing from a revolutionized conception of its constituents. That revolution would chiefly concern neither the nature of the universe, nor the nature of knowledge, but rather the nature of ourselves, what we are and can be. And the problems it presents touch deeply upon the direction of our conduct.

The history of human kind, as a kind, has been, to this date, in a fundamental sense Ptolemaic. Here we are; there is the world. Men have taken their general task to be that of making a decent life for themselves within it, or at least getting along in it, however uncomfortably, as best they could until they got out of it. The external world is immensely perilous—moderately tractable on some dimensions but largely hostile—and men succeed in their aims to the extent they manage, through wit or endurance, to whittle out of nature the necessities and comforts of life. Human outlook has been Ptolemaic in this underlying assumption: whatever was to be done, it must be done *with* the world, to it, making it more manageable, more tolerable, more hospitable. We adapt ourselves to climates and terrains, to cities and institutions—but adaptation has consisted essentially in using what one part of the environment provides to protect ourselves (or advance ourselves) against the rest of it. Such efforts as there have been to deal with the human condition by altering the nature of humans rather than their conditions have been aimed at making our selves relatively impervious to the slings of fortune, or by causing ourselves to believe that interaction with the natural world is ultimately of no consequence. Either (it has been suggested) all that counts is within, and external suffering therefore trivial; or all that counts is in another world entirely, and external suffering therefore but of passing concern. Genuine human accomplishment, meanwhile, however little or great, has taken place with little regard either for stoical withdrawals, or supernaturalist escapes, in their several varieties. With many remarkable successes, and uncountable miserable failures, men have looked from themselves to the "natural" world as though at a kind of thing as different from themselves, at root, as terrestrial clay was believed different from celestial crystal.

John Dewey's reinterpretation of knowledge was, to be sure, a powerful effort to bridge this gap. The revolution that looms, I submit, is the closing of it. Some philosophers— Democritus, and Epicurus, and Spinoza, as well as Dewey—long insisted that humans are essentially part of the natural world; but the fact remains that the conduct of men has not genuinely incorporated that conviction. "Human nature," it is tacitly and generally supposed, is one of the givens. We may develop or manipulate the potentialities it provides, with schoolrooms, or spankings, or psychological conditioning rational or irrational, but it is human clay in the end, not much better in the person of Bertrand Russell than in that of Parmenides. Now we enter a time in which that given is seriously to be questioned. Human beings, as a kind, will become the *subject* of experimentation and deliberately guided change, in a sense far deeper and more remarkable than as the individual recipients of torture, or training, or indoctrination. They will be the material out of which humans having quite different natures will be made; and the agents effecting change will be among the subjects in the process of being changed. The reflexivity of the situation is intellectually amusing; the possibilities for good and evil are so immense as to boggle the mind.

Copernican is the only word to catch the full spirit and impact of this change. Where previously human agency was directed chiefly toward what was "out there," to render it more tolerable or pleasing, the consequences of this revolution will be that the location of the satisfactory and the worthy will come to lie in the interaction of a world partly managed by human agents, with those agents having essential natures also under human control. As the movement of the spheres, after Copernicus, was seen to be a resultant of movements above and movements below, and knowledge, after Kant, was held to be a partnership of content without and capacities within; so now, the entire texture of what is human may come to be constituted by the warp of physical circumstance, and the woof of a deliberately self-changing humanity.

Events of great moment, impending in human history, often have early impact in limited spheres, or spheres tangential to their ultimate full thrust. These special manifestations—as with the early traces of the industrial revolution—are seen as troublesome, or perhaps hopeful, but are rarely appreciated as the foreshadowing of developments of which they are but the first vibrations. So, perhaps, is it here. It is first in the sphere of the fanciful—in Huxleyan dystopias and assorted works of science fiction—that the notions of controlled genetic mutation, and test-tube human breeding, and the like, are entertained. Quick on their tails come the horrors of actual genetic mutations flowing from accidental exposure to radioactivity, and from the ingestion of pharmaceuticals prepared with very different objectives. The prospect at first is quite frightening, and the possibilities of real and permanent benefit to mankind through deliberate self-change are very partially and very dimly glimpsed. Struck chiefly by the enormity of the dangers which open, as well as by the ignorance with which we face them, the understandable reaction on the part of many is to keep hands off—as though deliberate intervention in the evolution of the species were an act of profaneness, a "playing God" by men unworthy and incapable of the role.

But the course of history—barring some sudden nuclear or environmental catastrophe of now unimagined scope—is virtually certain to oblige humans to play that creative role. If we accept, as premise, that human population will continue to grow until meeting some horrendous "Malthusian check" or until rationally controlled and stabilized, we will either suffer unimaginable catastrophe, or we will sooner or later face up to the task of devising the instruments for that control and stabilization. Then, liking it or not, we will have to make decisions, some collective, some individual, regarding what we aim to do with humanity, with ourselves, and how we can tolerate doing it. Even the deliberate decision to take no controlling measures, when such measures are in fact in our power, is a decision which, if made, surely ought to be made with the greatest forethought. Perhaps we shall want to rely, for the future of the species, upon some sort of environmental lottery, or even upon a true lottery of our own devising. For my own part, it would seem that humanity deserves more from the use of its collective intelligence than sheer randomness.

In spheres in which previously randomness has always been taken for granted, how better it with deliberate control? No one can speak confidently now, of course. But so soon as this Copernican revolution is taken seriously, even as a distant but probable prospect, two philosophical tasks will confront us:

The first is that of deciding, or of developing criteria which may ultimately be used in deciding, what kinds of beings human beings are to be. In what directions are we to move in imposing rational control of human nature? No sooner do we say that some humans may reproduce and others not, than we must provide some principles upon which permission, or restriction, in this sphere, are to be grounded. The measures of control may be at first very limited, seeking only to eliminate those defects or negativities in the species upon which there is universal agreement. When that proves insufficient, as eventually it will, we shall face some much harder questions about what is to be encouraged, positively striven for, and at the expense of what. That is one serious practical task for philosophers—for whoever makes those decisions will be philosophers, whatever the source of their livelihood.

A second philosophical task, easier, I think, but not much easier, is that of devising the institutional machinery needed to reach those hard decisions. What kind of governance does a human community, undergoing a continuing process of deliberate self-change, need? What can it hope for? What trade-offs between ideal equalities and practical inequalities, between restraint in some spheres and liberty in others, are human beings prepared to accept or entertain? And what are the alternative modes of decision-making in this sphere, among which we might conceivably choose?

If you find these questions painfully general, and hard even to begin to answer, I am on your side. I would observe, however, that hard as they are, we are almost certain to confront them eventually, and when we do, they will be even harder than they now seem, for two reasons.

First, it is probable that we won't really face up to these painful decisions until we absolutely must, and that means we will put off the needed judgments until it is too late to apply *any* rational judgments without imposing great suffering—moral agony, or physical pain, or both—upon many. It is likely—not certain, but in view of recent history, very likely—that bold strategies upon critical matters will be collectively agreed upon only after the crisis is upon us. That will probably render them harder to devise, but impossible to postpone.

Second, the nature of the issues here is such that the framework within which the quest for direction is undertaken is itself in flux. That, I reemphasize, is the peculiarly reflexive, Copernican character of this revolution, if indeed it comes. The deliberate guidance of the evolution of the species, and the possible transformation of its members, is not simply another great step, or marvellous discovery. It will change every understanding of what human "steps" are or ought to be, and it may deeply alter our understanding of what "discovery" is for.

If I am anywhere near target in all of this speculation, it is right, I think, to view this contemplated revolution as Copernican in nature and sweep. About the future, of course, no knowledge is thoroughly reliable; we prognosticate with differing degrees of warrant. How much warrant there is for what has been suggested here I leave the reader to judge. But I note, in concluding, that if such a Copernican revolution does indeed lie before us, we are more likely to survive it, as a species, with success, if we begin to think about the trials it will present not after they are upon us, but now.

COMMENTARY

JOSEPH CROPSEY

THE CRUCIAL affirmation of Professor Cohen's stimulating paper consists in his characterization of Copernican revolution; and the decisive movement of the paper is toward extending the application of that conception of Copernican revolution beyond physics to the social life of man. The paper contains other things that are worthy of comment, but the crucial affirmation and the decisive movement raise such weighty questions that I think it will be best to concentrate attention on those in these brief remarks.

From the outset, Professor Cohen takes note of the fact that all revolutions are changes but not all changes are revolutions, and he wishes to make clear what distinguishes revolutions from other changes. He argues that revolutions are more far-reaching than, for example, reforms. Among revolutions, there are a few that exceed the others in scope and reach, and these are Copernican revolutions. In order to grasp Professor Cohen's characterization of those overmastering Copernican revolutions, it is necessary to begin with revolution simply. Revolutions are changes that, in his words, "move us . . . from one theory to another, from one state of development to another . . . [and that moreover] overturn an established system, scientific or political, and replace it with another having an essentially different nature. . . ." Examples are the Industrial Revolution, the discovery of non-Euclidean geometry, and the Revolution of 1688, as well as the "Marxist revolution."

It would have been distracting from the main thrust of the paper if Professor Cohen had developed to precision the meaning of the notion of "essential difference in nature" as between two theories or states of development. For want of such precision, there will be disagreement over the judgment that the Marxist movement is revolutionary while the "populist" movement, as spoken for by the Students for a Democratic Society, is not. But rather than pursuing this, I wish to call attention to Professor Cohen's application of the concept of revolution in a single sense to theories and to states of development indiscriminately. The consequences of doing so cannot be made to appear until one reflects on Professor Cohen's characterization of Copernican revolution. I believe it is fair to say that the kernel of that characterization lies in these words, which are used to describe the discoveries of Copernicus himself: "what other things seem to be, seem that way because we are in the condition that we are. . . ."

Let us remind ourselves of what exactly Copernicus was doing. He was explaining the apparent motion of the sun (among other bodies) and the apparent rest of the earthly observer by asserting the motion of the earthly observer and the motionlessness of the sun. To that extent, what he did was not different from what had been done often and for ages before, namely, to discredit mere experience and to demonstrate that experience has a meaning or a basis that can be made manifest only through reason. At most, Copernicus showed that how the sun seems to move is explained or partly explained by a movement of our own. This is very far from showing that what the sun seems to *be* seems that way because we are in our actual condition.

That this is not a mere quibble is shown by Professor Cohen's subsequent description of the Deweyan philosophy as the product of a true Copernican revolution. Dewey is shown in corrective reaction against Kant. Kant, with his famous assertion of the unknowability of the things as they are in themselves, of course concurred in the view that we must therefore know them in the only other way available, namely, as they are for us. But in so knowing them, we do not simply "know" them; we contribute something of our own to the knowing and to the knowledge. The object of knowledge and we ourselves as the knowers have a constitution or character or nature that can be described as "fixed," and that is crucial to the contribution made by the knower to the knowledge. Dewey asserts that neither the object— the "course of nature"—nor the subject—man—is a certain something, but that all is flux, especially man who nevertheless appears to remain the measure of all things. At least I am not aware of any suggestion coming from Dewey that there is a baboon-conceived world-order and a self-changing baboon self-conception, and so on for mice and other beasts, and that these yield important alternatives to the human or rationally conceived world-order or course of nature. In Dewey's free-ranging displacement of universal centers, his mind fails to reach the one change that would provide a decisive test of his construction, namely, the change from a rational subject to a subrational one as *the* viewer of the whole. Instead, Dewey contents himself with proposals for human self-change that were in principle quite well worn by his time, having been made familiar, though more profoundly, through Rousseau's teaching of the vast alterability of human nature and the crucial importance thereto of education.

While I do wish to impugn Dewey's originality, I wish also and more emphatically to impugn his soundness. Before doing that, however, it is necessary to recur to Professor Cohen's point, which agrees with Dewey's own self-appraisal, to the effect that the Deweyan formulation is a, or perhaps the, true Copernican revolution. I do not think so. Copernicus taught the world that a certain empirical fact cannot be understood on the empirical level. He did not maintain, and on the basis of his construction one cannot conclude, that after Copernicus understood what he understood, the world was changed an iota from what it had always been and must always be; nor that either the universe as such or man as such would undergo essential change through mutual influence. Only one change occurred, and that was in the content of men's minds. Not the object, not the subject, and not the relation between them is presented by Copernicus in a radically new light. I say this not with the intention of depreciating the great scientist but rather to praise his sobriety, which is in the highest tradition of science.

Dewey's teaching that all is in flux, and most of all man is in flux, suffers from the grave defect that a self-creator who cannot refer to a fixed criterion to govern the direction of his self-creation must necessarily be at best an anomaly and at worst a demented and unhappy thing of caprice. For reasons of his own, Dewey did not speak the language of existentialism but rather of "science"—experiment would show the way to man's self-creation and would somehow enable men to distinguish a desirable from an undesirable social outcome. If we learn anything at all from Copernicus, it is the inadequacy of mere observation or experiment to perform such tasks as Dewey, who was of course not a scientist, hoped the mind of man would be spared.

When Professor Cohen, in defining revolutions, speaks of essential difference between two *theories* or *states of development,* he does so in order to take account of the fact that some revolutions are in the realm of thought, as is the Copernican, and some in the realm of politics. If a Copernican revolution is one that locates in the observer something that explains what hitherto was attributed to the object, then one can understand how, for example, Hume's thought might deserve the name of Copernican-revolutionary. But by this criterion, how can any political change qualify to be so called? The theoretical formulation that lies behind a great political upheaval might qualify, but I do not see how the event itself can do so, according to the definition given.

This means that there is only one locus of Copernican revolution by Professor Cohen's definition, and that is in thought, in the mind; never in the phenomena or in the object. Certainly Hegel might explain the slave by reference to the master, and might define the nature of mastery through the nature of slavehood, but a political change that introduced or abolished slavery would not itself have the least influence on the definition of master and slave. Is it of any importance that this point be cleared up? Let us consider.

A legal system based on the publicly accepted belief that what seems to belong to the object is in fact to be explained by characteristics of the beholding subject would pose incalculable obstacles to the definition and then to the administration of justice. Is a crime the act of a criminal or the opinion of the witness? Or perhaps the passive act of the victim? Who are the real oppressed—the poor and exploited, as has always been taken for granted, or is it really a matter of perspective. The questions could be multiplied. It seems to me that the familiar distinction between theory and practice has survived for a reason that is more powerful than are the arguments for revolutionizing it. Professor Cohen has referred to the fact that, attached to the notion of revolution is the notion of the closure or completion of the turning about. I am put in mind also of the center or focus of a regular curve, some fixed point by reference to which the motion of a particle along the curve is alone intelligible. Speaking politically, the place of that fixed point is occupied by justice. Change and flux are endlessly enchanting. We must therefore be at pains to recall that change would have little meaning and less worth if there were not an unchanging and intelligible whole within which change can occur and be known.

SCIENCE AND SOCIETY: PRESENT

VI   The Twin Moralities of Science

STEPHEN E. TOULMIN

ANY scientist or philosopher whose fame lasts until his five hundredth birthday becomes the private property of those intellectual undertakers, the historians of thought. The victim's immediate relatives and friends are long dead, so he cannot defend himself and we are free to make of him what we please. We can embalm him with praise, or varnish him over with hagiographic gloss. We can cremate his reputation, and scatter it to the winds. We can use him as a cadaver, and subject him to ever more minute dissection. Or we can cast a more-or-less lifelike statue of him—its accuracy does not terribly matter and erect it as a ceremonial boundary-stone or benchmark, indicating the start, the climax, or the end of some historical phase or "era": though an "era," one must add, in history as particular *modern* historians see it.

Nicolas Copernicus has commonly received this last, sculptural or statuesque treatment. Comparing him with his immediate successors, Galileo and Kepler, we can discover little about his individual character and personality. His learned writings, too, are more often praised than read; their merits are more often guessed at than demonstrated. So the things that make Copernicus an object of interest today are things that tell us more about ourselves than they do about him. For Copernicus the man has been petrified—turned into the great historical milestone marking off the Old Dispensations in physics and astronomy from the New. Rightly or wrongly, he has come to symbolize, at the same time, the end of Scientific Antiquity and the beginning of Scientific

Modernity. Once we reach Copernicus (it seems) we can set those metaphysical old Greeks on one side, and begin dealing with scientific arguments in the modern familiar way.

I see no real alternative, here, to following this last example; but I want to follow it in my own eccentric way. For it is important to remember Nicolas Copernicus's *general* contribution to our intellectual traditions and enterprises, not just his narrow technical achievements. He was, after all, not just the man responsible for that technical transition in computative astronomy which initiated the so-called Copernican Revolution. More importantly, he is also the man who, after some twelve hundred years, staked out once again for the human mind the central claim on which any self-confident natural science depends; the claim that the human mind does have the capacity to develop consistent and comprehensive theories about the world in which we live, and can do so even on a scale that goes far beyond the direct reach of our experience.

This claim, which underlay so much in classical Greek science, had been explicitly or tacitly abandoned during the Alexandrian period. Speculation about the natural world then took second place to other, religious preoccupations; and speculation about the nature of the Heavens came to appear especially fruitless. Even Ptolemy himself was content to use a variety of different astronomical constructions, which were—physically speaking—quite inconsistent with one another. As a result, his picture of the planetary system was—or so Copernicus put it—less a lifelike portrait than a botched-up picture, as though a painter were to take the arms and legs of a figure, its body, head, and hands, from quite different subjects and run them together into one. In Copernicus's view, such a scientific amalgam was intellectually unacceptable. We must aim at a unified, coherent, and consistent picture: nothing less had the power to be (as he put it) "absolute and pleasing to the mind." And, with these words, we find Copernicus making a crucial declaration on behalf of the scientific intellect: one that has played its part in the development of natural science since his time. One precondition of all effective scientific speculation from Copernicus on has been this belief: the fundamental goal of science is one that human beings can indeed hope to reach.

My purpose, however, is not directly historical. Rather, it

is to ask what is becoming of the Copernican spirit in the science of our own time. For the contemporary situation in natural science has a curious feature, on which some of the other contributors to this symposium have also commented—this is, the quite unfamiliar sense of *moral ambiguity* among working scientists today, the novel doubt in the scientific profession itself about the fundamental value (even, about the fundamental justifiability) of the scientists' own purely intellectual activities. Often in the past painters, poets, or literary men rejected the values and achievements of science *from the outside*—through the satires of a Jonathan Swift, the engravings and poems of a William Blake, the plays and prefaces of a Bernard Shaw. But these doubts have rarely spread into the community of men who were pursuing scientific investigations for themselves. Nowadays by contrast, alongside a deepening suspicion of science from the outside, we can observe a fresh lack of conviction within parts of the scientific profession itself—a lack of conviction that has no exact counterpart in the last three hundred years. It is as though the psychological preconditions for science, as established by Copernicus more than four hundred years ago, have at last begun to crumble away, leaving us apparently at the end of the intellectual era which he inaugurated. And in one important respect at least (I shall be arguing) we are indeed now in a period of transition, which is compelling us to reconsider the intellectual values and the moral demands of the scientific enterprise.

Why should this be the case? It is easy to confuse ourselves about this point, through failing to look at the matter in a long enough perspective. In so many ways, the twentieth century situation appears to us unique; now, above all, we are, it seems, in a peculiarly scientific or technological age. Yet, I shall claim, this *quantitative* growth in the amount of scientific work being done is far less important than a *qualitative* change in the relations between the scientific enterprise itself and the rest of human life. From this point of view, the truth is more ironical than we have hitherto realized; yet, at the same time, it is also more hopeful.

From the seventeenth century on, the men who made it their business to invent and develop modern "natural philosophy" (i.e., science) devised, for the purposes of their work, some highly characteristic means, or instruments, for dealing with their problems. These instruments included both *intellectual* ones and also *sociological* ones. On the one hand, they developed certain procedures of investigation; on the other hand, certain institutional arrangements. And, if the scientists of the late twentieth century are at last in a position to realize the practical dreams for human welfare that formed so much of the original propaganda for science, credit for this fact must ultimately be given to the efficacy of those scientific procedures and institutions.

Procedurally, the key move was to devise a kind of intellectual "division of labor"; to select, out of all the available questions about Nature, certain abstract groups of issues that could be made the concern of particular specialized scientific disciplines. In the intellectual, as in the economic sphere, this division of labor has paid enormous dividends. Though Goethe and Blake might object that the abstraction which it involved introduces an element of artificiality into the results of science, it is not hard to demonstrate that much of the fruitfulness of scientific procedures has its source precisely here. Instead of aiming at a comprehensive, all-embracing "natural philosophy" from the start, the thing to do is to concentrate on the specific problems of biochemistry or neurophysiology, particle physics or systematics, and leave their *integration* for the future.

Alongside the development of these scientific "disciplines," there developed also the human self-images characteristic of individual scientists and of the professional collectives into which those scientists gradually came to organize themselves. As individuals, modern scientists came to see themselves as being disinterested, single-minded seekers after Truth, who confronted a Nature which they could study *from the outside,* as "rational onlookers." As collectives, likewise, professional scientists formed themselves into the modern counterparts of monastic orders or guilds, which accepted responsibility for cross-checking the achievements of their individual members and thus satisfying themselves of their intellectual merits. So science-the-rational-enterprise, represented by the totality of scientific "disciplines," became embodied in science-the-communal-activity, represented by the totality of corresponding scientific "professions." And my central thesis is this: the very *success* of the procedural and institutional means that scientists have devised during the last three hundred years—the very fact that these methods have at last put us in a position to realize Francis Bacon's social dreams—is in our own day doing more than anything else to transform the relations between scientists and the world they have to deal with.

Thus the very *success* of modern science is the greatest obstacle to its own *continuance.* The very means whose adoption led to this success have *unfitted* science, as we have known it over the last three hundred years, for dealing adequately with the novel situation created *as a consequence of* that success. The joint specialization of disciplines and professions—disciplines organized around self-isolated groups of questions asked "from the point of view of," say, molecular biology and professions organized around the self-validating groups of professional scientists who embody those disciplines—has been an indispensable element in the intellectual triumphs of the last three centuries of scientific work. But this same specialization has now become an obstacle to dealing with the problems that will face us all (scientists included) in the *next* three centuries, as a by-product of those triumphs.

Since 1650 (I am arguing) the very success of science has brought it to a state that is *maladapted* to the problems that will face us henceforth. We are, I believe, in a position not merely to identify the sources and character of this maladaptation, but also to personify them. For the whole history of modern science has been dominated by the examples and actions of two men, whose intellectual ambitions were not far apart, yet whose views about the *morality* of science rested on quite different principles.

On the one hand, we have Francis Bacon, to whom I have already referred in passing. Bacon himself, of course, was very little of a scientist: even less of one than Thomas Jefferson, for example—who may have been a populist by political conviction, but who was not above shocking the multitude by keeping dinosaur vertebrae in the attic of the White House. Rather, Francis Bacon was a Chief Justice, a historian, and political theorist. That charming seventeenth-century gossip, John Aubrey, reports Bacon's personal physician, William Harvey, as remarking, "He philosophized like a Lord Chancellor—I have cured him!" For all that, it was Bacon, rather than anyone else who first depicted at length, and in detail, the long-term human and social goals at which modern science has directed itself. And all those men, such as the late J. D. Bernal, who have discussed the social functions of science over the last fifty years, have taken Francis Bacon as their exemplar. Thus the title page of Bacon's *New Atlantis* quite naturally provided the dust jacket and frontispiece for Bernal's own book, *Science in History*.

What the first few generations of "new philosophers" never clearly saw, however, was just how far science would have to go, before there could be any question of realizing *any* of these long-term Baconian ambitions. Over the first two hundred and fifty years of modern science, scarcely anything can be shown in the way of "human welfare" that might justly be laid to the credit of science or the scientists. Throughout this time the scientific enterprise remained, not so much an instrument of social improvement, as (in Jacques Barzun's phrase) a "glorious entertainment." And if we are honest with ourselves, we must surely admit that this is *still* largely the case. Any academic scientist with a spark of self-awareness must surely have exclaimed, in some amazement, "To think that I can do the pure research I enjoy doing more than anything else in the world, *and get paid for it too!*" Ever since Samuel Pepys, the diarist, who was also the Secretary of King Charles II's Admiralty, persuaded the king to finance the newly established Royal Society of London, as a result, professional scientists have used Francis Bacon, not as a source of moral or intellectual inspiration, but as their public relations man. It is not just during the last fifteen or twenty years of "grantsmanship" that scientists have become adept at finding social pretexts for research projects whose authentic justification is purely intellectual. Rather, the scientific history of the entire last three hundred years is littered with unredeemed social IOUs. Although scientists have continued to dream—and to preach—about a future world in which their activities would be an instrument for improving the conditions of human life, and have repeatedly infected their fellow men with these same dreams, their actual work has been largely devoid of direct practical consequences.

So Francis Bacon has remained one of the great totem figures of modern science: the man whose picture, and social ideas, scientists have been accustomed to put forward in their external dealings with the rest of the world. Bacon has been their Minister for External Affairs. By contrast, the man scientists have actually emulated—the man whose lifestyle they have made their own and whose single-minded devotion and intellectual severity they have taken as their own ideal—has not been this worldly, successful, socially minded Judge. Instead, it has been that strange, dark, introverted, obsessive individualist whose curious, contorted psychohistory Frank Manuel has recently unraveled in his *Portrait of Isaac Newton.* Despite all propaganda about the "mission-oriented" character of scientific research, despite all the public emphasis on the long-term material benefits of science, the actual preoccupations of most working scientists have been strictly intellectual. They have learned to choose their research questions not on practical grounds, but with an eye to the conceptual shortcomings of their respective disciplines; and much of their success has been owing to that very fact. Indeed, if at that earlier stage they had done anything else—if, for instance, they had concentrated on research topics directly relevant to the practical needs of human beings, rather than to the intellectual deficiencies of their theories—they would have been acting like King Canute. The result would have been to slow down both the intellectual and practical progess of their work. For, until a solid foundation of scientific understanding had been laid down, by exploring theoretical questions in whatever direction they led, regardless of social benefits or technological spin-offs, practical fruits could not realistically be expected.

When I say that scientists have emulated the "life style" of Isaac Newton, I am not suggesting that they have modeled their personalities after his. (Given all that Frank Manuel has told us about Newton's infancy, that would be hard to do!) Rather, I have in mind the patterns of institutions, the forms of life, the "ways of being in the world" that have grown up around the Newtonian way of doing science: all those things that have helped forge the intellectual procedures and sociological forums characteristic of the modern scientific enterprise—those things that have given this union of speculative innovation with communal criticism the sustained effect it has in fact achieved over the last three hundred years. This Newtonian "way of being in the world" is the central thread around which the professional institutions of sciences have crystallized, and it is this too that permits John Ziman to compare these modern professions with the "monastic orders" of the Middle Ages.

Even today, most youthful apprentices to the scientific professions still adopt the same single-valued, single-minded rule of life as the religious novices of earlier ages. They devote themselves to their pursuit of better scientific understanding in a chosen discipline with the same total commitment that the young medieval monk gave to his pursuit of Sanctity. Any scientific apprentice with a serious concern for his own career, for example, knows that one prime professional necessity is to demonstrate that he is "sound," rather than "wild." This means resisting all peripheral temptations and distractions: no popularization, no publicity-seeking, no overt excursions into philosophy, or very obtrusive politics, or good works—not even any trespassing into other branches of science. The "sound" young man, who wishes to make his way in science, soon learns that he must follow the strict rule of his particular discipline, at any rate until he has "made it" into the National Academy or the Royal Society. A story is told about J. D. Bernal and his son Martin, who is also a talented scientist. They had had a stimulating evening, speculating about some oddly ill-understood topic—very possibly the theory of the liquid state on which Bernal spent much of his later life. It occurred to J. D. that it might be worth preserving the results of this conversation in the form of a joint paper. Martin's response was short and to the point: "That's all very well for *you*, Dad. *I've* got my career to consider!"

The collective commitments which young scientists share, and in terms of which they appraise each other's performances, display this same single-valued concern. If we listen to serious professional conversations among scientists, we soon discover that the tacit orders of merit in which they rank the members of their respective professions depend, similarly, upon purely Newtonian criteria. Who cares about social or political welfare, when intellectual or spiritual virtues are at issue? What young religious novice would not have preferred to be recognized by his peers for saintliness, rather than ending up in some worldly position, like being an archbishop or even a cardinal? And what young scientist would not prefer to be recognized by *his* peers for intellectual power and ingenuity, rather than ending up as something boring, like president of Harvard University or chairman of the President's Science Advisory Committee?

These Newtonian commitments, which show themselves in the guiding motives of working scientists, are reflected equally in the formal institutions into which modern science has been organized. It is natural enough that a professional association of nuclear physicists, for example, should elect its members and officers from among the men who have made distinguished intellectual contributions to nuclear physics. Yet the same, purely intellectual criteria are also relied on, even in the case of institutions whose goals are, on the face of it, Baconian ones—institutions, such as the National Academy of Sciences, which were originally established by legislative statute to give advice to government about practical and social issues. Indeed, it comes as a surprise to professional scientists if anyone calls this practice into question. Yet we should be asking, is it really the case that men whose talents have been tested in *disciplinary* forums alone will necessarily be best suited, also, to give advice about the implications of science for social *welfare* or national *policy?* It is at this point that the Newtonian morality of science finally begins to run up its own intrinsic limits.

Two general features, then, have characterized natural science during the centuries since Copernicus: first, the guiding epistemological vision of Nature, as presented to the detached scrutiny of a scientific onlooker who is not himself a *part* of Nature; second, the single-valued Newtonian morality that has shaped the activities both of individual scientists and of the scientific professions throughout the modern period. It is precisely those two characteristic features that have broken down in recent years and thus given rise to our current quandaries.

As to the epistemological feature, the farther we move along through the twentieth century, the more difficult it is for human scientists to continue viewing themselves as "rational external observers," studying a Nature of which they do not themselves form a part. That picture, which was for so long fundamental to the development of modern science, has broken down both in theory and in practice. A century and a half ago, for instance, so distinguished a scientist as J. J. Berzelius could still afford to be quite sceptical about the possibility of a scientific neurophysiology. In his eyes, the very notion of a nervous system—and so a mind—studying its own operations involved some kind of a vicious regress. Such a mind would be trapped in an intellectual Hall of Mirrors. There were thus intrinsic limits on the applications of scientific ideas. The world of physicochemical mechanisms could never be expected to embrace the brain, or the higher mental functions in which that brain is called into play. For us today, such a self-limitation is out of the question. Just *how* the physicochemical workings of the brain are related to those higher mental functions is still an immensely complicated and difficult problem; but there can be no question of regarding the whole idea of a "cognitive neurophysiology" as involving gross paradox, still less as a contradiction in terms. Even those who still place some limitation on the development of neurophysiology nevertheless regard this whole area of inquiry as among the most active and important elements in contemporary science. Indeed, from neurophysiology, all the way across a broad spectrum, to human ethology, primatology, and psycholinguistics, the thrust of scientific development at the present time is toward *reintegrating* Man into Nature. On the most fundamental level of theory, Man the creator of Science has become part of his own subject matter: The Knower and the Known are involved in one another, more radically than Descartes and Newton ever supposed.

This novel epistemological involvement between Man and Nature also shows up in other areas of twentieth-century natural science. Whether we turn to relativity or quantum theory, the involvement of the observer with his observation—the impossibility even of determining what counts as a "natural phenomenon" in a manner from which the observer himself is totally divorced— has become a commonplace of theoretical physics. In these respects, our whole theoretical analysis has become so *fine* that the older dichotomies and separations have become quite impossible. Meanwhile, on another level, human intervention in the world of Nature has, in point of practice, become so *gross* that there, too, our own activities have become part of our own fundamental problem. We can no longer handle or exploit the resources of Nature for our own practical purposes as though they were unlimited, and as though our drawing on them made no difference. On the contrary, this intervention begins to make a crucial difference, so that we can no longer even define the conditions of our practical problems without taking our own involvement in the relevant situations properly into account. Whether we consider *either* the development of contemporary scientific ideas *or* the handling of contemporary practical affairs, in both areas we are faced with brand new tasks. What we now have to do is to come to understand what it is to operate *from within* a World of Nature of which we Men are ourselves a crucial part. On the intellectual side, to "operate" means to engage in those intellectual activities by which we seek to *understand* that world from within. On the practical side, it is to engage in those projects by which we deal with, and take advantage of, the characteristics of the world in which we live. Either way, with Heisenberg and Einstein at one extreme, through cybernetics and information theory, to ecology and conservation problems at the other extreme, the message of contemporary science is the same. As we approach the last quarter of the twentieth century, our science is having to become *new,* not just in its content and concepts, but in its epistemological principles.

Against this background, I shall explain in what sense the present time in which we are living represents an age of "transition." The transition in question is a transition from a state of affairs in which the lives of individual scientists, and the collective institutions of science, could be—could *afford* to be, indeed, *had* to be—monastic, detached, cool, onlooking, apolitical, standing aside from society and human problems, to a new state of affairs which we do not yet fully comprehend. This new state of affairs will be one in which the monastic, Newtonian life-style is no longer adequate for the needs of science; no longer adequate, because science itself has achieved the first of the tasks it set itself three hundred years back (the intellectual task) and now must move on to the second, and more difficult, Baconian or practical task.

The single-valued morality of traditional science arose from the special needs of the Newtonian Era. So long as the intellectual goals of natural science could be pursued, and had to be pursued, in this monastic manner, the moral claim of Truth or

Understanding on the mind of the scientist was, for professional purposes, an *exclusive* claim. During that era, a man who wandered off and concerned himself with building steam engines, or inventing medicines, or worrying about the state of the sanitation, in this way, quite simply, *stopped being a scientist.* So long as this era lasted, the Baconian dream was something for external consumption only. It imposed no active moral demands on the day-to-day life of the scientist, or on the structure and activities of his institutions. For the time being, the pursuit of understanding was everything. Practical fruits could wait.

This state of affairs has come to an end. For better and also for worse, the Newtonian Era is over and Francis Bacon can no longer be denied. We may not be clear in our minds just *how* this change is going to affect the lives of individual scientists, or the activities of scientific institutions. In these respects, we face the new age with far too little done by way of preparatory thought and analysis. What is certain is that, from this point on, any *exclusive* claims for the Newtonian morality of science can no longer be sustained. If scientists do not begin to take their own Baconian morality more seriously in practice, indeed, their fellow citizens will be increasingly tempted to impose a greater social responsibility on them from the outside. In short, the time has come when our pious references to Francis Bacon must go beyond lip service, and—*even while remaining scientists*—we must start to redeem all those Baconian IOUs.

What will this mean for science? That question raises a whole spectrum of very large issues, some of which are discussed by other participants in this symposium. Certainly, I would claim that the institutional arrangements which natural scientists deal with the rest of society are ill adapted to the demands of our new society. During the years since World War II, the politics of science has been too much the politics of the pork barrel and too little the politics of liberalism and reform. So it is an ironical chance that several of the most significant agencies operating in this area have been swept away by the current United States Administration, just at the moment at which the legislative branch is in the process of setting up its own brand-new Office for Technology Assessment, since it is this O.T.A. which may yet develop the means of dealing with the Baconian problems that natural scientists have too long set aside or postponed.

I call this chance an "ironical" one for historical reasons. To many of us, it has long appeared that the executive branch of the government has a built-in power to monopolize the best in the way of scientific advice, and so to increase the ascendancy of the executive or the legislative branch which is typical of the present time. Now, the situation is suddenly and arbitrarily reversed. The legislative branch alone now possesses the beginnings of a machinery which may enable the results of science to be taken properly into account in the drafting of legislation and the formation of policy. Meanwhile, we are left with a collection of older, longer standing scientific academies, societies, and other institutions, whose very structures reflect the needs and habits of the previous Newtonian period. How then can we devise new channels of communication, new procedures of the discussion and exchange of ideas, by which the development of science and the needs of humanity can be brought into greater harmony? This is too broad a question to be left even to the National Academy of Sciences. It calls for a kind of reflection in which historical and sociological, political and philosophical insights will all be meshed together. I wish we had some better ways of answering those questions, but symposia like the present one at any rate contribute to arriving at them.

Finally, what will the life-style of scientists need to be in this new Baconian era? What will the new breed of men have to be like, who are both scientifically expert and humanly concerned? How will they escape from the influence of the older, single-valued attitudes that shaped the professional activities of science for so long? Those too are large questions, which I am here only interested in opening up. The one thing I am sure about is this: Newtonian monasticism will increasingly appear an unacceptable and oversimplified way of confronting the moral problems of the scientist's professional life. The twin moralities of science—Newtonian or internal, and Baconian or external—which came into existence together at the time that modern science was created, must both of them somehow be integrated into the future patterns of the scientific career.

Let me end on a note of nostalgia. When I think about the psychology of science—when I think, that is, about the considerations that have for so long drawn intellectually curious young men into the sciences—I am strongly inclined to believe that the Newtonian single-valued goal of the scientist's life has been one of its main attractions. In the Middle Ages, those men who found the complexities and compromises of more worldly modes of life distasteful, unmanageable, or overwhelming, could turn to the monastic life of the cloister for relief. So also more recently, by entering the glorious world of abstract thought, in which the pursuit of greater understanding was its own justification, the apprentice scientist has been similarly freed from moral conflicts at least in his professional life; and that is something that could bring comfort and exhilaration to the scientific novice. Alas! That monastic option is no longer open to the scientific novice of today; hence, the feelings of having been plunged into a new sea of moral quandaries. For three hundred years, one of the two scientific moralities that were born in the aftermath of Copernicus's work—the Baconian morality—lay dormant, while the other Newtonian morality achieved a monopoly of professional attention. Now the other twin is reawakening. What does this mean for the motivation of individual scientists? How far will the professional life of natural scientists be subject from now on to the conflicting moral claims from which it has so long been exempt? These things remain to be seen. For the moment, the walls of the scientific monastery are down, and the cold winds of moral ambiguity are blowing through its rooms. How much of the old structure and intellectual ideals will survive the debate on which we are just embarking, few of us would dare to guess.

COMMENTARY

HENRYK SKOLIMOWSKI

The Moral Dilemmas of Modern Science

IT WOULD BE a trite beginning to say that I find Stephen Toulmin's essay elegant and eloquent in form, and rich and rewarding in content, and that I substantially agree with the large body of ideas it contains. Indeed I accept almost entirely his *critical* analysis of present science and only wish he did go further in his critique. This is precisely what I intend to do. My critical remarks about science, and indirectly about Toulmin's essay, stem from my genuine concern about the future of science and of society, a concern which I no doubt share with Toulmin. Because I consider the situation so alarming, I allow myself to take a rather uncompromising stand. I wish to say at the outset that I am intellectually indebted to Stephen Toulmin and that I regard myself fortunate to have interacted with him on so many occasions, whether it was at Oxford or in London, at Brandeis or in Boston, in Amsterdam or at Ann Arbor.

Why do we reexamine the moral status of science? Why do we reevaluate its place in our civilization? What has happened to the alleged moral neutrality of science? What has happened to its Olympian indifference as the superior mode of dealing with the world? Stephen Toulmin says: "For the moment, the walls of the scientific monastery are down, and the cold winds of moral ambiguity are blowing through its rooms." And he adds, "How much of the old structure and intellectual ideals will survive the debate on which we are just embarking, few of us would dare to guess."

There have always been critics and hecklers of science. However, from the mount of its Olympian superiority—and because it felt that it was carrying the torch of progress and had a great historic mission—science in the past consistently ignored its critics and intimidated its hecklers. All of this is now changed. There is a deep crisis in the citadel of science itself. Here again I agree with Toulmin when he says:

> the contemporary situation in natural science has a curious fea-
> ture, . . . this is, the quite unfamiliar sense of *moral ambiguity*
> among working scientists today, the novel doubt in the scientific pro-
> fession itself about the fundamental value (even, about the fundamen-
> tal justifiability) of the scientists' own purely intellectual activities. . . .
> Nowadays by contrast, alongside a deepening suspicion of science
> from the outside, we can observe a fresh lack of conviction within
> parts of the scientific profession itself—a lack of conviction that has
> no exact counterpart in the last three hundred years.

The priests of science are losing their faith. Toulmin does not quite bring himself to say this, but this is what is implied by his remarks. So far I am in full agreement with Toulmin. But from this point on our paths diverge. For, after he perceives the unprecedented crisis, after he acknowledges that this crisis touches even the fundamental justifiability of the scientific enterprise, he seems to refuse to take a hard, critical look at the foundations of science, as if afraid to commit a blasphemy by casting a shadow on the goddess of science. Toulmin's analysis ends up in paying homage to the traditional deity of science while paying lip service to its present inadequacies. In essence Toulmin suggests that we have to switch from one idiom of science to another idiom of science; from the Newtonian perspective, to the Baconian perspective. I will argue, on the other hand, that we have to change not only an idiom *within* science, but we have to change the whole idiom *of* science. This will bring about an altogether different morality of science.

Let me first point out certain limitations of present science which indicate that its historical mission has gone sour. From this analysis it will clearly follow, I hope, that neither the Newtonian, nor the Baconian morality of science would do, for both represent the idiom of science which is distinctly waning.

The practice of science was meant to free the human mind from superstitions and prejudices characteristic of the prescientific cultures. The pursuit of knowledge in the way of science, being "objective," "value-free," and uncontaminated by the fictional elements, was meant to free the mind and *somehow* make the possessor of this mind a better human being. Now, in the process of liberating ourselves from one kind of tyranny, the tyranny of prescientific prejudices, we have become captives of new kinds of prejudices; we have become subjected to another conceptual tyranny: the world view of science the uniformity of which is imposed on us with a compulsion characteristic of traditional orthodox theologies.

Now we must clearly understand that the nature of knowledge determines the nature of the world around us. There is no understanding of the world without the intervening agent which is called knowledge. We perceive and understand what we are made to perceive and understand through the knowledge we acquire. The dominant position of science in our systems of learning assures the domination and perpetuation of the scientific world view which amounts to a vision of the world through the spectacles of science.

This problem has a more general aspect. Our common sense is no doubt a recapitulation of the physical world as formulated by Newtonian physics. Our language is profoundly influenced by the relationships which are taken for granted in science and in common sense. Now, our thinking is to a large degree determined by our language. The conceptual framework, which a given language represents, is a conceptual matrix which enables us to express certain relationships and does not enable us to express other relationships. Every civilization is under the influence of these subtle yet all encompassing conceptual determinants. Every conceptual framework is selective. It acts like a sieve and thus controls the flow of phenomena which will become the subject of our understanding and upon which our understanding and ultimately our action will be based. The conceptual framework of Western man in modern times has been profoundly affected by science. And so was his language and his thinking. We have no choice to accept or not to accept the knowledge and the world science offers to us, for the dogmas of science are thrust upon us from an early age. The point I am making is that science has not liberated our mind from all prejudices, but has provided a new set of preconceptions which is a new set of prejudices through which we are compelled to view the world for fear of being ostracized from a society that is governed by scientific reason. Equipped with the principles of our superior understanding and the principles of scientific rationality, we have disrupted natural cycles, and in fact the whole ecosystem, and we have constantly misinterpreted and distorted the world views of other people and their cultures by imposing on them our world view and our culture and trying to bring them into conformity with our own principles.

Furthermore, the evidence becomes more and more compelling that the objectivity and rationality of science are not only more or less arbitrary assumptions which allow us to play the game called "science," but that they are, at least in some respects, harmful dogmas. The objective approach advocated by science requires and demands a rather special attitude of the mind. The mind that perceives spirits behind every tree and every mountain (e.g., the mind of the American Indians) is not the most suitable instrument for scientific inquiry in our sense of the term "science." In order to enter the labyrinth of science and not get lost in its corridors, we have to train ourselves in the power of analysis and abstraction. We make sense of nature in scientific terms by the painstaking separation of some phenomena from other phenomena. We separate some phenomena from other phenomena, or separate some aspects of phenomena from other aspects and then pretend that these separated aspects are things-in-themselves, the ultimate components of nature. This separation is accomplished through the process of intellectual abstraction. The deeper we probe with the analytical tools of science and the more successful we become on science's terms, the more deeply we must steep ourselves in the process of abstraction.

The atomization of phenomena brought about by the process of ever-increasing abstraction is another name for *conceptual alienation.* For we must alienate phenomena (from other phenomena with which they are naturally connected) in order to bring them into the sharp focus required by science. The habit of scientific veracity is the habit of atomization, abstraction, dispassionateness, alienation. The scientific mind is the epitome of abstraction, atomization, and alienation.

We cannot change the structure of our mind from one occasion to another. Once our mind is molded in a certain way, it functions this way. Our minds are molded by science and the scientific world view. Our minds are structured to receive the world according to the requirements and dicta of science.

Conceptual alienation, which is a precondition for the successful practice of all quantitative science, is only one step removed from psychological alienation. The atomized mind, once it is set in motion, does not discriminate. It atomizes equally efficiently the phenomena of the physical world, the phenomena of the social world, and the relationships of the human world. Alienation in the human world is at least partly the result of the habit of scientific thinking with its abstractedness, separation, and atomization.

Thus, we cannot escape the conclusion that *conceptual alienation, developed under the auspices of modern science, is at the root of all alienation in social and human relations.* It is not an accident at all that alienation is the social disease which afflicts most severely those societies that are scientifically and technologically most developed. Alienation is a by-product of technological change. Alienation is the human residue of the process of mechanization, atomization, compartmentalization which science and technology push forward with a relentless thrust.

The ideal of the scientific mind which is geared to factuality, objectivity, and abstractness is one side of the coin of which the other side is moral indifference and ultimately moral anaesthesia. We did not have a problem with this mind and science at large as long as science did not start to dominate our entire civilization. The moral crisis of science came exactly at the point when science emerged triumphant.

What then are the objectionable moral consequences that the practice of science produces in the long run? Let us once more reflect on the universe of science. It is the universe of physical facts and objective relationships that occur among them. Is there a place for values in this universe? No. Except for one value: the value of objectivity. In enhancing and perpetuating the value specific to science, namely objectivity, we had to suppress other specifically human values. The gospel of the indifference of science is in fact an ethical stand. In the long run it produces moral anaesthesia. The immorality of science lies in the fact that it makes us forget that man is a moral phenomenon, that it desensitizes us as ethical agents, that it "purifies" man of moral concerns. Within rigorous curricula of modern scientific-rational education we are taught to be objective, not compassionate; we are taught to be efficient, not morally concerned. Our mind is prodigiously developed to seize physical relationships and connections, but is lamentably unprepared to seize and articulate human and moral relationships and connections. In the light of this analysis it is transparently clear that neither the Newtonian morality of science, nor the Baconian morality of science can serve as the possible foundation for the pursuit of science in the humane context. Both the Newtonian and the Baconian moralities stand for the same mental attitude: the attitude of apparent secularization which is a mask for the idolatry of Physical Fact; the attitude of apparent liberation from the constraints of natural phenomena which ends up in the violation of the limits of nature, and thereby produces even more severe constraints; the attitude of apparent tolerance and enlightenment which ultimately subjects everybody to the terrorizing uniformity of scientific reason. In short, the Newtonian and Baconian moralities stand for the attitude of the domination of nature (and other civilizations) via the instrumental and quantitative manipulation, coercion and compulsion. This has been the dominant attitude of the Western man in the post-Renaissance time which science, practiced in the Baconian-Newtonian key, has remarkably helped to develop. This attitude has not been uniquely formed by science. However, given the tendency of the occidental mind in the post-Renaissance period, science has become a most ruthless tool in the pursuit of our secular goals, and ultimately, through the marriage with technology, came to epitomize the occidental frame of mind.

In spite of his occasional statements to the contrary, Bacon *was* one of the prophets of the instrumental civilization which is based on the faith in Pure Reason, Fact, and Manipulation of nature. Bacon wrote in a memorable passage: "For the whole world works in the service of man; and there is nothing from which he does not derive use and fruit . . . insomuch that all things seem to be going about man's business and not their own." Entailed in such beliefs are the germs of the ecological doom. Not in the rudiments of the Judeo-Christian tradition, as Lynn White insists, but in the sentiments like Bacon's are concealed the seeds of the ecological disaster.

To accept the Baconian morality of science (as Stephen Toulmin urges us to) is tantamount to acknowledging that science *still* holds the key to human progress. This is the premise which Toulmin still cherishes, when, for example, he asserts: "Technology is ecologically damaging not because it is too *scientific,* but because it is too *unscientific*" (italics his). This kind of sentiment is, in my humble opinion, an expression of the simplemindedness that borders on the criminal.

The point is that we have lost the faith in science exactly because we ceased to believe that science still holds the key to (further) human progress. In such circumstances, switching the emphasis from one idiom of science to another idiom, while preserving the enterprise of science as it is and while ignoring the main cause of the present crisis of science, simply will not do. The crucial question within which all our problems could be condensed is: "Can science fundamentally change itself by its own means?" Stephen Toulmin and other protagonists of science answer this question affirmatively. My answer to this question is negative. It suffices to examine the tools and concepts of science *as* science to realize at once that science does not possess means to transform itself into something that could bring us into harmony with the cosmos and which would heal ecological wounds and close the disrupted cycles of nature. We must remember, furthermore, that the tools and concepts of science were not developed by science itself, but were spurred and inspired by larger ideals such as the secularization of our world view and material progress. The ideals of *secularization* and of *material progress* were *not* physical concepts or empirical data but were the projections of the mind which enabled us to explore a new conception of the world and a new mode of what we once considered a worthwhile life. We need new social and human ideals now in order to remake present science so that it enables us to explore a new postempiricist conception of the world, and a new (only dimly realized but clearly emerging) conception of a worthwhile life.

What I wish to propose is the return to the Copernican morality of science. And it gives me great pleasure to do this during this Copernican celebration. Copernicus, as we all know, was engaged in the description of the celestial spheres. But this was not meant to be a simple description of certain mechanical interactions. Copernicus was fascinated with the idea of the perfection of the circle and haunted by the imperfection of noncircular motions. His New Astronomy was meant to bring back a perfect order to the celestial spheres disfigured by Ptolemaic astronomy. It was the divine cosmos that Copernicus wanted to grasp and express. This cosmos entailed the aesthetic and the moral orders which were all fused together in the visible movements of the celestial spheres. Science was thus for Copernicus the unified knowledge of the mechanical, the moral, and the aesthetic, all of which must simultaneously be present in order to make sense of the divine cosmos. Science was for Copernicus a moral and thus a normative enterprise.

If the tradition of modern science is characterized by pervading empiricism, then Copernicus is not a part of this tradition for he was a Platonist and an apriorist *par excellence*. If the tradition of modern science signifies the avoidance of values and normative commitments in favor of mere descriptions, then Copernicus is not a part of this tradition, as he considered science to be a normative enterprise: a vehicle to reveal the beauty and harmony of the world rather than its mere physical aspects. If the tradition of modern science, particularly in its Baconian embodiment, ties science with utility and material progress, then Copernicus is not a part of this tradition for he saw the main value of his astronomy in its beauty and the beauty it reveals.

The impulse and the inspiration of the Copernican conception of science was his conception of the universe which he considered harmonious, beautiful and endowed with value. The Copernican morality of science has the same roots: the conviction that we live in a worthy universe, and that through science we are able to bring together the inherent value of the universe and the intrinsic values of man. Copernicus's terminology and various fragments of his *De revolutionibus orbium coelestium* eloquently testify that science for him was not a value-free set of descriptions, but the revelation of the harmony of the world. He talks about the symmetry of the world, about *mundi forma, forma perfectissima, admiranda, nexus harmoniae, harmonia totius mundi* which make the world *pulcerrimum templum*—the most beautiful temple. And here are some fragments from *De revolutionibus:*

> Among the many and various studies and arts which give the satisfaction and nourishment to the mind this, in my opinion, we should first of all devote ourselves to and most ardently cultivate which are in the realm of beautiful things and most worthy of knowing. Such are the sciences which are concerned with the celestial movements in the world, with the course of stars, their magnitude and distances, their rising and setting and the cause of other phenomena in heavens, and which finally explain the whole form of the world. And what is more beautiful than the heavens which contain everything that is beautiful? The names themselves: *Caelum* and *Mundus* are an evidence, of which one signifies purity and ornament and the other a work of a sculpture. It is because of its exceptional beauty that many philosophers called the heavens simply the visible deity. [Introduction 21]

> Nothing is more repugnant to the order of the whole and the form of the world than a thing that is out of its place. [Chap. I, 23]

> In the middle of all resides the sun. In this most beautiful temple, could we have a torch in a better and different place than the one

from which it can simultaneously illuminate everything? Because of this some call it the lantern of the world, others its reason, still others its ruler. [Chap. X, 38]

In this order we have thus found an admirable symmetry of the world and an established harmonious relationship between the movement and the magnitude of the spheres which could not have been found in any other way. [Chap. X, 39]

Now, after science became a predominantly mechanical enterprise, we tried to purge the Copernican science, and thus the Copernican morality of science of its moral and aesthetic connotations. We assumed, of course, that we knew better while Copernicus was somewhat misguided and marred his astronomy with irrelevant philosophizing. Giordano Bruno and Johannes Kepler fell victims of the same process of reinterpretation of their entire philosophico-scientific opus into simplistic categories of the mechanistic science, with the result that they were declared to be only "precursors of science," "belonging to the period of transition," "not fully liberated from the wrappings of the medieval theology." Most recent research clearly indicates that science for Copernicus, Bruno, and Kepler was far richer, more mysterious and interesting a phenomenon than we were led to believe by the orthodox interpreters of the history of science. One of the reasons for this richness and wholeness is that their science did not limit itself to the physical aspects of the universe, but also included the moral, the aesthetic, the transcendental. Herein lie the roots of the Copernican morality of science; or at least what I take to be the Copernican morality of science. I reject the notion that this conception of science signifies a return to medieval obscurantism. The notion of medieval obscurantism, it must be remembered, is the fabrication of the eighteenth-century enlightenment which was as brilliant as it was shallow. It may be claimed (whatever misconstructions the eighteenth century imposed on the Middle Ages) that in medieval times the dominant philosophy qua theology suppressed science. It may be equally validly claimed that in recent centuries science has suppressed philosophy and theology and a great deal of human content that was expressed by philosophy. And who is to say with certainty which of these suppressions is more detrimental to the human spirit in the long run? Rhetorics aside, I do not argue here for any return to scholastic science, or the scholastic constraints imposed on science, but I rather argue for a broader conception of science which will be based on a broader conception of knowledge or on the unification of the cognitive with the normative. This broader conception of science does not invalidate the laws of Newtonian mechanics, nor does it render the validity of all science subjective and depending on the whims or the fancy of particular individuals.

The life of the human species is normative and science as a normative enterprise will be subject to norms that are intersubjective. But this intersubjectivity will not be regulated by the canons of scientific objectivity, for if this were the case we would have undermined the whole idea of normative science, as based on new human ideals and as contrasted with present quantity-ridden science which is alienated from the human context. And let me also add that I do not attribute the process of the alienation of science from human concerns to any specific *political* ideology (which was the favorite strategy of the Nazi Germans, for example), but rather to the very process which enabled science to become triumphant: the process of abstraction, atomization, ever-increasing specialization—those very things which make science great (in its own specific universe) and which made it ultimately pernicious (in the human universe at large).

In conclusion, one of the fictions that was fabricated under the auspices of modern science and modern empiricism is that truth and goodness must necessarily be separated. In the universe in which man actually lives as man, in the universe that is both alive and compassionate, there is *no* separation of the moral from the cognitive, let alone pitting one against the other. The evolution of living organisms into ever more rich and ever more versatile forms has been accomplished by the accumulation of knowledge which is life enhancing. Such must be the function of knowledge that the future science will provide. We must make science produce knowledge that is humanly and thus morally enhancing. We must adjust science to human morality and not conversely. We must bring science back to the nexus of the moral universe.

This will require enormous changes in the nature and method of science. Science will have to become a normative enterprise directly related to the fulfillment of the innermost human needs. We shall have to return to the pre-Platonic ideal of the unity of truth and goodness.

Only then will the genuine morality of science be secured. Only then shall we be able to eliminate the spurious and pernicious dichotomy between the human morality and the alleged morality of science. Man is a spiritual agent and science that undermines his spirituality is not worthy of man. It is an unsound fancy to expect that man should serve the cause of science rather than science the cause of man.

VII　Biological and Social Theories — A New Opportunity for a Union of Systems

A. HUNTER DUPREE

ON the five hundredth birthday of Copernicus we come to announce that his day is at long last over. The earth will continue to orbit the sun; the sun is still a star; the stars still are suns. Yet the displacement of man from the center of the universe emerges as the most colossal misunderstanding in intellectual history. In the long run the great fruit of the Apollo flights will not be any practical return but will be the perception that the observer is at the center of the universe. The astronaut, after looking at Ptolemy's blue-green earth and Copernicus's sun from his own chosen path away from both, eventually returns to Earth; he as the observer takes the center of the universe with him.

The Plan

Talcott Parsons has elaborated a social system through a long and rich career. He has also placed it as a system into a general action system.[1] The plan here is to recognize both that Parsons's system has been already worked out in considerable depth and sophistication and that it is still undergoing development. Hence, instead of elucidating it for detailed criticism, I am accepting a simplified version of it. He has worked out four subsystems of action as follows (fig. 1):

A. Behavioral System
G. Personality System
I.　Social System
L. Cultural System

136

Each of these subsystems shows quadripartite divisions in four functions, which Parsons calls the four-function paradigm—(A) adaptation, (G) goal attainment, (I) integration, and (L) pattern maintenance. In November, 1970, Parsons began a fresh search for a connection between his subsystems of general action and biological systems at the molecular and genetic levels.[2] Parsons has often expressed himself on the exciting possibilities he sees in the application of cybernetic theory to the resolution of the gaps between the social and physical sciences. Because I had made a few remarks at the Bellagio conference which led to the volume, *The Twentieth Century Sciences,* edited by Gerald Holton, concerning the implications of the absence of the late Warren McCulloch and the convergence of systems theory as it emerged in widely separated settings among the disciplines, Parsons discussed these issues with me on December 23, 1970.

My plan, taking Parsons's system as given though incomplete, is to put it in a connected setting with DNA, seen as an information program, and to develop a measurement behavior by which the total system can understand itself, however incompletely. Although this plan is too ambitious for rigorous exposition, I shall concentrate on a description of the system as a whole, including a sketch of the physicochemical background, recognizing that many components may have to be revised later.

System of Action as a File Organization

The fourfold pattern of Parsons's systems and the fact that he himself assigns them to different levels suggests not only the arrangement of his systems into a file organization, but also opens the way for the extension of the fourfold boxes at both ends of the hierarchical structure. I shall call the fourfold boxes arranged in a hierarchical structure, for convenience, a file stack. To avoid determinisms and false analogies, I wish to deny at the outset that the file stack is either a mechanism or an organism. It is simply a file (in an office, not a workshop connotation of the work) which conforms to a normal theoretical file organization as worked out by Becker and Hayes (fig. 2).[3] Thus the file stack can be conceived of as a sequence of nested boxes. It is further convenient to separate the boxes into levels to stress their hierarchical ordering (fig. 3).

The items in any file cannot be known simultaneously. Hence the file has an input limited by the rate of entering items and an output limited by the rate of retrieving items. To make a request for information is to enter an item in the file. Each item can be described as (1) matter—that is, an artifact in the ordinary sense, but including documents; (2) energy—that is, the amount of energy expended to enter, store, and retrieve the item; and (3) information—that is, a signal which puts a sufficient constraint on random noise to be distinguishable from the noise by a value exceeding a chosen threshold.

Each file stack has its origin in the formation of a zygote by fertilization which lays down an information program in DNA, the deoxyribonucleic acid which has become the key substance of the age of molecular biology. Since the file stack will be observing itself, these are the genes of the observer, located on level 1 of the file stack, Parsons's Behavioral System (A) (fig. 1). Because the DNA program can express itself through RNA to synthesize proteins, the file stack begins to build on it immediately, ultimately to express itself in the behavioral organism. At the same time, according to the central dogma of molecular biology, protein cannot synthesize DNA. However many feedback loops may connect parts of the file stack, they cannot finally close the loop with the DNA program within the individual organism. Because of this stop-valve on the recirculation of information, the pattern of the file stack itself remains stable even when widely different sets of items are entered into it. I am making the assumption that (1) the genes are those of *Homo sapiens,* and (2) one of the characteristics of the file stack will be the ability to observe.

Level 2 becomes Parsons's Personality System (G) (fig. 1) which, purely for convenience, I shall call the Freud-Erikson level. Both these levels have a maximum entering and retrieval time in their circuitry equal to ontogeny of the individual, that is from the time the genetic program is formed at fertilization to the dissolution of the program, or death. Even if some of the genes in the form of one phosphate chain of the DNA double helix have passed by meiosis to another fertilization, a new individual and hence a new file stack is formed. Though the file stack we are describing here is obviously for the genetic program *Homo sapiens,* it would be possible to build file stacks from the DNA program of any other species.

Level 3 (fig. 1) is Parsons's Social System (I) and for convenience I shall call it the Parsons level. Here the file items consist of other file stacks and aggregates of file stacks, each tied to its own genetic program but mapped into our original file stack at the societal level in the form of behavior settings. In achieving this level the file stack now takes in items outside the skin of the observer, but these assemblages of behavior settings belong to him and his information system none-the-less.[4] It is at this level that most of the data called history are generated. The individuals are tied together by the templates of institutions—economic, political, religious, and familial. Here also are the templates of educational institutions and the networks which can move items laterally from one file stack to another. Hence while each new file stack has to be stocked afresh, it can enter aggregates built up in other file stacks. There is an efficiency loss in the process, but the file stack is no longer imprisoned in its own ontogeny, so that its systems may be summed with those of others.

Level 4 (fig. 1) becomes Parsons's Cultural System (L) which I shall call the Lewis Henry Morgan level. Like the Parsons level 3 this level contains assemblages of file stacks. Since, however, the items at this level are symbols, they are already programmed by definition for efficient storage and retrieval. The number of items transferred laterally is much fewer and the loss of information in transfer less marked and less given to fluctuation. Some elements of language, for instance, have maintained a high signal-to-noise ratio over a considerable fraction of phylogenetic history. The symbols which make up the cultural system of *Homo sapiens* can be found without exception in two forms. One form, internalized culture, is housed in living individuals, especially in their central nervous systems. The other form, material culture, is a symbol system which has been impressed on the external environment. While material culture must have ultimately come from an individual, he or she may be long dead. It is the two-way interplay between internalized and material culture which constrains all four boxes of the file stack at level 4. In summary (refer back again to figure 3), we see that each level provides a control mechanism for the level below it and the cultural level provides a cover for the file stack which comprehends all the systems and subsystems below it.

To provide an environment for the cultural level in the file stack it is necessary to explore the levels above level 4 very briefly and sketchily, even if it means going beyond Parsons's terminology and carefully developed systems. Since this is a design for a file organization, the designation of the categories is to an extent arbitrary and a mistake can be corrected. Above level 4 in the file stack, level 5 becomes the Earth Ecosystem, or the Linnaeus-Darwin level (fig. 4). Here the major divisions are the inorganic matrix, plants, animals, and the total population of *Homo sapiens*. The latter three sets of items are an assemblage of file stacks organized by DNA programs, hence biological organisms. These three we are calling the biosphere; one, *Homo sapiens* is set off here because the DNA program is that of the genes of the observer of himself around which we have built the file stack. It would be equally possible to build a system with the genetic program of the mouse as a base. It would only be necessary to put the mouse in the separate box, and *Homo sapiens* individuals would then be filed under animals.

Level 6, the Solar System (fig. 4), I have called the Copernicus-Newton level. The level has no items based on DNA filed in it. Only the circumstance that we have built a file stack which observes makes it possible to construct this level. Time consists of periodicity in the movement of satellites and thus is tied neither to the ontogeny nor phylogeny of *Homo sapiens*.

Level 7, the Galactic System (fig. 4), I have called the Einstein level and can scarcely do more than mention this cover of the file stack. It would, of course, always be possible to extend the hierarchy to $L_n + 1$. But to go on to levels 8, 9, and 10 is to go with Dante, beyond the stellar heaven and the primum mobile to the Empyrean.

A footnote to the description of the file stack might point out that it could be extended in a hierarchy downward as well as upward. The atomic, nuclear, and particle levels immediately suggest themselves on the basis of the accomplishments of physics. Let us call it the Bohr range in order not to get lost in detail. The absence of the DNA information program at these levels does not make impossible the entering and retrieving of information. Yet Bohr's principle of complementarity suggests that an observer is necessarily present in the very act of observing. As Gerald Holton has said,

> What Bohr was pointing to in 1927 was the curious realization that in the atomic domain, the only way the observer (including his equipment) can be uninvolved is if he observes nothing at all. As soon as he sets up the observation tools on his workbench, the system he has chosen to put under observation and his measuring instruments for doing the job form one inseparable whole.[5]

The observer mentioned here would be a file stack through levels 1 to 5 in order to have the bench, the tools, the paradigm, the eyes, and the nervous system to make the observation.

The Circuits of the File Stack

For the file stack to hang together one must provide channels to transfer items from one box to another. The channels must be paired for two-way flow in every case, making a total of six pairs of channels flowing laterally at each level. In those cases where the information flowing through one channel of the pair elicits information on the other channel of the pair, a feedback circuit is formed. The items transferred presumably contain information if they are distinguishable from noise, and they must involve the transfer of either energy or energy and matter. Flow may be continuous or discontinuous. The ratio of information to energy and matter in the transferred item may vary through a large range. Since the aggregate of items within each box are those most like one another as compared to the items in all the other boxes, they will be bound together more tightly by interior circuits than the total binding force of the lateral channels. Likewise by definition the cohesion of the total level will be stronger than the total pulls of the channels between levels. Since the channels are connecting stores of items in the various cells of the file stack, and they provide a network of two-way communication, they provide the possibility of equilibrium within the cells and levels while also providing the possibility of changing equilibrium among the various stores.

The channels between levels (fig. 5b) plotted out as lines from each of the four boxes of the next level make sixteen pairs of channels. Thus any level in the file stack will have a possible thirty-two pairs of channels, sixteen to the level above and sixteen to the level below. Again we should note that each passage of information along a channel will be made up of energy or matter and energy as well as the information content. The signals may be intermittent or continuous.

One major advantage of the simple linear circuitry we have used so far is that it can accommodate directly the pairs of channels already worked out by Parsons under the term "General Action Level Interchanges." But it should be immediately clear that in addition to the obvious oversimplification involved, such linear connections would produce either a static file in which there was no interchange, or if some intruding force began a circulation of entries and retrievals, the first minor fluctuation would be subject to positive feedback and soon wrench the file stack by oscillations. Hence to ensure continuing validity of the file categories by a flow of requests and retrievals from one box to another we need movement of information in at least some of the channels. To damp positive feedback oscillations we need a governor to supply a set of difference signals as negative feedback. Such a regulator is built into the file stack by Parsons's definition of "pattern maintenance and latent tension management" as one of the four functions of the "four-function paradigm."[6] In this view each level is treated as a single whole with the pattern maintenance boxes serving as a regulator. Parsons has identified these boxes in the four levels worked out by him (fig. 1) with the letter *l*.

David L. Clarke proposed a tentative general model which can be adapted here as the circuitry for a given level.[7] In his model the inputs are sums of inputs from levels above and below under the name environment. Each level has both an external environment above it in the file stack and an internal environment below it. Each level is capable of an indefinite internal subdividing of the items filed within it. As the model moves from a first state to a second state in time, the equilibrium of items is determined by result of a table of probabilities into which the inputs are entered and from which the outputs emerge.

The exceptional level of the file stack is level 1, the behavioral organism. Here the regulator is the genetic information program coded in the DNA and, in accordance with the central dogma, there is no input into the DNA molecule from anywhere but previous DNA molecules. The molecular biologists have been creative and ironic in the choice of terms for additions to the general language, never more so than in labeling this fundamental proposition a "dogma." Gunther Stent has noted:

> an essential feature of the central dogma [the only feature we need here] is the one-way plan of information,
>
> $$DNA \rightarrow RNA \rightarrow protein,$$
>
> a flow which is never reversed.[8]

Hence a linear diagram of level 1 would correspond to figure 5c. We shall return to this level and this arrangement when later we begin the task of placing the DNA program of our observing file stack into a set of coordinates.

Having delineated the circuits, provided for change in the state of the file-stack system, and introduced pattern maintenance regulators at each level, we can control oscillations toward stable states, provide adaptive response, and introduce regularities by placing constraints on the possible states of the system.

To go beyond the tentative general model of Clarke, we might possibly conceive of the file stack as a set of perceptrons and adopt the computative geometry of Minsky and Papert.[9] Such a procedure would certainly be desirable, but it is beyond my powers at present to arrange the coefficients rigorously. Therefore I shall fall back on a simpler, looser model, environment-organism-environment circuit of the psychologist Roger Barker.

Without pushing the model too far, and always attempting to exorcise the anthropomorphic connotations of the word organism, we can look upon Barker's "organism" as one of the levels of the file stack. The environment means all the interaction with the levels above. The central processes are the interactions with the levels below, creating the internal environment—the term echoes Claude Bernard. Barker's perceptors and effectors represent the channels of input and output, or entering and retrieval. By linking a series of these ecobehavioral circuits in order and with the definitions of the file stack, a picture emerges of the behavior of the levels of the system and we see that the stack as an information processor can perceive in itself certain regularities. In other words it can measure itself.

Measurements

In a slightly different context and with a somewhat different procedure I have attempted to formulate elsewhere a conception of measurement on the basis of Barker's ecobehavioral circuit. The system can be made to work on the channels of the file stack. What needs to be done is to make some remarks about its application in this new context.

In the first place, the perceptor cannot take in the total contents of a file category simultaneously, and it must expend energy for entering and retrieving. Also some time must elapse in the process. Therefore the only things which can be measured are the inputs and outputs—not the contents of the file itself. Hence the diagrams of the file stack are pictures of levels of black boxes, or—to use Stafford Beer's term—esoteric boxes. Of course one can take any file category and subdivide it, creating a denser retrieval of information by nesting subdivisions of boxes as in Figure 3 in structures built on geometric progressions of two. This reductionist procedure will eventually get down to the Bohr range, where complementarity requires the presence of the observer and his instruments to make a measurement. Hence no amount of subdivision can fill the boxes per se.

The one regulator which enters the file stack in a form that is invariant is the DNA program at level 1. From there the copying principle involved in mitosis as the individual develops injects the progression of 2^n into the whole file stack. The doubling sequence—1, 2, 4, 8, 16, 32, 64 . . .—derives from the phosphate chains which make up the double helix. The genetic code injects the number 3, which when coupled with the doubling sequence produces a sequence of $3(2^n)$—3, 6, 12, 24, 48. . . . The twenty amino acids available for building fall in a sequence of $5(2^n)$—5, 10, 20, 40. . . . At the level of the gene, 3 gets into the picture as the Mendelian ratio, which is comprehended in the 2^n progression as we have seen it in the file stack.

One need not claim mystical powers for numbers to observe the early and ubiquitous presence of both geometric and arithmetic progressions. A few sequences are likely to be often repeated, and hence strongly reinforced in the circuits which provide the raw material for measurement. If only arithmetic progressions were available the DNA information system would not transmit information in the way it does. On the other hand, if regulators did not cut down the rate of doubling, no stability would be possible. An embryo starts off gaily 1, 2, 4, 8, 16 cells, but soon the rate begins to change in some cells until at maturity the rate of increase sinks to zero. This alone juxtaposes a geometric progression of 2^n with an arithmetic progression.

$$
\begin{array}{llcccccccccccccccc}
n+1 & 1 & 2 & 3 & 4 & 5 & 6 & 7 & 8 & 9 & 10 & 11 & 12 & 13 & 14 & 15 & 16 \\
2^n & 1 & 2 & & 4 & & & & 8 & & & & & & & & 16 \\
3(2^n) & & & 3 & & & 6 & & & & & & 12 \\
5(2^n) & & & & & 5 & & & & & 10
\end{array}
$$

In order to fill in the gaps left after the first two lines above, 3, 6, 12, and 5, 10, 20 . . . are the first two to suggest themselves. The sequences beginning $7(2^n)$, $11(2^n)$, and $13(2^n)$, appear less strongly in the file stack, but sabbatarians, dice players, and card players know they are there.

The construction of the file stack itself based on Parsons's fourfold paradigm and adapted in this paper to a normal file organization naturally gives a strong prominence to the progression 2^n. The binary digit (bit) in information theory has created an atmosphere in which a measuring system on the base two has again become important. Furthermore genetically controlled bilateral symmetry in the human body forms the basis of the pace as a measure, since only a biped would count in that way. The 5, 10, 20 sequence of fingers and toes is bilaterally symmetrical. The cubit, arm, fathom (two arms) is likewise genetically set up in a 2^n progression by bilateral symmetry.

The measuring done by a person of no special skill and measuring done by a craftsman (observing organisms both with file stacks of their own) use the 2^n progressions with human body referents at level 1 to erect measuring grids which draw standardization and enforcement from society (level 3) and symbols from culture (level 4). Activity in the channels at each level appears to most perceptors as smooth curves of distribution. The system selects a unit near the peak of the curve, places boundaries on its oscillation, and then incorporates it in a grid composed of ratios among integers.

By putting the geometric and arithmetic progressions which abounded in their file stack together, however, mathematicians by the seventeenth century had it in their power to make a grid from a single unit with identical proportions. Let us call it the Napier relation. Once this came into play the discrete integers of the everyman-craftsman measuring system could be grouped again into smooth curves by the calculus. The Napier relation with the superstructure of calculus built upon it, injected into some file stacks which were suitably tuned what we may call scientists' measure. This grid has had its greatest successes at the high and low levels of the file stack, culminating in the theory of relativity in level 7 and quantum mechanics in the Bohr range.

One of the most significant accomplishments of the measuring activity of the file stack was the ability to measure DNA in its genetic setting in level 1. It is significant that Crick and especially James D. Watson approached the DNA molecule by using measurements made by others who had mastered the far ranges of scientists' measure and then thought "simply," essentially with the aid of a set of everyman's tinkertoys.

The Coordinates of the File Stack

With the ability to see DNA as an information program comes the possibility of fixing a file stack in two sets of coordinates. All else is relative to the interior of the stack, but the DNA molecule has a connection with a world outside because it has been produced by a previous DNA molecule by replication. The gene program establishes equilibria and regularity of signals and compares them, but each file stack occupies an external environment which we have not yet defined.

The file itself constructs a sort of vertical axis through its pattern maintenance boxes at each level (fig. 6). Since the distance between levels is information distance rather than physical distance I wish to turn to time for the definition of the horizontal coordinate. The conventional time coordinate is usually represented by a single line with equidistant markings representing a period such as the ticking of a clock, or the rotation of the earth. Such a time line can be generated as a powerful signal on level 6 and be passed down through the file stack. But for our purposes such a measure is arbitrary through the whole set of levels controlled by DNA. Likewise, time drawn from vibrations of the physical spectrum, which I have called radiation time, is arbitrary through the whole set of levels of the file stack controlled by DNA.

Therefore an adequate definition of the time coordinate must add to physical time a biological time which is tied to DNA. The first species of biological time I have called phylogenetic time. In its pure form this axis runs from the first self-replicating DNA to the fertilization which lays down the program for the file stack. The unit of phylogenetic time is the generation which spans from one meiosis, or division of reproductive cells, to the next. Level 5 has as an outer limit the total duration of phylogenetic time. Levels 3 and 4, while still referable to phylogenetic time and its unit of the generation, are restricted in their operation by the formation of the gene program in DNA known as *Homo sapiens*. In between the first replication of DNA and the formation of the species *Homo sapiens* are many patterns of DNA programs with the branch points where new species are formed, but the elaboration of that pattern may be left to evolutionary biologists.

Levels 1 and 2 are governed by yet another species of biological time, what I call ontogenetic time. The duration here is set by the particular program laid down by the observer genes and runs not from meiosis to meiosis but from fertilization to the dissolution of the file stack. The information moves from one cell to another by DNA, but by the kind of division called mitosis rather than meiosis. Thus the coordinates can be established by the limitation of the processing time of the various levels.

The horizontal axis, more like the conventional time axis, is inadequately defined by x values in physical time, which fix the whole DNA system in a set of coordinates but can measure it off only in arbitrary units. Therefore we must go on to fix the file stack relative to the vectors formed by the action of DNA itself. To look back at figure 5c, we can see again the crucial importance of the information input to the DNA program of the observer-file stack acting as a regulator to level 1 but insulated from all feedback.

The sperm and the egg which meet to create the DNA program of the observer-file stack (fig. 7) each bear a half-program which had split off by meiosis from a cell existing in other ontogenetic time realms. When one counts back from those two splits *through* the fertilization which produced the mother or father to the meioses which produced those earlier fertilizations, one has defined a unit of phylogenetic time. This is an open system at both ends, bounded in the past by the sum of genetic experience carried in the DNA and in the future only by the ability to produce individuals who can reproduce.

The Darwinian Trajectory

One of the great intellectual accomplishments of all time was the construction in the nineteenth century of the concept of evolution out of information entirely subsumed in a system not unlike the file stack we have here described. I am calling that construction here the Darwinian trajectory in order to avoid calling it the Darwinian tree or the evolutionary tree. The most famous diagram of the Darwinian trajectory is that drawn by Darwin himself in *On the Origin of Species.*[10] Darwin actually accomplished the tremendous feat of making up the chart on phylogenetic time using a unit of 1000 generations. But he then ruined the effect by likening the diagram to a tree and changing the vertical axis to geological, hence physical, time. Therefore the term Darwinian trajectory has to be used to denote a construction of modern biology in which the coordinates of physical, phylogenetic, and ontogenetic time are carefully distinguished. In figure 6 I am indicating the Darwinian trajectory by the simplest possible scheme.

In a world with G. G. Simpson, Ernst Mayr, G. Ledyard Stebbins, and the Philip Handler volume,[11] I do not dare launch on a full description of the Darwinian trajectory. I can only point out here some characteristics relevant to this particular discussion. I wish especially to point out the many characteristics that the Darwinian trajectory shares with the file stack.

Since both are based on DNA some of the same numerical relations prevail. The replication of DNA sets up the 2^n (1, 2, 4, 8 ...) geometric progression. Yet pure geometric progressions of DNA replications would be completely uncontrolled. The presence of arithmetic progression in the inorganic matrix (i.e., all other molecules of matter and energy than those controlled by self-replicating DNA) set up a particular form of the Napier relation, which Darwin recognized without knowing anything about Napier or calculus, making it over into the Napier-Malthus relation. Darwin reconciled the arithmetic progression of the environment and the geometric progression of the organism in phylogenetic time by means of natural selection.

The flow of the Darwinian trajectory—set off in one direction and held in course by the stop-valve of the central dogma—is made up of information, matter, and energy. The emphasis within the channels also varies. They make up a network which carries information in two directions, and the presence of regulators is part of the adaptation which Darwin called natural selection. The neural networks of the more complex organisms make admirable servomechanisms which hold both the interior and exterior environment in equilibrium with the organism.

The characteristics just outlined mean that the Darwinian trajectory can be measured by generations, the same kind of units and proportions as we have already seen in the file stack. Roger Barker's environment-organism-environment circuit can here be used without the emendations we had to adopt for its use in the file stack, and here its very looseness makes it superior to the perceptron as a figure of speech.

The ultimate outcome of the play between organism and physical environment can lead to change. The differentiation of the gene flow by natural selection in phylogenetic time eventually populated the environment with a variety of organisms, so that the environment of the Darwinian trajectory is made up of behavior settings relative to a particular organism which contain both physical objects and other organisms. Some of these have the same genetic program as the first organism, falling within the same species boundaries. Some organisms in the behavior setting of the first organism are differently coded in their DNA programs. When a behavior setting becomes overpopulated with different species it sorts itself out into niches which are themselves arranged in hierarchical structures.

Despite the dazzling beauty of the Darwinian trajectory, we must remember that since we are still using the same measuring system that we described in the file stack, we are only recording channels which connect esoteric boxes. Analysis enriches the network but always leaves as a core an esoteric box.

The Intersection

We now close in on the object of our pilgrimage as we contemplate the problem: Where does the Darwinian trajectory intersect with the file stack built on genes of the observing individual with its *Homo sapiens* program?

The answer is immediately apparent. The intersection occurs at level 5 of the file stack—the Earth Ecosystem. The population of *Homo sapiens* in the file stack is surrounded by plants, animals, and the inorganic matrix in a dynamic equilibrium. The Darwinian trajectory exhibits exactly the same elements. I have called this level the Linnaeus-Darwin level because of the problem, which once seemed insuperable, of mapping the elements of a hierarchical structure made of the biosphere elements (Linnaeus) onto a set of geographical coordinates supplied by the earth itself, and made apparent by the ecological niches highlighted by the concept of natural selection (Darwin). It is now not too hard to map the level both ways in either the file stack or the Darwinian trajectory, a double relationship hard for nineteenth-century men to see. The Gray-Agassiz debate before the American Academy of Arts and Sciences[12] is a neat illustration of the clash of these two views.

The angle at which the two systems intersect is so sharp that it produces a stereoscopic view of the Darwinian trajectory. This three-dimensional view, when fed back into the file stack, markedly changed the values not only at level 5, but down through the lower levels to level 1. Perhaps we have accomplished something, but is the appearance of depth, the opening of long vistas back to primordial DNA, a true escape from Bishop Berkeley's prison of idealism? We no longer have to believe that biology is only a file stack. But to get over from the file stack to the Darwinian trajectory still requires a leap of faith. Curt Stern put this clearly when he said:

> Could not one of the pioneering agriculturalists in the Middle East have counted the numbers of different plants in a segregating field? Could he have observed 3:1 ratios (perhaps helped in his arithmetic by belonging to a people who used the duodecimal rather than the decimal system)? . . .
>
> These are idle questions, but I raise them to confess my belief in the reality of the external world however inexhaustible in character and however incompletely represented at any stage of its recogni-

tion. Given enough time, genetics, whether born in ancient history or in the most recent centuries, would have reached a state very similar to the present.[13]

The Placement of the Seed

To go on some steps farther is possible because we have left some obvious inconsistencies in the picture drawn in figure 6. In the first place, the DNA information program for *Homo sapiens* is filed as an item in both levels 1 and 5. The entry of the particular strand of the Darwinian trajectory which stems from the speciation of *Homo sapiens* leads straight into the population at level 5. The dilemma is not easily solved by choosing one level or the other, because the two-way channels run all the way through the stack, summing to a single vector from 5 to 1 and another from 1 to 5. Without a genetic input at *both* levels, the file stack would not have the characteristics we have specified.

Another look at figure 6 gives us a clue to the trouble. The axis which we constructed from the genes of the individual observer back to the primordial DNA is an axis only because, Descartes-like, we constructed it that way. If we start at primordial DNA and move toward the file stack, we see immediately that it is by definition the phylogeny of *Homo sapiens* and hence falls within the Darwinian trajectory. This idea was another one very hard to get used to in the nineteenth century, for there seemed to be something extraordinary about *Homo sapiens* which resisted relegating him to level 5 along with the rest of the mob in the Darwinian trajectory.

Scientists have a special reluctance about moving the observer into the population of *Homo sapiens* as a whole. Since the seventeenth century they have tried to find a place for themselves outside both the file stack and the Darwinian trajectory so that they could see how the whole thing works objectively. The scientific revolution divided mind from matter without having a clear conception of the organism in which mind resided. The late Warren McCulloch, in recalling the time when the physicist Boltwood recounted an outline of the history of science in a single evening, has remarked that Galileo

> split natural science down the middle by refusing to admit mind or anima, as an explanation of physical events. We agreed that this was

the way to make physics, but that it left mind—anima—a sort of self-sufficient item capable of wandering around the world having ideas, even perceptions, of a world devoid of anima.[14]

He might have added that the observing mind outside the physical system could only see its own position in the biological organism, either as an observer performing pure Cartesian cognition and getting results on biological systems identical to the results on physical systems, or by falling back on some form of vitalism for a biology with roots outside the tradition of the physical sciences. To a greater extent than is often admitted, the biology of the seventeenth to the twentieth century was a combination of the two approaches.[15]

The inability to distinguish ultimately among scientist, craftsman, and everyman means that the term observer refers simply to any file stack based on *Homo sapiens* DNA which gets started. As long as its adaptation circuit (a function in Parsons's four-function paradigm) works well enough to keep it operating at all, it is observing enough to meet the definition of observer. The conclusion is inescapable that we must select a DNA program among the population of *Homo sapiens* at level 5 and designate it the individual program of level 1. Far from being a disaster, this move opens up a number of advantages for our scheme.

The steps are as follows:

1. Topologically deform the file stack so that the genes of level 1 are located within the DNA population of level 5, which is also now within the Darwinian trajectory, as in figure 8.

2. Group the communication channels running from level 1 to level 5 into summed vectors of the four functions of Parsons's four-function paradigm.

3. Place levels 6 and 7 outside the Earth Ecosystem to the top and the Bohr range outside it on the bottom, which makes the physicochemical spectrum a straight line with DNA in its place in the molecular range.

The arrangement created by this deformation has many advantages. All four of Parsons's subsystems now connect with the Earth Ecosystem and thus partake of the environment of the Darwinian trajectory. The behavioral organism finds its place within the biological population of the species. One can visualize much more clearly the adaptation function, the integration function, and the system goal attainment function of the four-function paradigm in addition to the pattern-maintenance function that has heretofore guided us in constructing the file stack.

Now the flaw in the conventional view that Copernicus displaced man from the center of the universe can be seen quite clearly. Level 6, the Copernicus-Newton level, is part of a spectrum which gives plenty of scope for physics from the very large to the very small. Without violating the laws of mechanics or thermodynamics, we have taken the whole of biology and the social sciences out of the straight-line physical spectrum and given them a cybernetic form in which the DNA domain becomes itself a feedback loop. All of the levels of action thus fall into an equilibrium which makes the organization of the total observer possible. The physical spectrum includes the whole self-replicating DNA domain, but only the DNA domain provides the instrument for observing the physical spectrum. For our purposes man is still at the center of the universe because we have built the file on his DNA program, not because he resides in a particular location.

Yet two major incompletenesses still hover over the reformulated system.

1. Because of the one-way valve of the central dogma in the DNA of the behavioral organism, there is a preponderance of counterclockwise currents in the diagram of figure 8, giving the functions a pronounced counterclockwise spin. Since there is an energy loss as one passes around through the various levels (now spokes on the wheel) and since the countercurrent comes only from feedback, the functions will reach level 5 only in weakened form, and ecological, cultural, and societal responses will have trouble getting back to the behavioral organism.

2. Parsons has long been impressed by an idea of Alfred Emerson that "the 'system of cultural symbolic meaning' played a role analogous (in the proper biological sense of analogy) to that played by genes in biological heredity."[16] He further has assigned a predominance of each to the four functions to the subsystems of action—adaptive function to the behavioral organism; goal attainment to personality; integration to the social system; pattern maintenance to the cultural system. With the counterclockwise spin mentioned above the cultural subsystem is a long way from

the genes and has available little but reflected energy, inadequate for pattern maintenance on the levels close to the behavioral organism (the locus of the DNA program to be sure, but also of the systems it constructs in ontogenetic development).

To close the gaps left by the above two considerations let us look carefully at the point of jointure of level 1 and level 5. We selected a DNA program in the level 5 population to be the DNA program for level 1 also. Looking back at figure 7 we can designate an individual as the selected gene program. But the same gene program can be mapped in two ways.

1. By assuming that the gene program of the individual is the start signal for the behavioral organism we move from fertilization to the end of the program. If there are any mutations in the genes they may pass on to the children but cannot affect the phenotype (behavioral organism) as it lives out its life.

2. By mapping gene programs as a part of the level 5 population we must connect them with the other elements in the gene population. This matrix is measured in phylogenetic time. The gene population matrix is not closed, and programs, including mutations, are recombined in accordance with Mendelian ratios. It has gone through a sequence of states beginning with primordial DNA, and the successive selections—Darwin's natural selection—through interaction of the matrix with the environment have laid down the gene program in a pattern available to the individual which is not random, but a summation of previous experience.

Symbol as Gene

To start a clockwise countercurrent in the loop in figure 8 from the gene program at level 5 to connect to the cultural subsystem on level 4 we must first decide to use phylogenetic time. If the gene program at level 5 were not *Homo sapiens* but some other organism, we could conceive of the probability matrix of which it was a part being held stable by the DNA programs which make up the genes. The environment into which this organism moves is also a matrix which has been selected and to a certain extent ordered by the population of organisms (echoes of L. J. Henderson's *Fitness of the Environment*). These two systems make their mutual selections by exchanging information, and the adaptation of the population to its niche will proceed, incorporating some of the mutations which arise in the genetic matrix. However, each unit of the gene population lives in a phenotype, which impinges on the environmental matrix by molar behavior of the total organism, not by individual bits of information flying around like the molecules of a gas. Clearly some molar behavior is a direct result of the genetic program, but even in quite simple animals some information available to it comes as feedback from the environment. Since the environment is partially ordered by the phylogeny of the organism, this behavior based on the environment-organism-environment circuit is reinforced over many generations. The molar behavior involved—courtship rites, fear-and-aggression patterns to protect territory, evasive tactics against predators—although phenotypic is a part of the total information program available to each organism.

If such patterns of molar behavior can find a mechanism to transcend ontogenetic time and be passed down in phylogenetic time, they will rotate clockwise on figure 8, hence moving from level 5 to level 4. When dealing with *Homo sapiens* the ability to shape a nongenetic mechanism for transmitting a molar behavior pattern in phylogenetic time is highly developed in the form of symbol. Thus an information program supplementary to the gene program is available to *Homo sapiens,* and while the fit is not perfect, symbol sets in effect provide a line of codons extending beyond the end of the DNA chain. They are not as immune from environmental feedback as mutations are, but given relative stability in the environmental niche, they can achieve sufficient stability greatly to extend the pattern maintenance repertoire of *Homo sapiens* as a species. For our system, it is the flow clockwise by the supplementary symbol system flowing through levels 3, 2, and 1, that the chance of equilibrium in the DNA sector for man is achieved.[17]

The Shape of the Individual in the System

The term "individual," despite or because of its many overtones from common language, has a place in the system. Looking back at Figure 8, one can see the deformed file stack which has bent to start and end with the same gene program. Although there is matter and energy flowing in and around this stack from the environment, its organization can be traced *both* from the gene which lays down the program for the behavioral organism *and* from the matrix of genes and symbols in the population in its setting in the biosphere. To unroll it again we can see it in Figure 9. The gene-sets at either end are of course the same genes.

At this point a difference between this formulation and that of Parsons should be emphasized. At least in some descriptions of his action systems, Parsons refers to a descending hierarchy of controlling factors descending from "High Information" above the cultural system, and a "Hierarchy of Conditioning Factors" ascending from a "High Energy" level.[18] My view, since the whole stack is divided into hierarchical levels which exchange filed information, results in a two-way flow of information which is fueled at both ends by both energy and information. Since through much of the file stack this information is contained in the brain-mind system of the individual, it consists of very small amounts of energy transfers and inputs transduced by the senses from physical vibrations to neural impulses. The high energy metabolic system is also two-way in terms of energy flow. Culture and society place the individual where his metabolism can get its chemical inputs from the environment, and thus metabolism carries the process down to the cellular level, finally as ATP, whence the energy is released back up the stack in the form of heat and work.

Hence to say that culture "controls" downward from a level that is specialized for pattern maintenance (as L level in LIGA sequence) is to specify only one of four pathways. It is true that cultural symbols will only indirectly affect the behavioral organism. But the constraints put on the organism of *Homo sapiens* by his beliefs and actions are felt even at the cellular level. Similarly, the operation of the behavioral organism by repeatedly sending signals through social action can affect symbolization at the cultural level.

It might be argued that differentiation in the social sense is impossible below the level of "identity," and that Parsons's four-function paradigm operates on the behavioral organism only by analogy. Such an interpretation unnecessarily takes sides in the internal warfare among psychologists, for example, between Piaget and B. F. Skinner. The organismic, structural bent of this interpretation might seem to dictate a vote for Piaget, but the accomplishments of the behaviorists should not be thrown out lightly, particularly when the tradition from Kurt Lewin to Roger Barker offers a possibility of getting room for the strongest parts of both views.

Hence a straightforward way of treating the behavioral organism might be adopted from biology. The behavioral organism might be considered a set of subhierarchies comprised of the molecular level, the cellular level, and the level of the organism. The fight between the behaviorists and the structuralists might then be narrowed to the question of whether and at what level the organism achieved consciousness, and through it, identity. At the personality level (regardless of one's attitude toward Freud or any particular theory, Parsons's insertion of the personality level is a major accomplishment) identity clearly exists and consciousness integrates the organism to what the psychologist would recognize as molar behavior. Somewhere at the level of the organism the individual becomes a plural federation of organs and tissue. It may not be differentiated along the lines of Parsons's action system, but differentiation is a necessity for the achievement of *homeostasis,* an equilibrium below the threshold of consciousness. Biologists might wish to revise this formulation, but it is a matter of detail as compared to the crucial question of whether the double mapping of the genes at both ends of the involved individual can work, and if so, on what layout.

If we can for the moment assume the genes at the base of the individual, we can summarize that the balance of exchange between all the levels below the social—i.e., personality and behavioral organism—is tied together more strongly by vertical lines of integration than by direct exchanges with the environment. The individual swims in a physical environment, and a wandering cosmic ray occasionally collides with a molecule, or the sun and weather beat on the skin. But the important currents of environmental input, perception and metabolic intake, are organized in terms of individual identity and give the individual a tough outer boundary within which systems maintain a clearly defined pattern. Ontogenetic time has a particular relevance here because the genetic program which extends from fertilization to the dissolution when the interior equilibrium ceases to maintain itself, is the unit of duration of the individual itself. And up through personality, the strong boundaries prevent a direct flow of information beyond the ontogenetically bounded individual.

At the levels of society and culture, however, the boundaries delimiting the individual, while still perceptible, are weak as compared to the open matrices operating in phylogenetic time. The social and cultural systems get internalized as that part of the general matrices that each individual takes in and makes a part of his own hierarchy. Internalized culture stocks the memory bank of the individual's brain-mind system, while the same system in interaction with others serves as a servomechanism responding to the social matrix around him. Thus, a part of culture and society operates within the individual, provides him with guides and norms, and it also dies with him.

The Hookup of Individuals into Action Systems

To get beyond the temporal cross section of figure 8, it is necessary to turn the diagram around 90° and see how the individuals hook up into an ongoing chain. The first continuing system, and the one that is crucial to the two-ended monad we have been describing, is genetic. The second system will be the social (level 3) where both the individuals in a demographic grid and culture as a set of templates interact. Third we shall consider the cultural system.

Figure 9 makes an approach to the genetic connections. The gene program at the bottom represents ontogenetic time, that is, a time whose unit begins with fertilization and ends with the dissolution of the organism. This time sequence can be measured in physical time by a limit set by the vital statistics on the population of the species. But an individual's genes cannot receive feedback signals from its own file stack because of the fundamental dogma of DNA. Hence the dead end at the death of the organism becomes a fundamental limit, as it has been in popular thought since prehistoric times. However, by way of fertilization the gene is open backward to the summation of the species program, and back of that through a series of binary decisions to primordial DNA. This accumulation we have designated Σ_G. The Monads M_1 through M_3 are connected in phylogenetic time by their gene programs. Even if M_1 develops a fully stocked file stack with many internalized symbol-sets and social values and norms, the process can go no farther biologically without grouping the monad at least into a dyad, or in common language, in *Homo sapiens* (and many other species) a sexual mating must take place for the process to continue.

Darwin, who went over so much of this ground with fruitful results, spent years experimenting on different forms of flowering plants to test the generalization "Nature abhors perpetual self-fertilization."[19] No amount of special cases from primitive plants and animals, and no amount of threat of clonal reproduction in man, as Watson has predicted, can change the fundamental importance of making the genetic program open to the future as well as the past by the self-replication of DNA in meiosis, and the recombination, with the possibility of incorporating mutation, in fertilization. When two monads get together (an action at the social level conditioned by the cultural environment, by the way) they open up the future for a new $\Sigma_G + (m_1, m_2)$. If fertilization takes place, the dyad becomes a tryad and a basic unit for the social level, the family. If all does not go well, natural selection has been at work to affect the matrix of the gene pool.

In figure 9, the M_1, M_2, and M_3 genes at the top of the diagram are really the same genes as at the bottom, but are repeated at the top to indicate the curvature of the action system upon themselves so that the genes will be next to (but not touching) the cultural level of the individual file stack. The genes cannot touch because, subject to the fundamental dogma of DNA, it cannot receive information from the file stack at any level. The difference between the genes at level 5 and those at level 1 lie in two differing circuits into which a given gene program is wired. A major consequence of this double circuitry is the famous dicta that (1) acquired characteristics cannot be inherited, and (2) adaptation is achieved by the selection of genes, including mutations, for perpetuation in the next generation.

The Horizontal Linkages at Levels 3 and 4

As we have seen, the currents within the file stack of the individual are encased in strong vertical boundaries at the level of the behavioral organism and personality. When one begins to form a collectivity of monads, however, the dominant currents run horizontally by the interaction of the individuals. As individuals they are still windowless to the intervening environment and to other individuals of the species except through perception, but they now can establish horizontal channels by the use of symbols generated at the cultural level. It is the pattern of these connections to which we have given the name templates.

Templates, by defining the terminals and exchange points of information flow, put bounds on the roles played by the individual monads. Conversely, the monads as they fit into templates become servomechanisms for controlling the horizontal information flow at level 3. A template, by having individual servomechanisms replaced, cannot only keep going beyond the lifetime of an individual; it can evade the iron law of organic decay à la Spengler. Yet the influence of ontogenetic time at the social level is great because the demographic matrix of individuals is continually receiving additions and subtractions. Each time an individual dies or drops out of a template, a vacancy chain begins by which other individuals move up into new slots until a new recruit finds lodgment. Society thus renews itself, but at a cost. Because of the feedback mechanisms provided by those marvelously complex brain-mind systems operating in roles in the polity and economy, very great multiples of efficiency can be obtained in energy control as compared to a biological population under genetic control alone.

Since a social system is made of both templates and individuals, the telling of the story from the bottom, the individual monad, is only a part of the whole. The individual against the collectivity is one of the great tensions of history, a tension that is joined at the level of society. Yet we must also try to account for that powerful horizontal stream which flows at level 4, culture.

Even if we define culture as a set of symbols, we note in figure 5 that the boundaries of the individual monad extend weakly all the way up through the cultural level. Hence some symbols that make up the horizontal flow of information by culture (Σ_C) are internalized in individual monads. Internalized culture is of course present in even the most deprived of individuals, and in the form of language is heavily dependent on very long tides of summations in the history of the species. Hence internalized culture is not sufficient to account for all the information flow (Σ_C) because the capacity of the demographic matrix at any given time is much more limited than the total stock of culture available.

The factor which gives culture its strong horizontal force in time and its cumulative power is that in each generation some of the symbols internalized in individuals get impressed on some of the matter in the environment. Hence material culture is more than just a set of artifacts. Each set of artifacts presupposes an actor who made it, and behind the actor his whole file stack. Thus over time a correspondence develops in the arrays of symbol-sets inside individuals and those outside the individual impressed upon the environment.

The material culture components of the culture stream are a part of the environment, but their information content is greater than that of material which has not received the impress of human action. In the coding and decoding of this information, not only can individuals communicate with one another, but they can pile up a memory bank in the external environment as well as internally. Templates can then be made from symbol patterns stored in the memory bank and retrieved for appropriate social situations.

To accomplish these exchanges between culture and society it may be necessary to interpose an intermediate subhierarchy on the boundary between levels 3 and 4, a special set of templates which we may call cultural institutions. Their role in society (and they are manned by individual servomechanisms like other social institutions) is to care for the memory bank and provide for the transmission of cultural symbol-patterns.

Descartes and Vico

The question cannot be avoided of where to locate the observer in the system. Since only *Homo sapiens* are included, it is hard to imagine how the observer could stand outside the biosphere like Voltaire's Micromegas. Yet the location of the observer within the circle of the system is also perplexing, because the several levels participate in the observation. Within the DNA domain the information system looks at itself; outside the domain the principle of complementarity brings the observer from inside the domain to the farthest reaches of the physicochemical spectrum.

One way to locate the observer is by the method of Descartes. "Cogito ergo sum" locates the observer in the file stack on level 1 in the box "cognitive capacity" in Parsons's subsystem. From that point by a series of axioms he achieves an analytical geometry of great clarity. It is admirably suited to dividing up the boxes of the file stack into smaller and smaller subdivisions, and it is also well suited to erecting a grid that is convergent with nodes on the physicochemical spectrum. Descartes's procedure, however, by reducing the observer to a point, creates the illusion that he is outside the whole physicochemical system, including the DNA subsystem, and looking at it as an outside observer. No matter how fine his network of analysis, his geometry does not delineate man or the system of which he is a part. It merely locates any point on a set of arbitrary coordinates. Both the brilliant success and the disastrous misunderstanding of the Copernican tradition are tied up in this conception of the observer. The successes of this choice of the locus of the observer shine brightly in the history of modern physical science. But the choice is hard on both biology and the social sciences.

An alternative to Descartes was already seen in the eighteenth century by Giambattista Vico. In the terms of this paper he chose as his starting point an item of material culture—an artifact. It might be a ruin, or a document, or a myth. He could know an artifact because he could make one. Since he knew himself he could imagine the maker from the artifact. Thus he reconstructed the file stack and hence the DNA circle by the combined evidence of the artifact and himself. This meant that he moved around the circle counterclockwise until he reached his artifact. Then he began to move clockwise toward the present, achieving a sense of phylogenetic time, especially by using symbol as artifact. He achieved the certainty of the reality of the historical past, not the clarity of the Cartesian geometry. In Vico, the system observed itself from the inside, achieving thereby the axes of ontogenetic and phylogenetic time. A child of Vico, who has chosen as his artifacts the children of Descartes, wrote this essay.

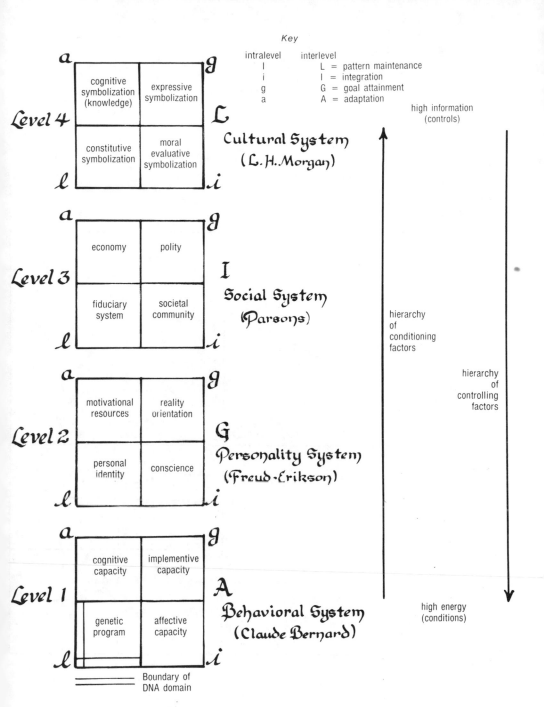

Fig. 1. General Action System of Talcott Parsons

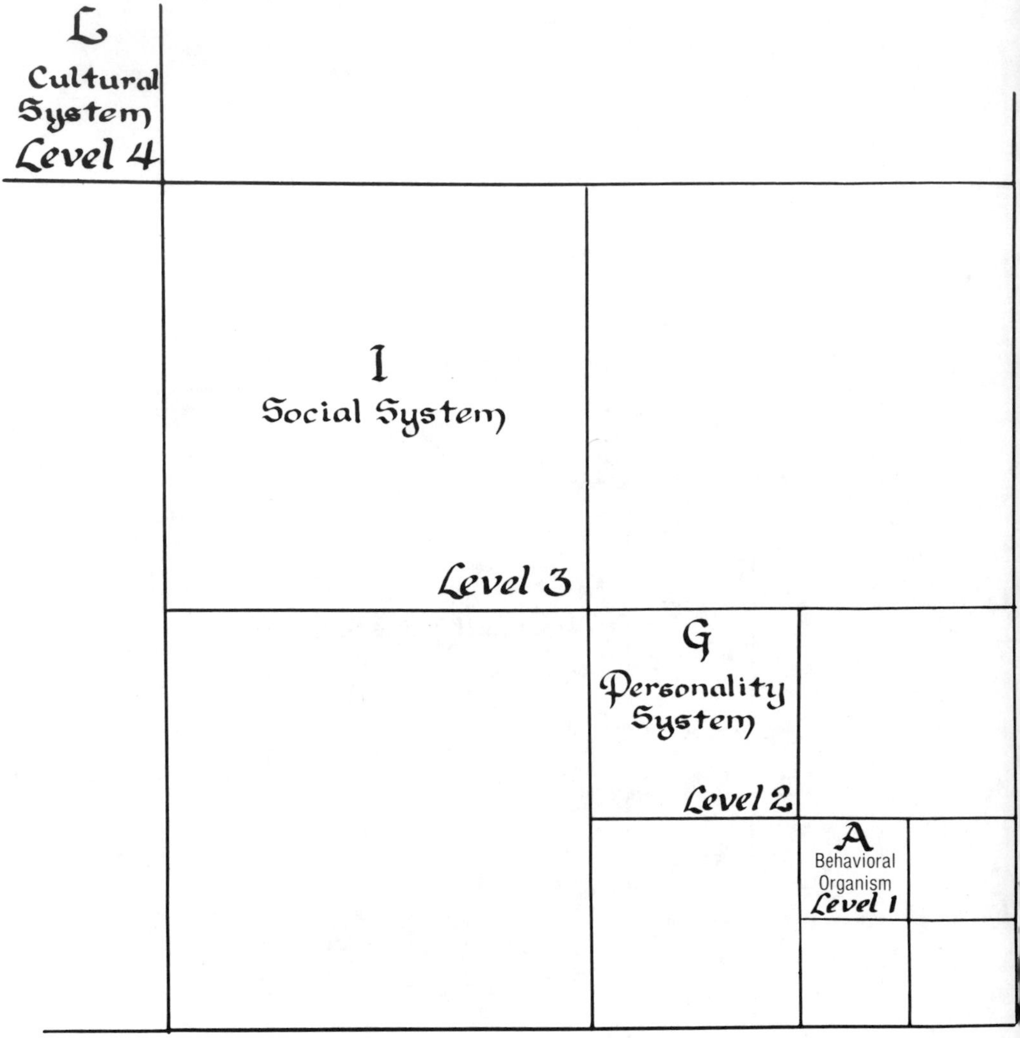

Fig. 2. Parsons's Action Systems in a normal file organization

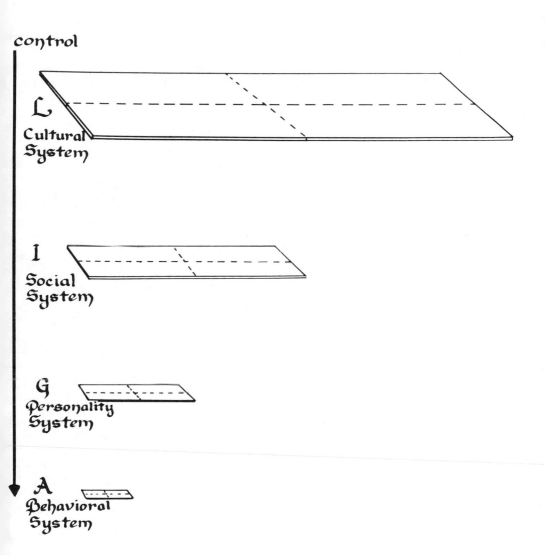

Fig. 3. Nested boxes of file organization arranged into levels

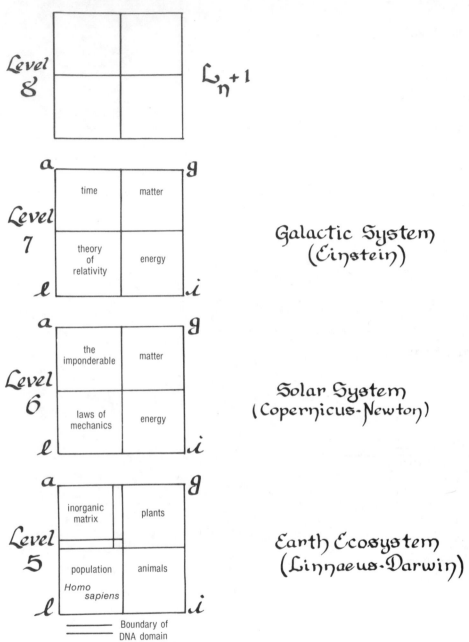

Fig. 4. Levels superimposed on Parsons's Action System (Levels 5 to 7, the physical environment)

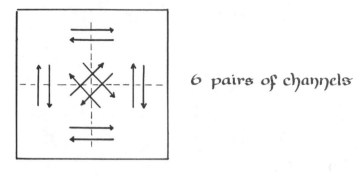

6 pairs of channels

a. Lateral flow channels within a level

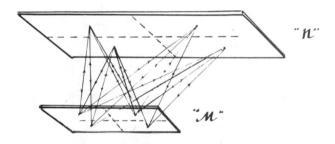

"N"

"M"

b. Interlevel channels connecting each of the 4 boxes at level "M"
with each at level "N"

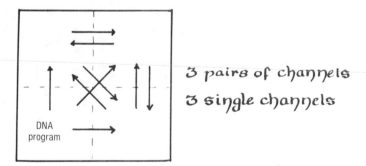

DNA
program

3 pairs of channels
3 single channels

c. Lateral channels in level 1

Fig. 5. Flow channels

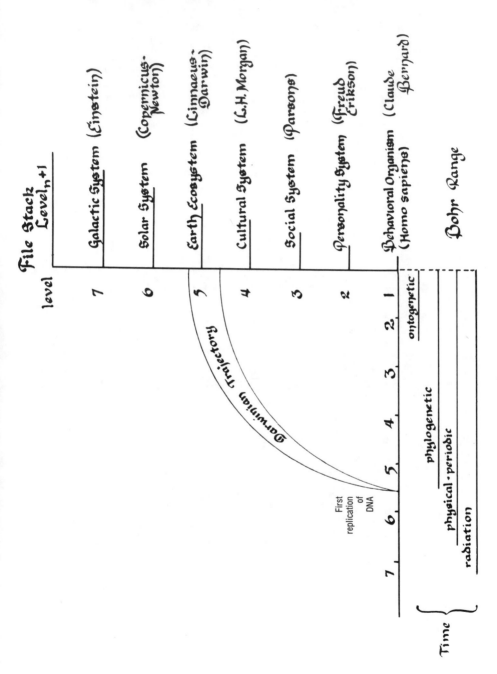

Fig. 6. Arrangement of levels combining Parsons's Action System with an evolutionary setting (including some eponymous clues to in-

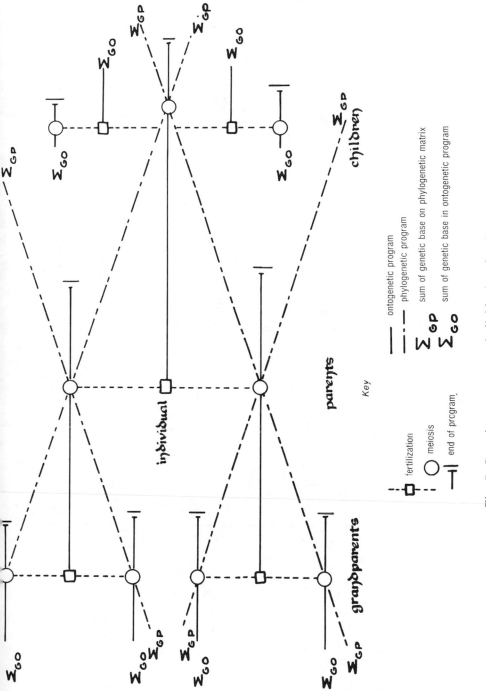

Fig. 7. Genetic programs as individual and collective matrices

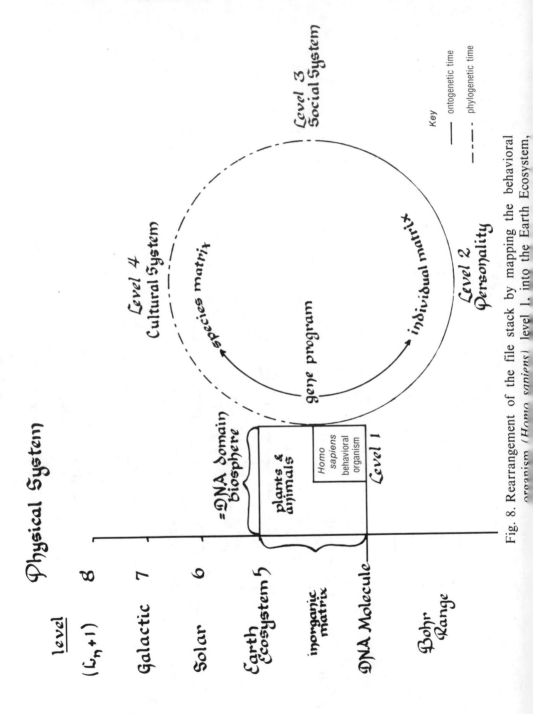

Fig. 8. Rearrangement of the file stack by mapping the behavioral organism (*Homo sapiens*) level 1, into the Earth Ecosystem,

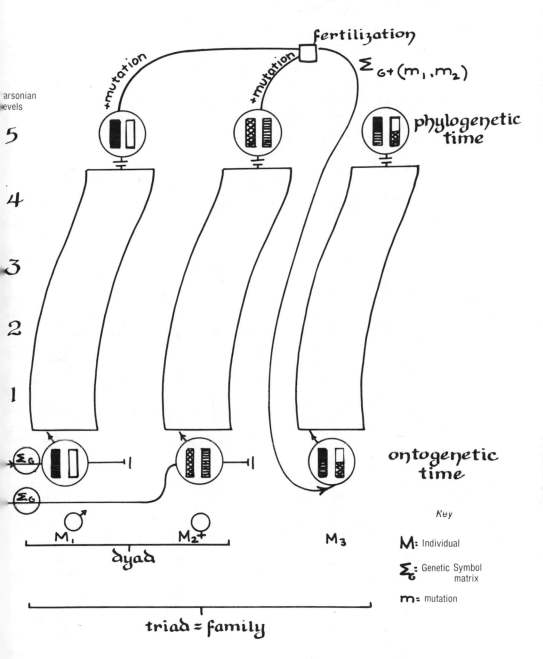

Fig. 9. Genetic connections of family of three with Action Systems

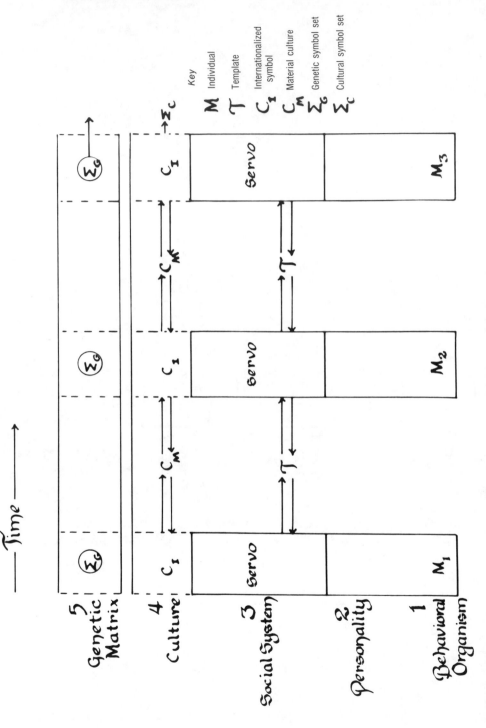

Fig. 10. Genetic and cultural matrices

NOTES

1. A brief introduction to Parsons's ideas may be found in Talcott Parsons, "On Building Social System Theory: A Personal History," *Daedalus,* Fall 1970, pp. 826–81, especially pp. 844–85; also in Gerald Holton, ed., *The Twentieth-Century Sciences: Studies in the Biography of Ideas* (New York: Norton, 1972) pp. 99–154.

2. The immediate stimulus for this line of speculation was from the following two papers: Curt Stern, "The Continuity of Genetics," *Daedalus,* Fall 1970, pp. 882–908; and Gunther S. Stent, "DNA," *Daedalus,* Fall 1970, pp. 909–37.

3. Joseph Becker and Robert M. Hayes, *Information Storage and Retrieval* (New York: Wiley, 1963) p. 387.

4. Roger Barker, *Ecological Psychology* (Stanford, Calif., Stanford University Press, 1969), is my source for the concept of behavior setting.

5. Gerald Holton, "The Roots of Complementarity," *Daedalus,* Fall 1970, pp. 1018–19.

6. Parsons, *Daedalus,* Fall 1970, p. 844.

7. David L. Clarke, *Analytical Archaeology* (London: Methuen & Co. Ltd., 1968), pp. 73–74.

8. Stent, *Daedalus,* Fall 1970, p. 926.

9. Marvin Minsky and Seymour Papert, *Perceptrons: An Introduction to Computational Geometry* (Cambridge, Mass.: The MIT Press, 1969).

10. Facsimile of first ed. (1859), Introduction by Ernst Mayr (Cambridge, Mass.: Harvard University Press, 1964), between pp. 116 and 117.

11. Philip Handler, ed., *Biology and the Future of Man* (New York: Oxford University Press, 1970).

12. A. Hunter Dupree, "The First Darwinian Debate in America: Gray Versus Agassiz," *Daedalus,* vol. 88, no. 3 (1959) pp. 560–72.

13. Curt Stern, "The Continuity of Genetics," *Daedalus,* Fall 1970, p. 907.

14. Warren S. McCulloch, "The Beginnings of Cybernetics," working paper prepared for the *Daedalus* conference at Bellagio, Italy, in September, 1969.

15. Michel Foucault, *The Order of Things: An Archaeology of the Human Sciences* (New York: Pantheon Books, 1970), deals with the intellectual space occupied by classification in the Classical Age.

16. Parsons, *Daedalus,* Fall 1970, p. 850.

17. Gregory Bateson has discussed these subjects in a way relevant to the present argument in *Steps to an Ecology of Mind* (New York: Ballantine, 1972), pp. 411–25.

18. For example, Talcott Parsons, "Societies: Evolutionary and Comparative Perspectives," in *Foundations of Modern Sociology Series,* ed. Alex Inkles (Englewood Cliffs, N.J.: Prentice Hall, Inc., 1966), p. 28.

19. Charles Darwin, *The Different Forms of Flowers on Plants of the Same Species* (New York: D. Appleton & Co., 1877).

COMMENTARY

GERALD M. PLATT

Another Wedding of Biology and Sociology: Can Molecular Genetics and Parsonian Theory of Action Live Together?

PROFESSOR DUPREE has written a complex and exciting essay. His attempt to integrate molecular biology with Parsonian sociology brings to this type of merger a new perspective: one which is not found in the literature even where previous attempts to bring the disciplines together exist. The forms of integration up to now have been two, analogies in terms of evolutionary and functional perspectives. In no sense does Dupree crassly employ either of these models.

The evolutionary form is the familiar Darwinian-Spencerian model.[1] In this framework the Darwinian phylogenetic scale, in one form or another, was superimposed upon cultural development. The Spencerian model was unilinear in terms of cultural and social development and assumed that certain types of societies were more highly advanced, as illustrated, for example, in the adaptability of technological developments, and finally implying that Western cultures were more advanced and better adapted "species" than less technologically developed societies. Such a framework was obviously culture-bound and biased toward the Western world.

Anthropological ethnography of the 1930s and 1940s forced a withdrawal from this position. By demonstrating that preliterate societies frequently had very complex cultural patterns which were well adapted to their circumstances, the cultural invidiousness in terms of evolutionary adaptability was withdrawn from, and in its place, the polar position of cultural relativity was substituted for some time.

A second mode of integrating sociology with biology came into view about the same time as the evolutionary perspective was losing ground. The position was in part grounded in the work of the French sociologist, Emile Durkheim, and then in the works of the anthropologists, Ratcliffe-Brown and Bronislaw Malinowski, it reached its height. A fundamental proposition of the approach was that of the teleology, the purpose or the function of social activities, groups, institutions, and so on in society; the framework fostered reductionism in terms of the purposes of social activities and groups. This approach dominated Harvard and Columbia University sociology until the 1960s.

This analogy with biological functioning had its difficulties. It was possible to tell of individuals what homeostatic processes consisted of and what was organically vital in *Homo sapiens,* but such specifications were more difficult, if not impossible, regarding institutions, societies, and cultures. Attempts to modify and salvage the theory in strictly functional terms failed.[2]

Another criticism hurled at this approach was that in its focus on how societies or cultures homeostatically sustained themselves, it was ideologically conservative. Parsons's work especially came under attack from neo-Marxists[3] because as part of this "structural-functional" movement (as it came to be called), he once wrote that his primary interest was how systems were ordered and how they maintained themselves; he would save analysis of change for some other time.[4]

Parsonian theory was hardly the most crass among these

structural-functional approaches,[5] but Parsons was the most prominent and ingenious in this camp, and thus his work became the object of the greatest derision.[6] Parsonian theory has sustained both the evolutionary and functional perspectives, but in radically altered forms. In this sense then, Parsonian theory is still grounded in biology and thus, to a degree, lends itself to the wedding Dupree is attempting.

Dupree's Foci

It is not surprising then that in light of Parsons's prominence in the field of sociology and his orientation, Dupree has leaned upon Parsons's work in uniting biology and sociology. There are other reasons, too, why Dupree has chosen to rely on "the theory of action," but these are more accidental than substantive and can be put to one side.

Dupree has developed a unique way of integrating sociology and biology. His attempt is more sophisticated than anything previously put forth in this genre, and certainly it is not a simple superimposition of biological theory on sociological phenomena.

The uniqueness of the attempt stems from several sources. First, there is Dupree's own genius at seeing integrating features between the disciplines in a way never before recognized. But also important are the recent developments in molecular biology and the movement away from the earlier structural functional position in the field of sociology to a point of view Parsons has recently called "the theory of action." Action theory relies heavily on cybernetic theory and information processing similar to that found in molecular biology.

While all of these circumstances permit new levels of sociological and biological codification, they also present difficulties. Dupree perhaps has gone too far; that is, he has carried his analysis beyond plausible inference and integration, accomplishing the integration at the cost of violating some fundamental propositions of one or the other of the fields. Indeed this is a problem we will soon point to in the articulation of molecular biology and the theory of action, for the integration is not possible because of endemic incompatibilities of the two fields. This is true in spite of the many ways in which the two disciplines resonate with each other.

Before we can proceed to these issues, it is necessary to simplify, as best as possible, what Dupree is attempting in his complex and difficult paper. Dupree's paper, however, is not only complex, it is also diffuse. By diffuse I mean that in trying to break new ground, he has covered a welter of topics, some more fully elaborated than others. And as a result, the foci of his work do not easily emerge.

In spite of the diffuseness and complexity, there appear to be two related central topics. The first of these is the development of a taxonomic analysis of a sociological, biological, and physical system, its overview and integration. He accomplished this with the aid of three analytic tools: the file stack that integrates a great many levels but is fundamentally built upon Parsonian theory and molecular biology, which is derivative from the observer gene in the organism; the second tool is "message or item transfers" (information, energy, matter transfers), which *link* the system and which again are found in more or less analogous form in both the sociological and biological frameworks; and finally, he employs the Darwinian trajectory (suggested by Dupree as another way of looking at Darwinian phylogenctic development) built upon phylogenetic time, which is subsequently deformed to bring together the primordial gene of the trajectory with the observer gene of the file stack.

The second focus of Dupree's work is epistemological and addresses itself to the problem of how the scientist objectively views his world and the phenomena in it. The answer to this lies in the link between the earth-ecosystem and the organism, making it possible to combine observation and flow of information from the highest to the lowest levels and indeed to also reverse the process. Scientists are thus able to see objectively not because they are *separate* from the system[7] but because they *are part* of the system. "Artifacts," events, phenomena, processes, etc., in the world are products of the combinations of the file stack, and in turn, the observer can objectively see from within, because the object permits the reconstruction of the file stack's effects upon it. Thus, it is only by being part of the file that we know it.

There are throughout Dupree's construction and analysis of these points some inconsistencies of definition and some vagueness. For example, in one point, observation is seen as no more than measurement of the flow of information between levels in the field. (See Dupree's comments on measurement of inputs and outputs.) At another point the observer DNA can "see" the total file in a manner which cuts across levels and indeed "sees" by combining levels. (See comment on the observer gene in the DNA program.) Further, the qualities attributed to the observer gene are more than those of the DNA program—they must be humanoid because they imply the literal capacity to physical vision, hearing, smell, and so on. And finally, the deformation of the file stack plays havoc with Parsons's conception of organism and its environments. The organism is, in Dupree's terms, directly linked to the culture and not mediated by personality. There is even some suggestion that the organism is analytically no longer independent of the elements, organic and inorganic, in the ecosystem. But with these issues, vagueness of formulation and boundary of definition play the largest part in the difficulties.

Most important among his difficulties, however, are the reformulations of Parsonian theory which Dupree attempts to achieve in the integration. I have here in mind especially the connection of the "first replication of DNA" (previously Dupree used the term "the primordial DNA") with the observer DNA in the behavioral organism at level 1 in the file stack (see figs. 6 and 8); this later became the "genetic symbol matrix" and the "individual" (see figs. 7, 9, 10). To accomplish this, Dupree had to seriously violate Parsons's conception of cybernetic hierarchy of control which gives priority to *cultural factors in the action process.*

We must take time to very briefly elaborate Parsonian theory of action and its evolution over his long career.

A Very Brief Overview of Parsonian Theory and Its Development

Parsonian theory, unlike most of contemporary sociology, is systematic and simple. By systematic, I do not mean what is generally suggested in social scientific literature—that is, comprehensive, although it is that also. Rather by systematic is meant deductive, deduced from a few simple propositions. Further, by systematic is meant that *all aspects*—theoretic and empiric—of analysis are derivable by deduction from the simple model.

By simple, I do not mean that the language is simple because by now it is conventional wisdom that Parsons is "hard to read." Rather, by simple is meant that once you know the fundamentals of the model, the key to the system is available for its easy unlocking.[8]

I will elaborate on the systematic aspect of the theory following a short digression.

By contrast to Parsonian theory, most sociological theory is ad hoc or problem-oriented. Most sociologists, given a particular issue, will formulate a set of hypothetical propositions about the phenomena so that a general orientational framework can be established within which that problem can be researched. The end product of the investigation may alter the original formulation, but once the work is done, so is the issue; it has been resolved.

Rarely do sociologists attempt to accumulate this information into a more general framework or even to deduce a related problem by which the conceptions and evidence of the first are extended, elaborated, and pushed forward. More likely the same researcher will go on to another problem or another area, more or less oblivious to what has just been accomplished in the previous work. If this may in some degree caricature specific sociological workers, it does not do violence to the field as a whole. Thus, while a sociological ancestor may investigate a particular problem, for example social classes in society, a subsequent investigator may nominally be researching the same phenomenon, but there is only a vague relationship between the old and new work in its conceptualization, variables, and operationalizations. In short then, the very nature of the social organization of sociology has inhibited its cumulative growth and systematization. Professor Shils put this problem cogently when he wrote:

> It is just this point that the great difficulty of present-day sociology lies. To take up where one's predecessor left off entails the acceptance of the delineation of the variables studied by the predecessor in exactly the form in which they were put by the predecessor. The point of leaving off in sociology is difficult to locate because the vague interpretations of data always exceed the data.

The categories involved in interpretation, that is, the theoretical construction, are always broader and vaguer than the categories in which the data were collected or observations made. The latter vary too much from one investigator to another, even though the investigators believe themselves to be working on the same problem. For good reasons or for poor, each subsequent investigator "improves" on his predecessor's working delineation of the relevant variables. What in fact happens is that a slightly different variable replaces the one previously studied; they are both thought to represent the same thing when in fact they do not. As a result the cumulative and progressive development of sociological knowledge is rendered more difficult.[9]

In the field of modern sociology, Parsons's work stands separable from this description. The theory itself has evolved in the hands of its creator for more than forty years.

Parsons has suggested a certain continuity in his work since the first major publication, *The Structure of Social Action* (1937).[10] In reality, however, the "breakthrough" to systematization came in 1951 with the publication of *Toward a General Theory of Action.*[11] It was at that time that Parsons and his co-workers developed what were called the "pattern variables" and, although there has been a reduction from five to four variables and considerable theoretical elaboration, this can be considered the marking point of the modern form of the theory.

The pattern variables were a way of redefining Töennies "gemeinschaft-gesellschaft" dichotomy of traditional and modern societies into a framework which could account for traditional and modern forms of action systems extant within the context of modern institutions. Much of this had grown out of Parsons's early studies of the medical profession whose behavioral patterns he could not at that time explain in terms of commodity-money modern market mechanisms.[12]

Deductive justification for these patterns were given, but primarily they were inferred from observation.[13] In time, however, Parsons realized that the whole categorical scheme demanded better theoretic justification. Several papers built on this point in the late 1960s and then in the early 1970s; such justification was forthcoming.[14] And by this time Parsons had abandoned the "structural functional" orientation and was now referring to his work as "action theory." The emphasis had shifted from functional operation to interaction and information flow among acting units in a system.

In legitimating this framework, Parsons only made a few simple assumptions. The first of these was that any "living" system (a system in interaction with other systems) must be separable and yet interacting with its surrounding environment and other systems. It must give and receive informational messages within itself and between itself and other systems; this process was itself the character of action. The information would be processed, used, and adjusted accordingly. Any system that did not meet these criteria and was not bounded from its environment was not a "living" acting system.

On the basis of this simple set of assumptions, Parsons suggested that all living systems are confronted with a finite set of problems or dimensions with which they must cope. One of these is the boundary problem—that is, getting and giving resources or information across their boundary.

The second of the problems was that of the processing or use of resources once they are acquired. Among these problems are such as the allocating of resources or information and determing whether the system is doing what it should be doing to ensure its survival, or more simply, the evaluation and adjustment of the system in light of the processes it performed.

The third dimension is that of time. All of these processes could not occur simultaneously. Thus, for example, getting information or resources should tend to occur prior to their use for, let us say, achieving goals of the system or their allocation or integration internally to the system. And finally, there is the time needed to evaluate and adjust as to whether things are going "correctly" or well with the system. If this all sounds a little anthropomorphic, it is not intended to, and one must keep in mind the processing of information, its achievement, use, allocation, and evaluation is at the heart of the whole of the operation.

If then, these are the problems confronting any living system, we have two axes which are independent of each other and thus bisect each other, producing four quadrants or four "functional" problems for any system. These are the earlier and later processes. From this then arises the system which Dupree has used within the middle range of his file stack.

However, because Parsons wanted to relate the temporal axis to traditional sociology, he has distinguished these as means and ends. But he has suggested a clockwise spin in terms of temporal processing, while a counterclockwise spin in terms of "cybernetic hierarchy" of control over the system.

Thus, the receipt of resources is achieved in the upper left-hand quadrant (A) while the evaluation of the processes of the system operates from the lower left-hand quadrant (L). Although all of the processes are vital to the system, they are not equally influential to the continuance and substantive expressions of the system. How the system *will* and *should* operate is more importantly related to (if not dependent upon) the informational messages emanating from the lower left-hand quadrant than those from the upper left-hand cell.

There is then a rank order of importance of the operative processes for the system, running counterclockwise from the lower left cell to the upper left cell. As Parsons put it, the lower left cell gives information (direction similar to the DNA central dogma) to the system and the upper left cell gives energy (motive force) to the system. The (G) cell is in the energy direction while the (I) cell is in the informational direction, but again ranking between the polar extremes of information and energy as described by the processes in the (L) and (A) cells. It is this ranking of importance to the system of the processes of the functional cells that Parsons

	t_1	t_2
External	A (Adaptation)	G (Goal-attainment)
Internal	L (Latent-pattern maintenance)	I (Integration)

refers to as the "cybernetic hierarchy of control" and while it is a proposition at the general theoretical level, it also holds for all empirical phenomena to which the action paradigm can be applied. Thus, for any substantive phenomena, and for Parsons, this means that for any living and acting system, the hierarchy of control obtains. *It is this proposition within the theory of action that is closest to the central dogma of molecular biology.*

Also, because there was the desire to articulate the whole of the framework to previous sociological analysis, Parsons has given each of the processes a "functional" description, but function presently means process more than purpose or teleology. A = adaptation, G = goal attainment, I = integration, L = latent-pattern maintenance. It is not necessary for me to describe these further because Dupree has already done so in his paper. More important is to note that each of these "functions" is by analogy given a *substantive* or empiric focus for different levels of phenomena. Thus, for example, Dupree illustrates that Parsons distinguished four levels of society coincident with each of the functions, i.e., organism (A), personality (G), societal community (I), culture (L). In turn there is the nesting series so that each of these levels again manifests these same processes. Parsons suggests therefore that for societal community (I), the economy subsystem serves the adaptive function (A), the polity serves the goal attainment function (G), the legal or normative system the integrative function (I), and the family and education the latent-pattern maintenance function (L). More examples of the application of the paradigm could be given but it is not necessary to elaborate these here.

Because Parsons moved away from "structural functionalism," he began to develop ways of "informationally" connecting or integrating systems; these he called media of exchange. Using money in the economy (A) as already analyzed in classical economic theory, he began to develop informational media for the other subsystems of the societal level. Almost simultaneously, papers on "power" in the political system (G) and "influence" in the societal community (I) as informational media of exchange were published.[15] Finally, value-commitments as the media in the fiduciary system (L) of society were forthcoming.[16] Each of these media were primary *informational flows,* and as Parsons has pointed out frequently, money in its exchange value is a way of transmitting messages. Again the media were integrated and derivative from the fundamental AGIL paradigm and this makes the theory of action appear analogous to message or code transfers in cybernetic theory and molecular biology.

But since the latent-pattern maintenance process (L) stands at the pinnacle of the information or message hierarchy of control of the system and strains to maintain the purposes of the system, Parsons's theoretical work soon began to be seen as having a built-in ideological conservatism; that is, the system had a built-in inertia. Parsons responded to such accusations by developing a theory of change; or put more generally, he asked, when and how do systems change? Again the four processes provided the framework for the solution.

In brief, the theory of change provides that change occurs when the processes involved in the *attainment of the goals of the system become ineffective or cumbersome.* Under these circumstances, the system differentiates (G) into two better, more effectively functioning systems. But such a differentiation survives only if the other processes also change. The systems must undergo adaptive upgrading (A), a "process by which a wider range of resources is made available to social units, so that their functioning can be freed from some of the restrictions on its predecessors." The first two process changes present integrative difficulties and these can be resolved only when there is "inclusion (I) of the new units, structures and mechanisms with the normative framework of the societal community." And finally there is the problem of value-generalization (L) to complete the change: "if various units in the society are to gain appropriate legitimation and modes of orientation for their new patterns of action," the value must become more general.[17]

This is hardly the whole of the Parsonian framework. For example, it bypasses a welter of topical papers Parsons has written, most systematically derivative from the basic theoretic paradigm, and I have hardly given enough theoretic elaboration needed to illustrate the Parsonian framework in detail. However, it does give the sense of the systematization and the theoretical derivations therefrom. In short the unity and exhaustiveness of the theoretic endeavor has been illustrated.

Professor Parsons is now more than seventy years of age. He has had a prodigious career and has made enormous contributions to the social sciences. Given the length of his career and the quantity and quality of his work, it would be presumptuous of me or of anyone else to speak of *the* influences upon his intellectual development, especially since he has already spoken for himself.[18]   One influence which stands out in the content of his work and to which he points with a great deal of pride is his biological studies as an undergraduate at Amherst College and later the influence of the famous work of W. B. Cannon, *The Wisdom of the Body*.[19]

There are then some very substantial common bases upon which Dupree could integrate Parsonian sociology with biology, but not the least of these is that they have a common origin in biological theory. However, there are several other grounds upon which Dupree can justifiably integrate Parsonian theory with molecular biology. Parsonian theory is systematic in the deductive and derivative sense. As a result, it is analytically exhaustive in its categories, and Dupree is not faced with the problem of deciding where the theory begins and ends, nor is he faced with the problem of inclusiveness which would certainly be an issue with almost all other sociological frameworks. The Parsonian paradigm of bounded acting units, of levels of analysis in a continuous nesting series, and of process in terms of media of informational exchange, with special relation of the DNA central dogma to the processes of latent-pattern maintenance information, is compatible with the structure of molecular biology, if not specifically, at least in general analytic terms. Therefore, it is hardly startling that Dupree finds much in this sociology that shares the lines of contemporary biological thought, but the most important of which we are about to discuss.

Sociological Theory: Parsonian and Other

With these thoughts in mind we can now return to Dupree's foci and his wedding of molecular biology with the theory of action. Our main concern will be with the file stack (see fig. 6), certain of the conditions and processes attributed to it, its theoretic and empiric validity, and the theoretic transgressions, especially to Parsonian sociology produced by its construction and deformation as illustrated in figure 8.

In order to construct the file stack of seven or more levels, Dupree begins with the belief that there are discrete phenomena of analysis attributable to each of the levels. Within the middle range of the Dupree file (levels 1, 2, 3, and 4) Parsonian theory plays a particularly important role. In the theory of action, in accord with the four functional paradigm, Parsons has suggested that while these empirical phenomena may not be totally coincident with the functions, they are in their dominant processes expressive of each of the functions of the AGIL paradigm.

Thus for Parsons societal culture primarily operates in terms of latent-pattern maintenance functions (L) while society as a separable level operates in terms of integrative functions (I), personality is primarily a goal function (G), and the organism is principally adaptive (A). But this then means that while all of these processes are vital to society, culture is more influential and determinative than the others; *it is culture which primarily shapes the processes and the patterns of the society.*

Let us focus on culture (L) in contrast to society [i.e., the societal community (I)] to highlight this issue. In his work Parsons insists that "I have found this distinction to be completely indispensable." And he writes,

> By contrast with the cultural system which is specifically concerned with meaning [cognitive (A), expressive (G), moral-normative (I) and constitutive (L)], the social system is a way of organizing human actions which is concerned with linking meaning to conditions of concrete behavior in the environmentally given world.[20]

Here the distinction is clear: culture is *meaning* and society is *meaningful action* adjusted to the environment; analytically and empirically Parsons is saying that the levels *must* and *should be separated.* But even more important, he is saying what we have already noted; culture is determinative of meaning and the shaping of behavior to the situation is the adjusted process the societies must make in terms of normatively real conditions to express that meaning. Such an assertion is in accord with his basic proposition of his hierarchy of control.

Some sociologists would not accept this distinction regarding distinguishable levels of analysis, templates of culture or institutions, and the servomechanistic conception of latent-pattern maintenance.[21] And certainly they would not accept the belief of ranges of social phenomena similar to, let us say, a Bohr range, a Newtonian range, or Einsteinian range. Some sociologists, even those who more or less are sympathetic to Parsonian theory, do not accept the theory in a manner similar to his presentation, as for example, the latent-pattern maintenance system operating as it does, or that change follows the description offered by Parsons.[22]

Also I have in mind some prominent sociological theorists who would reduce all variables of action to the individual psychology or personality level, conceiving of the other levels such as culture as either unreal or irrelevant.[23] The line of thought would be that it does not matter what meanings exist in society; all that counts is how people really act, regardless of what morality, historical or contemporary, might exist in a society. This line goes further: these sociologists would question the worth of referring to a normative cultural heritage and instead would attempt to learn how people "really" behave, to uncover the "real rules" in real situations, or perhaps they would aim at discovering the situation ad hoc or the improvised rules of behavior rather than look to the servomechanistic conception of normative cultural dictates.[24]

I raise this issue not polemically, but rather to indicate that the theoretic stability of sociology and molecular biology are not on a par. There are many sociologies but there is only one molecular biology.[25] Dupree must realize the instability of sociological theory upon which he builds. And yet because he needs the four middle range levels to construct his file stack, he accepts what is generally termed in sociology and anthropology a cultural theoretical approach; but were he to accept some of the other sociological frameworks he could not talk of discriminable levels and he might be better or more easily able to integrate the genetic underpinnings of social behavior.

However, by accepting the cultural theoretical perspective (of which Parsonian theory is only one version; Alfred Kroeber's work is another), Dupree must accept the proposition of *hierarchy*

of control; that is, cultural messages as a latent-pattern maintenance servomechanistic device are of first order importance in
shaping societal processes and behaviors, as with the messages in
DNA which are basic in shaping the phenotypic structure of
individuals.

The Downgrading of Culture and the Hierarchy of Control in Dupree's Integration; The Loss of the Sociological Central Dogma

In short then, Dupree must live with culture, in its pattern
maintenance, servomechanistic, and determinative conception because he has accepted the theory and because he needs it and the
other three levels of society to develop his file stack. But almost
immediately Dupree's work modifies one of the sociological "central dogmas" of Parsonian theory by downgrading the importance
of culture as proposed in the theory of action. This downgrading is
not independent of the theoretical and empirical security of the
two fields Dupree is attempting to integrate.

Early in Dupree's work it becomes evident that there is a
certain priority of determinative motive power to action in the file
stack, emanating from the organism to the higher levels of the file
in which the DNA program exists. Dupree on this point writes:

> Each file stack has its *origin* in the formation of a zygote by
> fertilization which lays down an information program in DNA,
> which I shall call the observer genes. Hence level 1 of the file stack is
> Parsons's Behavioral Organism (A) (fig. 1). Because the DNA pro-
> gram can express itself through RNA to synthesize proteins, the file
> stack *begins to build on it immediately, ultimately to express itself
> in the behavioral organism.* At the same time, according to the
> central dogma of molecular biology, protein cannot synthesize DNA.
> However many feedback loops may connect parts of the file stack,
> they cannot finally close the loop with the DNA program. . . .
> (italics mine)

The priority of the DNA program in the file stack is a
theme repeated throughout the paper. This is to a degree justifi-
able. The central dogma of molecular biology is that DNA →
RNA → Protein, and this process is irreversible once the protein is
synthesized. Dupree cannot get around this theoretically substanti-

ated dogma, and in fitting it into the sociological-biological union, he must have his model conform to this known and substantiated fact. Between the more solidly based biology and the more tenuous sociology, social theory gets sacrificed.

The theme thus reaches a crescendo subsequent to the deformation of the file stack (fig. 8) when he writes of the circulating movement in the deformed file:

> Because of the one-way valve of the central dogma in the DNA of the behavioral organism, there is a preponderance of counterclockwise currents in the diagram of Figure 8, giving the functions a pronounced counterclockwise spin. Since there is an energy loss as one passes around through the various levels (now spokes on the wheel) and since the countercurrent comes only from feedback, the functions will reach level 5 only in a weakened form, and ecological, cultural, and societal responses will have trouble getting back to the behavior organism.

Dupree is quite aware that this violates a fundamental tenet of Parsonian theory—the one we have already alluded to. In Parsonian theory the currents should tend to predominate in just the opposite direction, that is, in a clockwise spin from culture (L) to the lower levels. Parsons as noted refers to this as his cybernetic hierarchy of control which insists that the dominance of the controlling informational flow in human societies stem from the meaningful symbol systems extant at the cultural level and that energy come from the lower file levels, such as the organism (level 1) and the personality (level 2), combining with each other to produce meaningfully directed social action in environmental situations.

Dupree tries to ameliorate this difficulty which he has obviously created by suggesting a two-way flow, clockwise and counterclockwise, but again organized in biological rather than cultural terms; the ontogenetic DNA programs he suggests stress the counterclockwise spin and the phylogenetic DNA programs stress the clockwise spin in societal behavior, social action and institutional organization. But in both of these instances of explanation, there is by sociological standards a *reductionism* away from the cultural analysis which he has accepted for his file, and Dupree recognizes this difficulty. Thus, he writes somewhat glibly,

"*If* such a pattern generated by the primordial DNA program can find a mechanism to transcend ontogenetic time and be passed down in phylogenetic time, they will rotate clockwise on figure 8, hence moving from level 5 to level 4." If that were the case, the place of culture in societal determination would be restored, but that *if* is a very big one; it smacks of Lysenkoian logic and is hardly connected theoretically to the whole of the thrust of his now molecular biological analysis.

But Dupree does not stop here in his effort to restore the "rightful" place of culture in the analysis. He now makes the attempt to leap from the phylogenetic DNA program to Parsons's frequently mentioned but hardly developed conception of cultural "symbol as gene." But this argument is not strong in Parsons's work and neither is it developed in Dupree's. Rather in an ad hoc way, Dupree tacks on the following: "Thus an information program supplementary 'symbol as gene' is available to *Homo sapiens,* and *while the fit is not perfect, it in effect provides a line of codons extending beyond the end of the DNA chain.*" (Italics mine).

Dupree's effort at dealing with the symbol system as countermotive force is not convincing and does not go beyond what little Parsons himself has said on this topic. We are left with more confidence in the central dogma of molecular biology; the material seems harder. And yet we also know that culture does shape behavioral expressions and that gene information in homo sapiens tends to be general and thus available to cultural shaping in a way that specificity of gene messages in lower forms is not.

The result of Dupree's formulation is that culture and society, the higher levels in the file, tend to lose their motive force as moral and normative mandates or as internalizations in personality in the socialization process. One of the corner stones of Parsonian theory or any cultural theory could be put the following way: the cultural variation among humans far exceeds their genetic variability and therefore culture dominates man's behavior. Dupree's theoretic formulation of the social levels of the file, which is fundamentally based upon one or another of the DNA programs, undermines such a position.

What I have tried to show up to this point is that Dupree

had to choose. He could have selected some sociological theoretical frameworks which would have been easier to integrate with biology. But instead he selected Parsonian theory because the discrete levels were compatible with his file-stack construction. However, he promptly proceeded to undermine the central dogma of the cultural emphasis of the theory of action and some of the known facts of social science concerning cross-cultural, societal behavioral variations. Dupree thus can't have it both ways and develop a unified theory of biology and sociology. If he accepts Parsons's version of cultural theory, he then can't promptly relegate the fundamental dogma of cultural determinism to a position of secondary importance in the analysis.[26]

Empirical Difficulties for the File Stack and for Parsonian Theory

I have to this point stressed the theoretic difficulties and incompatibilities that Dupree has encountered in wedding biology to sociology. There are also empirical problems and inequities between the two fields and these contribute to the theoretical problems already noted.

One problem Dupree realizes he faces and inherits from Parsons's work is that of fitting empirical phenomena or evidence to the file stack. Dupree describes the file stack not as a "mechanism" or an "organism" but "simply a file (in an office. . .) ." Items in the file "can be described as matter, energy and information." And finally he notes in describing measurement of the file that they are no more than "black boxes or . . . esoteric boxes."

While Dupree's main purpose is not what goes into the boxes, the use of the metaphor is of interest because perhaps inadvertently, he has stumbled upon one of the problems which has plagued Parsonian theory. In simple terms, the problem is that of what goes into those boxes and how does one decide what to place in them. This problem is somewhat akin to the one already noted by Shils in the sense that sociological variables are often differently operationalized, dependent upon the whim of the sociologist.

But more specifically, the problem which plagues Parsonian theory is that while he has defined the functional processes for the boxes and has offered substantive illustrations, *no precise rules*

of transformation from the theory to the empirical world have
been established in his work. In employing Parsonian theory for
empirical work, one uses a "rule of thumb" when, for example,
developing instruments to test the work, e.g., a questionnaire.
Decisions of this kind are often made in an empirical vacuum of
relevant information and thus several alternative formulations are
possible. Deciding upon one rule implicitly means specifying rules
of transformation and casting aside other possible rules. But to
this point, in no sense do the rules that are established unequivo-
cally lead to substantiation or negation of the theory no matter
which way the data finally fall. In Parsonian theory, the degree of
isomorphism between the rule of transformation and the theory
has never been established and negative findings could have been
produced by a mistaken transformation procedure while positive
findings only give us increased confidence that the theory *might*
be correct; but still there is *no literal* way to interpret the findings
as standing in the place of theory.

Dupree by accepting the nesting series and by offering no
more precise way of discussing the content of the boxes or even
that which is closer to his interest, the informational flow, has
inherited the unresolved problems from Parsons's work. The same
level of difficulty obviously does not beset molecular biology. Not
all the empirical problems confronting that field have been resolved, but the empirical foundations are solid. As Stent has
written, the major breakthrough is over; what remains is the
yeoman cleaning-up action.[27]

Thus again, Dupree has inadvertently or unconsciously
moved from strength. Between the empirically and solidly grounded molecular biology and the empirically unstable sociology, he
has tended to favor the former in interpreting processes and causes
of the development and structuring of society. Dupree is a rational
man of science and he is moved by the want of sound evidence,
but at the cost again of social science.

Some Direction of Analysis

Quite obviously, I do not wish to fault Professor Dupree
on the theoretic direction he has taken. The number of intriguing
propositions he has offered in his work were too many to note
within the context of a brief paper. But one connection especially

intrigues me and thus I will comment upon it. By connecting the Darwinian trajectory to the file stack, Dupree integrates phylogenetic time (which can be considered historical time) with ontogenetic time (which can be considered biographical time). In the history of science and thought, there have been efforts to connect the etiology of past with the etiology of present; recall the old saw that "ontogeny recapitulates phylogeny" as one of these. But Dupree's foci are on "observing" and "knowing" the past and the present and their relationship to each other. And his framework offers interesting possibilities for investigation.

Freud once wrote about the growing field of ego psychological theory, that is the oncoming integration of sociology and psychoanalysis, that he could not set sail for open water; in his own work, he had hugged the shoreline but he wished others well in their voyage. In spite of the brilliance of his work, perhaps Professor Dupree has found himself in the obversely analogous position—he may have voyaged too far and too quickly.

There are two directions of analysis which I believe should be encouraged before embarking on such an adventure. The first, which is not unrelated to the second, is the need for more detailed analysis within the Parsonian framework in terms of the degree of fit between social and biological phenomena; in short finer analogies between Parsonian theory and molecular biology. The second is to bring Parsonian theory closer to that of biology which demands more empirical substantiation.

It has already been suggested that one of the softer and more problematic spots of the paper has been the reductionism regarding culture and the association of the cultural symbol with the gene. On this matter simple questions need to be asked about such an analogy—questions which, if answered, would reduce the biological biases and upgrade the cultural factors in the integration. For example, in what precise ways are the gene and the symbol analogous? What are the similarities and differences of their messages? How are such messages sent and how are they decoded? What are the cultural analogies to variation and mutation? How do cultural symbols supersede and overcome biology? Questions such as these go unanswered and yet it is detailed thought on such issues which can validate the type of integration that Dupree is attempting.

I do not wish to imply that there has been a total absence of such an effort on the part of social scientists. Parsons has done some of this which we have noted, and Clifford Geertz has gone farther in associating the gene with the symbol.

On this topic Geertz has written:

> So far as culture patterns, i.e. systems or complexes of symbols are concerned, the generic trait which is of first importance for us here is that they are extrinsic sources of information (Geertz, 1964a). By 'extrinsic', I mean only that—unlike genes, for example—they lie outside the boundaries of the individual organism as such in that intersubjective world of common understandings into which all human individuals are born, in which they pursue their separate careers, and which they leave persisting behind them after they die (Schutz, 1962). By 'sources of information', I mean only that—like genes—they provide a blueprint or template in terms of which processes external to themselves can be given a definite form (Horowitz, 1956). As the order of bases in a strand of DNA forms a coded program, a set of instructions, or a recipe, for the synthesization of the structurally complex proteins which shape public behavior. Though the sort of information and the mode of its transmission are vastly different in the two cases, this comparison of gene and symbol is more than a strained analogy of the familiar 'social heredity' sort. It is actually a substantial relationship, for it is precisely the fact that genetically programmed processes are so highly generalized in men, as compared with lower animals, that culturally programmed ones are so important, only because human behavior is so loosely determined by intrinsic sources of information that extrinsic sources are so vital (Geertz, 1962). To build a dam a beaver needs only an appropriate site and the proper materials—his mode of procedure is shaped by his physiology. But man, whose genes are silent on the building trades, needs also a conception of what it is to build a dam, a conception he can get only from some symbolic source—a blueprint, a textbook, or a string of speech by someone who already knows how dams are built, or, of course, from manipulating graphic or linguistic elements in such a way as to attain for himself a conception of what dams are and how they are built.[28]

It is obvious that Geertz takes this analogy farther but it still stands on "soft" foundations by rigorous theoretic standards. But until such theoretic work is accomplished, the central dogmas of molecular biology and Parsonian sociology will not mesh and the latter will be, as Dupree has done, sacrificed for the more firmly theoretical discipline of molecular biology.

This then brings us to the whole question of empirical substantiation of the whole of the Parsonian framework. To this point "harder" empirical work in the Parsonian paradigm has been historical and developmental. Of the previous generation of sociologists, the work of two students stands out: Neil Smelser and Robert Bellah. Smelser studied the differentiation of the family-firm into the separated but integrated modern institutions of family and firm as functionally more specific units.[29] The work of Robert Bellah followed this same differentiation model and dealt with the modernization of Japan. Smelser's work is now an accepted classic within the context of the Parsonian "evolutionary" model of development. Bellah, without negating Parsonian categories, has retreated from his analysis of modernization of Japan through differentiation which he employed in his *Tokugawa Religion.*[30]

A whole generation of younger sociologists is now attempting to operationalize aspects of Parsonian theory in terms other than historical data—especially focusing on contemporary society. My own association with Professor Parsons has moved in both directions. Our theoretical-substantive analysis of *The American University* has only recently appeared.[31] The empirical study of a national sample of academics will follow on its heels.[32] In the sense in which I have defined systematic, the empirical volume will still only be so in an ad hoc way. Some topics will directly substantiate theory but some will only have a vague relationship to theory. But there is much more need of this type of work, empirical work which begins to clarify theoretical issues and gets to the problems of just how one goes about operationalizing Parsonian theory. When this is accomplished, Professor Dupree's model will either be altered, be on firmer ground, or there will be evidence enough to suggest that such a wedding, as novel as Dupree's attempt is, is just not possible or even fruitful.

Until then, Parsons for sociology and now Dupree for his wonderful attempt at integrating the disciplines has certainly accomplished a novel form for looking at the entire spectrum of the social, natural, and physical world. But Francis Crick, as if to underline the fundamental necessity of empirical work for theory, once said:

> I have always had great difficulty in publishing theoretical ideas in a vacuum of evidence. In the long run we do not want to *guess* the genetic code, we want to *know* what it is. . . . Whether theory can help by suggesting the general structure there is little doubt that its discovery would greatly help the experimental work. Failing that, the main use of theory may be to suggest novel forms of evidence and to sharpen critical judgement.[33]

By Crick's standards, at least Dupree has done this much for us.

ACKNOWLEDGMENT

I would like to thank Professor Carl P. Swanson, Department of Botany, University of Massachusetts, Amherst, for his time and effort in helping me grasp some of the finer distinctions in the field of modern biology. I would also like to thank Rita Kirshstein for her help in formulating and executing this essay.

NOTES

1. In sociology, this type of framework was also found in the works of Comte, Hegel, and Marx, just to mention a few others.
2. Robert K. Merton, "Manifest and Latent Functions," in Robert K. Merton, *Social Theory and Social Structure* (New York: The Free Press, 1968), pp. 73–138.
3. Tom Bottomore, "Out of This World," *The New York Review of Books* XIII, no. 8 (November 6, 1969): 34–39; Lewis Coser, "Social Conflict and the Theory of Social Change," *British Journal of Sociology* VIII, no. 3 (September 1957): 197–207; Ralf Dahrendorf, "Out of Utopia: Toward a Reorientation of Sociological Analysis," *American Journal of Sociology* 64 (August 1962): 115–27.
4. Talcott Parsons, *The Social System* (Glencoe, Ill.: The Free Press, 1951).
5. For example, compare it to the doctrinaire statements on the absolute functional necessity that had emanated from Malinowski's

A Scientific Theory of Culture and Other Essays (Chapel Hill, N.C.: University of North Carolina Press, 1944); or from the work of Parsons's own students who had overreacted to what he meant by "functional necessity." See D. F. Aberle et al., "The Functional Prerequisites of Society," *Ethics* 60: 100–111; Kingsley Davis, *Human Society* (New York: The Macmillan Co., 1949); and Robin M. Williams, *American Society* (New York: Alfred A. Knopf, 1960).

6. In the most recent "radical" attack on Parsons's work, Alvin Gouldner in his *The Coming Crisis of Western Sociology* spent more than 200 pages scathing Parsons's work but before doing so, he asserted that Parsons was *the* sociological theorist in the twentieth century and that it was necessary to get "past" his work. Gouldner, *The Coming Crisis of Western Sociology* (New York: Basic Books, 1970).

7. In sociology, this is a position taken by Karl Mannheim, *Ideology and Utopia* (New York: Harcourt, Brace, 1936).

8. Talcott Parsons, "Some Problems of General Theory in Sociology," in John C. McKinney and Edward A. Tiryakian, eds., *Theoretical Sociology: Perspectives and Developments* (New York: Appleton-Century-Crofts, 1970), pp. 27–68; Talcott Parsons and Gerald M. Platt, *The American University* (Cambridge, Mass.: Harvard University Press, 1973), chap. 1.

9. Edward Shils, "Tradition, Ecology, and Institution in the History of Sociology," *Daedalus*, Fall 1970, p. 819.

10. Talcott Parsons, *The Structure of Social Action* (New York: McGraw-Hill, 1937).

11. Talcott Parsons and Edward Shils, *Toward a General Theory of Action* (Cambridge, Mass.: Harvard University Press, 1951).

12. Talcott Parsons, "Illness and the Role of the Physician: A Sociological Perspective," *American Journal of Orthopsychiatry* 21: 452–60.

13. Talcott Parsons, Robert F. Bales, and Edward A. Shils, *Working Papers in the Theory of Action* (New York: The Free Press, 1954).

14. Talcott Parsons, "Some Problems of General Theory in Sociology," pp. 27–68.

15. Talcott Parsons, "On the Concept of Political Power" and "On the Concept of Influence," both in Parsons, *Politics and Social Structure* (New York: The Free Press, 1969).

16. Talcott Parsons, "On the Concept of Value-Commitments," *Sociological Inquiry* 38, no. 2 (Spring 1968): 135–60.

17. Talcott Parsons, *The System of Modern Societies* (Englewood Cliffs, N. J.: Prentice-Hall, 1971).

18. Talcott Parsons, "On Building Social System Theory: A Personal History," *Daedalus*, Fall 1970; and his earlier biographical sketch in *Alpha Kappa Deltan,* 1959.

19. Professor Parsons's pride in his biological training is most vividly illustrated in a give and take at the Bellagio Conference when his biological training was questioned and a famous zoologist and geneticist stood up on his behalf in this matter. Parsons relates this in a footnote (no. 16, p. 86, *Daedalus*) when he writes: "In the discussion at the Bellagio Conference, when this essay was first presented, a question was raised about the seriousness of this exposure to biology. I was greatly pleased when Professor Curt Stern said: 'May I make one very short point in regard to Amherst that not everyone might know. At Amherst, biology was taught at a very advanced, even a graduate level, although Amherst did not give doctors' degrees. These were highly distinguished people, and probably their influence was greater than it would have been had Professor Parsons gone to another college with good but less distinguished professors.' "

20. Talcott Parsons, "Culture and Social System Revisited." *Social Science Quarterly*, September 1972, pp. 253–66.

21. Harold Garfinkel, *Studies in Ethnomethodology* (Englewood Cliffs, N. J.: Prentice-Hall, Inc., 1967); Alfred Schutz, *Collected Papers I: The Problem of Social Reality* (The Hague: Martinus Nijhoff, 1962).

22. Wilbert E. Moore, *Social Change* (Englewood Cliffs, N. J.: Prentice-Hall, 1963).

23. George C. Homans, *The Human Group* (New York: Harcourt, Brace and World, 1950); also by Homans, *Social Behavior: Its Elementary Forms* (New York: Harcourt, Brace and World, 1961).

24. Howard S. Becker, *The Outsiders: Studies in the Sociology of Deviance* (New York: The Free Press, 1963); Erving Goffman, *The Presentation of Self in Everyday Life* (Garden City, N.Y.: Doubleday Anchor Books, 1959).

25. Gunther S. Stent, "DNA," *Daedalus,* Fall 1970, pp. 909–937.

26. Some of the quotes from Dupree's work that are cited in this paper come from his earlier version presented in Ann Arbor. He has made an effort to revise his paper to ameliorate criticisms made at the conference, but his fundamental biological reductionist position still exists even in the new version of his work.

27. Stent, "DNA," pp. 909–937.

28. Clifford Geertz, "Religion as a Cultural System," in Michael Ban-

ton, ed., *Anthropological Approaches to the Study of Religion* (London: Tavistock Publications, 1966), pp. 6–7.

29. Neil J. Smelser, *Social Change in the Industrial Revolution* (Chicago: University of Chicago Press, 1959).

30. Robert N. Bellah, *Tokugawa Religion* (Glencoe, Ill.: The Free Press, 1957); see also by Bellah, "Continuity and Change in Japanese Society," in Bernard Barber and Alex Inkeles, eds., *Stability and Social Change* (Boston: Little, Brown and Company, 1971).

31. Talcott Parsons and Gerald M. Platt, *The American University* (Cambridge, Mass.: Harvard University Press, 1973).

32. Gerald M. Platt and Talcott Parsons, *The American University, Vol. II: A National Survey of Faculty,* forthcoming.

33. Robert Olby, "Francis Crick, DNA and the Central Dogma," *Daedalus,* Fall 1970, pp. 965, 978.

VIII   Panel: Science and Education

SCIENCE EDUCATION
AND THE NEW HERETICS

JOHN M. FOWLER

WE ARE honoring at this time a historical figure whom many think of as a heretic, an antiestablishment hero. This identification is not really accurate. Copernicus was very much a member of the establishment—the Church. He was important enough in the hierarchy to be called to Rome to help revise the calendar. His work, however, began a movement that eventually split science from the Church and caused it to grow into a rival establishment.

During the five hundred years since Copernicus, science has gained dominance in the intellectual sphere. In the decades since World War II that dominance has become so complete that to many "Science" is now "The Establishment." In all honesty, those of us who consider ourselves scientists and science educators must agree that there is some reality to this view; that at least during the golden age of the sixties, science had the kind of unquestioning support, political power, and influence that we associate with The Establishment. Golden ages, unfortunately, never last and, when they pass, all paths from the top lead downward. In the late sixties and continuing into the seventies we have witnessed growing attacks on this establishment. We are listening to the increasingly loud voices of the new heretics.

Heretical attacks on the Scientific Establishment come, we might say, from both the left and the right; from the mystics—the antirationalists—and from the practical men of business and politics. There has, of course, always been some opposition from these quarters. At first the opposition was from religion, but now much of religion has made an uncertain peace with science. There are still some religiously motivated confrontations, such as the one being mounted against biology texts by the "Creationists" in California. But most of the heresy now comes from outside of religion.

The mystics whom I am calling the "new heretics" have, as one of their most compelling spokesmen, the historian Theodore Roszak, author of *The Making of a Counter Culture* and, more recently, *Where the Wasteland Ends.* In the former work, Roszak both brands Science as the Establishment and seeks to begin its undermining. The attack is put most savagely in chapter 7, which he titles "The Myth of Objective Consciousness." He first shows us the depth of his heresy:

> If the preceding chapters have served their purpose, they will have shown how some of the leading mentors of our counter-culture have, in a variety of ways, called into question the validity of the conventional scientific world view, and in so doing have set about undermining the foundations of technocracy.

After exploring the "objective consciousness" and defending the label of "myth," he then nails his theses to the door of our Cathedral (if you will let me move my analogy from Copernicus to Luther):

> While the art and literature of our time tell us with ever more desperation that the disease from which our age is dying is that of alienation, the sciences, in their relentless pursuit of objectivity, raise alienation to its apotheosis as our *only* means of achieving a valid relationship to reality. Objective consciousness *is* alienated life promoted to honorific status as the scientific method.

The rest of this book seeks to suggest other ways to know the world and, in fact, to replace the question "how shall we know" by "how shall we live."

The success of the mystic heretics in leading the young and even the not so young away from science is difficult to measure. The relative space occupied in bookstore shelves by astrology and astronomy provides some visible evidence; the enrollment decline that follows the dropping of science distribution requirements gives another dimension.

The attack from the right is not a new one either; many of

us who are *not* nostalgic for the fifties remember Charlie Wilson, Eisenhower's Secretary of Defense, and his contemptuous reference to scientists who were "trying to find out why grass is green." But the practical Wilson's disdain for the esoteric searchings of science was in that decade overruled by a public still grateful for the efficacy of wartime application of science. Today's detractors, with a "What have you done for us lately?" attitude, are more numerous and closer to the purse strings.

The attitude is put most tersely in a quotation attributed to the late President Johnson: "We've spent millions in medical research, isn't it time now to zero in on disease." For a view from the present administration we need look no further than the March issue of *Science* magazine where a "high official" was quoted as saying: "I don't believe science and technology are ends in themselves. Science isn't a superior thing of itself which we have to keep on a pedestal. It has its primary impact in relation to other things like trade and economy. . . ."[1]

The attack of the practical men on the establishment of science has produced the most telling effects; the falling curves of research support and the accompanying unemployment, reduced graduate student enrollment, and so forth. The attack from the other quarter is more insidious and may have a delayed effect. It infects the youth; they have yet to rise to power either within the larger society or the academic community. Science has not yet faced the full force of this heresy. It haunts our future.

Science Education

Heresy can be met in many ways—it can be overpowered, bought off, or absorbed, and, in modified form, made a part of future dogma. I am going to suggest that science education can learn from the heretics. But first let us take a look at science education, the "propaganda arm" of science, and try to assess the strengths and weaknesses and identify trends there which have a bearing on the way it can meet the challenges of the heretics.

Let me narrow my field of view to reasonable limits; I will talk only about science education as a part of higher education, choosing such a limited scope as most fitted to my time limits, to this audience, and to my own experience. I do not choose it because of educational priorities—they may well be quite different. I want to remark on two challenges, two related areas in which I believe change is and will continue to take place. These changes are brought about by historic pressures and are only affected in a limited way by the attacks of the heretics. Nonetheless, by occupying much of the attention of concerned science educators they will have a major effect on the success of our response to the heresies.

The first of these is the shift in the traditional role of postsecondary science education. Throughout most of its modern history science education has seen as its central task the training of science's next generation. If you add to this central aim the in-house activity of providing a foundation for other scientists and engineers you have encompassed not only the central aim but almost all of the target. Only geology and biology have traditionally attracted reasonable numbers from that population of students I will call "nonscientists" (this term, a very common one, sums up quite well the traditional view of these students; it tells us not who they are but who they are not—they are not of us).

The task, happily accepted, to produce more and better scientists received its greatest boost in the sixties when the National Science Foundation (NSF) poured many millions of dollars into curriculum reform and course development. The reform in secondary schools came first: Physical Science Study Committee (PSSC), Biological Sciences Curriculum Study (BSCS), Chem-Study, and so forth—only mathematics, the "loner" among the sciences, was able to begin in elementary school and revise and reform all the way up the educational ladder. The secondary school reform was followed by the establishment of the College Science Commissions which tried to assist and encourage undergraduate reform. The later sixties saw several major elementary science programs come into existence.

Although all of these programs have had effect on the scientific literacy of the general school population, the interest and experience of their developers, the tests they set, and so forth, all biased these courses in the direction of foundation building; they best served to help the budding scientist begin to build toward his professional goals.

The combination of reverses which struck the sciences, especially the physical sciences, at the turn of the decade have in them the clear message that concentration on the production of more professional scientists is not any longer a broad enough goal to occupy all of science education. Even the NSF millions did not increase the percentage of natural science graduates. From 1960 to 1970, the number of natural science graduates (engineering and mathematics excluded) was equal to 2.2 percent of the total number of students who had graduated from high school four years earlier. The individual years varied only within the narrow range of 2.15—2.25 percent.[2] The firmness of these percentages seems to point to a certain immutability of talent and desire. Since the growth curve for high school graduates is flattening out, one can predict only modest growth, at most perhaps 2 to 3 percent per year, of science majors in the 1970s compared to the doubling of science graduates between 1958 and 1970.

Science departments, however, have expanded greatly during that period, fed both by growth in student numbers and in support for research. These faculties are now faced with less to do of a professional training nature and with pressure from the practical men who hold the purse strings and who want more tangible productivity from the faculties they support. The response has been, at least in some of the more desperate hard science departments, to reach for the nonscience student.

I am most familiar with physics, but biology and earth science show much the same response. In the past three or four years there has been an enormous increase in the variety of courses offered by physics departments. These courses take many forms: "Science and Society" courses of the type which could be served by Schroeer's book, *Physics and Its Fifth Dimension: Society* or by Friedlander's, *The Conduct of Science.* In such courses the science component of public policy issues are examined in greater or less detail, as is the interface between science and society.

A second approach, very popular in biology, is the "Environmental Course"; the best seller, *Ecology, Pollution, Environment,* by Turk, Turk, and Wittes, serves some of these courses and is joined by a host of other "environmental readers." In physics the most popular theme has been "energy" and a number of texts, mine included, are at some stage of publication.

A third type of course which takes advantage of special interests and talents of scientists are the "Physics and/of _____" courses. There are several "Physics of Music" courses in existence; I know of a "Science and Literature" course. There are other examples which are familiar to you. One should also mention here the History and Philosophy of Science courses. Finally, there is a most welcome growth of interest in providing a more pertinent physics course for the premedical students, and the present list of two or three introductory physics texts with an emphasis on medical and biological application and example will soon be expanded by several manuscripts in the works.

Added to experimentation with courses is curricular experimentation; the inflexibility of the science major is being reduced by cutting requirements, providing optional tracks, and more room for electives. There are several interdisciplinary curricula in testing stages; Environmental Studies Programs, for instance, at such places as the University of Wisconsin, Green Bay, Governors State College in Illinois, Evergreen College in the state of Washington. The new major in human biology at Stanford University is apparently an outstanding and successful example of an interdisciplinary major; it is reported to be the most popular science major at Stanford, and its lower level courses are popular as distribution fulfilling electives for nonscience students.

Many science educators hope that these curricular changes and interdisciplinary majors will woo back the kind of student who would have been a science major three or four years ago, but who has deserted to other fields more recently.

A New Kind of Teaching

As science education moves away from what it does best—training future scientists—it is, I believe, also moving away from a style of teaching that has been honed by years of experience to one in which new needs must be met in new ways. Science education has been totally dominated by content. The table of contents of all the introductory science texts are organized around groups of topics which must be mastered in sequence. The only skill of the type that I would label a "process" skill has been problem solving, and even that has become so routinized that it is not very transferable; it does not, I believe, as it is usually handled, adequately prepare a student to evaluate a natural situation, identify the questions in it, phrase the questions in a manner which allows them to be answered, and then seek the data from which answers can be wrested.

Content-based education has worked for the training of scientists. As befits an Establishment, a science education is dogmatic, and necessarily so. It will probably continue to be. If new progress is to be made, then science students must be forced to climb quickly to the top of their discipline. There is not time for a leisurely examination of the foundations on which that structure is built, no time for each student to discover these basic principles for himself.

But as science education accepts more and more responsibility for the nonscience student, as it begins to search for ways to prepare all students for lives in which science-derived knowledge and technology will play a large role, this emphasis on content will have to be modified. There is neither the time nor the patience on the part of the nonscience students to build the edifice stone by stone.

It is true, I believe, that the most useful contributions to the hoped-for solutions of society's problems will be made by people trained in a discipline, who grow from that base, and add to it knowledge and techniques from their experience. I don't think that many scientific contributions to the solutions of society's problems will be made by generalists. But there are other needs; the citizenry of the future needs to feel more at home with science, to know how a scientist works, to recognize a scientific argument, even learn some of the basic strategies of science, such as estimation, the handling of uncertainty and error, and to acquire habits of observation and questioning.

We set many of these goals for our students now, at least in the memoranda to our colleagues in support of distribution requirements. But we have little experience with instructional techniques aimed deliberately at their achievement, or with evaluatory methods to measure the student achievement. Let us take just one example. I would hope that my course on "Energy and the Environment" at Maryland would enable a student to read the advertisements attacking automobile emission controls and to be able to evaluate the strengths and weaknesses of the arguments therein. I cannot accomplish this by dissecting the advertisement myself and asking them to memorize my criticisms; that would not give them a general approach, would not prepare them to handle the next such advertisement. I must discuss the basic issues in the energy crisis, give them examples of scientific and unscientific approaches, try to inculcate a habit of skepticism and searching, and then hope that they will apply knowledge and strategies in new areas.

For the nonscience student, for whom I claim content must take a lower priority than process, the laboratory takes on a new importance. The laboratory *is* process-oriented; it is here that the student should be able to see most clearly how questions are profitably asked of nature. Unfortunately, laboratory courses often manage to mask this essential excitement of science. If we could release them from the pressures of content goals, from "coverage," and give priority to the process goals, we could turn this albatross of science education into one of its strengths.

I feel very strongly that science education in the next few years will have to work very hard to articulate these "process" goals, and then work even harder to discover ways to achieve them and to test for this achievement. To be successful we must seek help wherever we can find it; from our colleagues in the humanities who have always set such "process" goals before their students and from psychology and education, where progress is being made in both the art of stating objectives and in the science of precision testing. Is it too much to hope that examples of process-oriented pedagogy and tests to evaluate student achievement of process goals will begin to appear in our science education journals?

Can We Learn from Heresy?

I have suggested that the education of science majors will show more variety in the 1970s and that the nonscience student will form a larger part of our audience. In carrying through these changes, can we learn from our critics? Few of us are willing to grant the extreme viewpoints contained in either the heresy from the left or the right. The establishment which can swallow and digest heresy, however, remains strong. I think there are adaptable ideas presented to us from both sides.

The students who come to us from outside of science are likely to be indoctrinated to some extent with the mystic, anti-objectivity heresy espoused by Roszak; we had better be prepared to meet it. We can not only meet it, but accomplish our goals with honesty and success by incorporating much of the heresy into our teaching. We must emphasize the humaneness of science, admit that we never really manage to separate, in Roszak's terms, our "in-here" from the "out-there." We must place the objectivity of science in context, show that its range is limited, example the important human areas in which it doesn't work. We must show it as a strategy which can answer a certain kind of question and help the students recognize the kinds of questions it can answer and to recognize that in other areas it not only cannot answer questions, but that the idea of questioning itself may be inappropriate.

If I were teaching a biology course, I would use (to make some of these points) the experiment reported on by the popular media in which a polygraph elicits all sorts of interesting responses from plants. Most of you have probably heard results reported: the plant shows excitement when threatened, can even pick a "plant murderer" from a lineup. Students ask me if I believe in this, and I tell them of my skepticism and of the basis for it. I don't discount it out of hand—that smacks of dogma, of persecution of someone outside of the establishment. This "experiment" or the earlier Velikovsky *Worlds in Collision* controversy, provide opportunities for discussion of the meaning of and the strengths and weaknesses of the objective approach.

But we must even go beyond the admission that the objective approach is limited in its application. We should admit that subjectivity and sometimes even mysticism play a role in science. Copernicus put forward as a major advantage of his new system that it was "pleasing to the mind," not that it worked better than the Ptolemaic system (it didn't). It was accepted for this reason long before the careful collection of data substantiated it.

Kepler, of course, was a mystic driven by a belief in number magic. There is evidence that Bruno was burned at the stake as much for his mysticism as for anything else. Even Oersted was a mystic and credits his discovery of electromagnetism to the influence of the romantic nature philosophers of the early nineteenth century who emphasized the basic unity of the universe— "of course, electricity and magnetism were related."

We would do well to give our students such examples; even to admit the subjectivity in our own experimentation. When I was doing coincidence counting experiments in nuclear physics, of course I cheered for the "spin-up" coincidences over the "spin-downs"; I just tried to arrange the experiment so that my cheers didn't affect it. Science is a human activity; it must be presented as one and not as an austere set of abstract rules which we find cleverly engraved in the secret places of nature.

We can learn from the practical men also. For too long we have sent our students down a narrowing tunnel and called it education. We have concentrated our energy on producing Ph.D.'s who would only consider a career which allowed them to continue the specialization of their theses. Industry no longer wants these narrow specialists and the politicians won't pay us to produce them.

Programs at the graduate level to change the pattern are being talked about here and there. The problems of change there are great and I will not attempt to discuss them. I would like to suggest some changes at the undergraduate level which might come from swallowing and digesting the heresy of practical men. We must now prepare a greater number of our students for employment at the bachelor degree level. We should target these students on careers; industry does not want aborted graduate students.

I have been engaged the past few weeks as a consultant to the Physics Department at the new Baltimore Campus of the University of Maryland and have been asked to help them formulate a plan for their future development. In that capacity I have been talking to many of the people who hire technical personnel. I find that they want students who have made application of their knowledge. They want graduates who have solved real problems on the computer, who have carried out project-like experiments with the kind of instruments they may use in industry. Most of all they want students who have had to set up their own problems, obtain their own data, make their own decisions about necessary precision, and so forth.

We can design undergraduate programs to produce such students without sacrificing the solid core of science education. We can introduce courses on the applications of our discipline—in physics, a course which shows how physics is applied in medicine, geophysics, astronomy, and environmental monitoring. We can also replace one or more of the elegant pregraduate senior-level courses with course seminars organized around the theme of modeling and computer solution. We can develop project laboratories. We should strongly consider the advantages of student internship in real jobs and begin to work out some of the troublesome details of such arrangements. In spite of present and projected curves of falling employment opportunities for Ph.D. degree graduates, we must believe that the increasing technological sophistication of industry will need more and more people with science-based training to make it run, and we should seek the advice of the practical men in preparing our students for such careers.

In slanting some undergraduate programs in a practical direction we must not, however, forget that science is a "liberal arts" discipline. We must not substitute a narrowing tunnel leading to a specific industrial job for the tunnel which has led to graduate school. What I see in my imperfect vision is still a basic education in science and the other liberal arts. I would replace a few of the specialized pregraduate courses by courses, broad in their approach, but matched to an environment in which discovered knowledge is fitted to new problems rather than matched exclusively to an environment in which the generation of new knowledge is the goal. I may be wrong here; by any name it may still be engineering.

Conclusion

You must have sensed by now that I am more concerned with the criticism from the Roszaks, Goodmans, and Elluls than with the criticism from the Charlie Wilsons, even though the latter is backed up by budget control. I think, therefore, that the major task before us is to humanize our teaching of science. We must remain part of The Establishment, but not blind to its faults. Where dogma and cant are necessary, we can use them and explain why we do; where they can be discarded, they should be. We should not be so much "defenders of the faith" as missionaries. As Paul Goodman says in our defense after criticizing much of science in his essay, "Can Technology Be Humane?": "Like Christianity or Communism, the scientific way of life has never been tried."

If science education can illuminate the meaning of the "scientific way of life" and convince the next generations of students that men can live this life and still remain human, the voice of heresy will lose some of its power, and science will continue its necessary and proper role in human development.

NOTES

1. *Science* 179 (30 March 1973): 1311.
2. *USOE Digest of Educational Statistics*, 1970, p. 49.

A COMPARATIVE STUDY OF TWO NINETEENTH-CENTURY EDUCATIONAL SYSTEMS

L. PEARCE WILLIAMS

AS A HISTORIAN I am a bit uneasy discussing the present state of science and society. I am more accustomed to dealing with science and society in the past. I shall describe science in the past, but with the clear objective of using that past to point some morals for the present and perhaps for the future.

In my research I am basically interested in why people do science and what they expect to find in their study of nature. Part of that research is biographical, but part of it also is institutional. I am interested in trying to trace out the ideas that have occurred to men such as Faraday and Ampère (my current subject), but I am also interested in the institutions that men such as Faraday, and Ampere, and hosts of others like them have tried to create in the hopes of passing on the scientific torch to future generations. And, therefore, I am interested in scientific education.

May I begin by pointing out a rather astonishing fact; that is, the history of scientific education is a *terra incognita*. It has not been touched upon or researched at all. There are no works at all, sophisticated or unsophisticated, dealing with it in any kind of depth whatsoever. If anybody feels desperate in casting about for a thesis topic or a life work, may I point out that this topic will keep one busy for at least fifty or sixty years if done properly. I intend here not to investigate this subject in any depth, but simply to skim the surface of what seems to me to be some of the important aspects of the history of scientific education in terms of our assessment of modern science.

Let me begin with what may or may not be a true prediction. Certainly we are today in trouble in terms of scientific education. You cannot pick up a journal, a newspaper, anything that deals with science in modern society without constantly today running across the theme that science is declining; engineering is declining; the combination of science and engineering is in a precipitate decline. Science and engineering are today not popular subjects. Perhaps one should say that science is through, that the decline will continue, that science will continue to remain unpopular, and that we are entering a postscientific stage in society where science, if not totally irrelevant, will have a less important role than it has hitherto. That I cannot predict. What I can ask, however, is Why in the past have people studied science and how in the past were institutions of education erected that created a stimulus for the study of science and brought science to the peak that it reached sometime in the mid-twentieth century?

Again, as a historian I would like to have at least one foot firmly planted in some kind of documentary evidence; therefore, take the examples of France and Germany in the nineteenth century. When the century opened, science was almost a French monopoly. If one looks at the indicators of scientific progress and scientific productivity, they all point, with a few hesitant exceptions, to France. The most important scientific work, by the general consensus of practicing scientists, was done in France. The growth of scientific literature was a French phenomenon. Specialized journals tended to be French. When they were not French, they existed only because of the existence of French models. Nicholson's *Journal of Philosophy,* for example, in England could not have existed had Nicholson not appropriated French articles on science and translated them and published them in his own journal. And all you have to do is open a copy of the early editions of Nicholson's *Journal* to see the enormous difference in the scale of science being done. Those articles that are indigenous to England usually are articles describing miraculous births of rabbits to country women or some meteorological display in the wilds of Scotland as opposed to the good sound physics and chemistry that was translated from the *Annales de chimie* or from a *Memoir* of the Academy of Sciences. The same is true in Germany. *Gilbert's Annalen* and *Crell's Annalen,* two of the earliest German scientific journals, were parasitic upon French science and could not have existed had they not again preyed upon the material provided by the French. And quite obviously, in terms of scientific education, the French were leagues ahead of anyone else in the world. The French were the first to introduce science into the secondary school curriculum, an experiment which was brief but which left some lasting marks on the French scene. The École polytechnique was certainly the greatest scientific school of its time and remained so for at least one generation. At the end of the century, French science is nowhere to be found, at least in the physical sciences. French physics is almost a contradiction in terms by 1880. There is not an important theoretical French scientist from the death of Laplace in 1827 until the writings of Henri Poincaré in the late 1880s.

What happened? How did French science decline? Why did French science decline? May I answer indirectly by comparing the decline of French science with the rise of German science. Because, once again, if one compares the beginning of the century with the end of the century, it is a startling comparison. In 1800 there is no important German scientist with the exception of Gauss, who was then considered to be a mathematician. By the end of the nineteenth century all physical science is clearly dominated by the Germans. This means not only mathematics, but mathematical physics, physical chemistry, all those studies which we would lump today under the rubric of the exact sciences.

How can one explain these two striking events? Briefly I can only isolate one aspect even within the educational system itself. And I would like to suggest that if you look at the educational systems you see, in fact, two broadly diverging philosophies of education and two evolutions of educational systems that widely diverge by the end of the century.

The French were the first, and Napoleon probably the first ruler, to recognize the practical effects of science. Napoleon clearly saw that the wedding of science and industry was imminent and that wedding would be fertile in terms of the power of the state. That was rather prescient of him, because in 1800 or 1810 when Napoleon was at the peak of his power that was not at all clear. It had been, of course, a dream since the time of Francis Bacon, but as late as 1834 in the debates of the Chamber of Deputies under the July monarchy, when the government was trying to cut Gay-Lussac's governmental sinecure, François Arago could rise in that chamber and point only to the lightning rod as an example of applied science that he felt would impress his listeners. To realize in 1810 that science would feed industry and would make industries powerful was at least partially Napoleon's foresight. But because he saw science in almost an entirely practical way, the institutions that Napoleon created tended to focus upon that practical effect. The Université de France, which was the institutional embodiment of Napoleon's ideas on science, was a university intended to create young men who would have practical knowledge literally at their fingertips. They were to be filled with practical scientific knowledge, even though this involved the study of higher mathematics. People who came out of the École Polytechnique were mathematically far more sophisticated than any other graduates of any other university system in the world. But the intent of the education was practical. And to ensure this practicality, the state set up a confusing barrage of tests—tests for the teachers, tests for students to get into the scientific schools, tests for students to get out of scientific schools. This system of examinations guaranteed that the graduate of a French scientific faculty would indeed have at his fingertips the very latest scientific knowledge.

Practical scientific knowledge failed, however, to keep science alive. And practice, practicality, practicality without a dimension that includes either the metaphysical or the theological, such a stimulus is insufficient to feed the life of science. I would argue that one of the reasons, and I am not prepared, unfortunately, to assess all of them in any convincing way, but one of the reasons for the decline, if not the death, of French science in the nineteenth century is that the appeal to practicality simply is not enough to inspire youth to pursue a scientific career. It was much more romantic and it was obviously much more fulfilling to leave science and to take advantage of one's intellectual talents in other fields.

If French scientific education neglected the metaphysical

and the theological, German scientific education did not. And one of the striking differences between the systems of education lies precisely in that point.

German science, first of all, was taught within the Philo-sophische Facultät. Science was not, as was the case in the École polytechnique, separated from other studies but it was plunged into a philosophical atmosphere. One cannot pick up a single treatise in German science in the nineteenth century without running headlong into philosophical problems. Scientists in Germany in the nineteenth century, willy-nilly, whether they wanted to or not, whether they later scorned philosophy as a waste of time, whether they saw in Hegelian metaphysics the vaporings of romantic fools or not, nevertheless, were deeply immersed in metaphysics and fought their way out of metaphysics only by coming to grips with it. And in the process of fighting their way clear of metaphysics, in the process of fighting their way clear of the theological dimension that accompanied metaphysics in the nineteenth century, they were forced to construct for themselves far more than the narrow world view that passed for scientific culture in France. They not only had to come to grips with philosophical problems such as the nature of space and time, force and mass, matter itself, but they had to come to grips with the question of the existence of a *physis,* of a world out there whose laws were imminent and the discovery of whose existence gave order and meaning to the world.

Thus, the spectacular rise of German science in the nine-teenth century was at least in part the result of the fact that science provided for the German scholar far more than an instru-ment for material well-being or for practical application. Science for the German scholar in the nineteenth century was, indeed, a way of life. It provided a *Weltanshauung* into which one could also inject, I suppose, a little *Zeitgeist.* But it provided a whole way of looking not just at science, but at the world, at man, and at man's place in it. It was, if you will, a religion, and it was a religion whose successes were to be found in the prodigious scientific achievements of the time.

We are in a position today where religion is a distasteful word. Theology is raised somewhat sheepishly as a possible subject of study. Despair, really, is the dominant theme. In scientific education today we have clearly followed the French. Those of you who have been through a scientific education recognize that the French mode is precisely what we follow here. You start with simple problem sets. You go on to more complicated problem sets. And after nine years of problem sets you ultimately face nature itself and that is the biggest problem set of all. We are, in T. S. Kuhn's words, all problem-solvers today. And increasingly, my reading of the modern mood is that the problems are no longer worth solving. Or to put it another way, the solutions are worse than the problems or the cure is worse than the disease. We have no ideology of science because science, we are told, is value-free, and to have an ideology of science is to imply that science is more than the objective discovery of nature. In short, the dimension that we have carefully bred out of science through our educational system is the dimension of metaphysics, of theology, and of value commitment.

It is at that point that science today faces probably the greatest crisis in the history of science, namely, its ultimate demise because it can no longer win converts. I do not know what the answer is. I have clearly implied what the answer ought to be. I am not sure it is possible. The answer ought to be, it seems to me, somehow, the creation, or the recreation, or the revival, or the rediscovery, of that religion of science, of that sense that science was, if you will, a means of approaching the whole; a mystical contact by which more than the part could be seen, but the relation of the part of the whole could be made manifest. I do not know if we can do that. I will only suggest that if we do not recover that excitement and that viewpoint, then I suspect science will not recover and that the postscientific world has therefore already been ushered in.

"SHOW ME A SCIENTIST WHO'S HELPED POOR FOLKS AND I'LL KISS HER HAND"

THOMAS J. COTTLE

MOST of the apartments on Taylor Street, behind the South End of Boston, were once fashionable houses owned by wealthy families. Years ago when these well-to-do white families moved out and poor black families, many of whom had migrated from the South, moved in, the houses were converted into apartments, sometimes with as many as four dwelling units on each floor. Turrets and towers that remain on the corners of the roofs of these apartments and the unusual designs that were cut into the stone facades of the buildings remind the present residents that once the wealthy lived here, once artisans built buildings here, once life was rich here. Today, Taylor Street is one of many streets in a poor neighborhood, a neighborhood hidden from the main currents of Boston by a developing South End area. Still, from some of the rooftop towers on Taylor Street, the children who play up there against their parents' wishes claim they can see all of Boston, particularly the taller buildings, and even the buildings across the Charles River belonging to the Massachusetts Institute of Technology.

They report, for example, sighting what looks to them to be weather predicting equipment atop one of the MIT buildings. Several of these children have decided, moreover, to explore the Institute, see it up close, and prowl about the buildings. The idea of scientists working so close to where they live intrigues them, and because I come from MIT to visit with them, it is only natural that they would tell me of their plans and turn our conversations to the work of scientists.

One of the boys who climb to the top of an apartment house on Taylor Street to spy on MIT is ten-year-old Keith Downey. He is a thin young man with large round eyes and a high forehead. He never goes anywhere without one of his hats, a fact which disturbs his parents who practically have to pull them off his head when he enters his home.

"I've got my thinking hats," he told me one afternoon. "I think science with them on, particularly my gray one. I can go up there to the tower and look at where those scientists work and just about get myself to imagine what they're doing."

"Which is what?" I asked him.

"Seems to me, they spend a lot of time working on rockets. They want to explore the planets, maybe even have people stay up there in space for a long, long, time so they don't ever have to come back down for fuel and all the other stuff they need, their supplies, like. They could work up there, I'll bet, for whole centuries, have their families live up there, have their babies born too, and when people die dump the bodies into space. Up there you don't need to bury people you know. You just throw them out of one of your spaceships. It's easier than the way we have it here. Like, when my grandfather died, my father said he didn't know whether it was cheaper to have him alive or bury him. People who buried him and did all they had to do at the chapel on Caldwell Avenue charged my father so much money, he said he couldn't pay them. He told my mother one night they might have to throw my grandfather into the ocean. She got real angry 'cause it was *her* father that died. They didn't know me and one of my sisters were listening to them fighting. They got the money though, from some loan company, which my father says is no better than stealing since he'll be paying them back for the rest of his life. Not really maybe, but for a long time. But I figured that if scientists really wanted to do a good thing for my family and folks like us, they should figure out a way to get rid of people when they die so that it doesn't cost us so much money to take care of all the dead corpses."

When Keith Downey walks with me on the streets of his neighborhood, he gives the appearance of being oblivious to almost everything. He jumps curbstones, peers in windows of grocery stores and restaurants hunting for his friends, but he does not seem especially sensitive to the environment. When I go with him into the downtown areas, however, and especially to college campuses, he begins to look at and touch everything that he can. We will be walking together, and Keith will rub his hand along a building or fence feeling the metal or stone or brick. He may pick away at chips of mortar and rub them between his fingers, or hit a pipe to listen for its special sound. When we visit MIT, his look is one of awe. In the Nutrition Laboratories, for example, where one of our recent visits took us, a graduate student explained to him the nature of some research on artificial food substances. Keith was deeply interested in this work.

Later, on the trolley going home he said to me, "You see how much scientists do for people. That food laboratory we saw there has to be an important place. When they get done with their work, there won't be a single person in this country going to starve any more. Now the President of the United States, he has all the power and all the money, but he doesn't have the brains like those folks at MIT. They're the ones who'll do the work so that pretty soon, like that one man said, a person can swallow a couple of pills and have all the food he needs that day. Or maybe that week too. That's the day, man, I want to see. Come into the kitchen and tell my mom, 'give me the breakfast pill, mom.' She'll hold it out for me and I won't have to come home again 'til supper, especially if I can stick my lunch pill in my pocket too. That man at MIT, he's got the right idea. Never go hungry, and never have to waste all that time sitting around at the table, listening to all your baby brothers and sisters screaming in your ear while you're trying to get something to eat. Scientists, man. There can be *nobody* on the earth doing better things than they are. What else did we see there, Tom?"

I reminded Keith of the Medical Laboratory we had visited where serological experiments were just getting underway along with some examinations of heart muscle. This latter work required the use of dogs who later on would be sacrificed. Keith had wanted to see the insertion of a canula into an artery. We watched together, robed in white laboratory gowns, and only once did he wince. The rest of the time, almost an hour, he stood transfixed, fascinated. Here again a graduate student explained the research to him.

But now, alighting from the trolley and walking the nine blocks to his home, his fantasies began to play on his recently acquired knowledge. "Here's what scientists are going to be able to do, maybe even have it done before I die. They'll have like a brain, or a heart, that's dying. The person isn't dying, see; just his brain or his heart is. Or maybe his stomach will be rotting away. I heard of a lady whose stomach did that," he interrupted himself. "The doctors said it was because she never ate the right foods. So they cut her stomach out and then she didn't have one. But see, when all those scientists we've been visiting get done, they'll just look you over, you know, x-ray you and whatever they do, take your heart out, and put a new one in."

"They sort of do that now, don't they, Keith?"

"Yeah, but those folks are dying. Scientists and doctors have to work a little more, just a little bit more; then they'll really be able to keep people from dying. If one part wears out, they'll put another part in. It's just like fixing a car. You got a bad nose, they'll take it off and put somebody else's nose on you." This idea struck him as particularly amusing and for almost half a block he was convulsed with laughter. "You know what I wish scientists would do," he said when at last he composed himself. "I wish they'd build cars that could go as fast as rockets only not go up into the air. I get kind of scared thinking about going into those rocket ships, although maybe I'll be an astronaut. But I wish they'd make cars go faster and fix it so no one would ever crack up. Then I wish they'd make pills to fix you up when you're sick." He was letting his mind wander over ideas he had played with during our visit to the Institute, almost hoping he would hit on some scientific invention that would wholly please him.

"I got it," he shouted out finally as we crossed over to St. Marks Street. "Dentists! Scientists are going to have to fix your teeth with pills and drugs that don't hurt. Maybe they could put you to sleep before you even leave your house going to see them. Get rid of that pain, man, and you've done something for everyone. Everyone that goes to the dentist, that is."

"The lives of scientists really intrigue you, don't they, Keith?"

"Better believe it, man." He was aglow with excitement reflecting on our excursion, as well as on the view he gets from his special tower on Taylor Street. "I sure would like to be one someday. Working in those laboratories like we saw, inventing things that no one has ever heard of, maybe get famous, discover something that would help people. Get rid of dentists," he smiled, "help babies get born right, make people strong.

"We didn't even talk yet about energy," he went on. "My science teacher tells us we got this, what they call an energy crisis. If we run out of oil and gas, he says, we aren't going to have any cars, or heat in the houses, you know. The way I see it, I mean, not me, exactly, but we talk about it at school, scientists, maybe me and my friends someday, are just going to have to find the answers, otherwise a whole lot of people are going to die. If you don't have the right food you die. If you don't get warm enough you die too. And if you can't drink the water 'cause it's all polluted with stuff, you're going to die too. So scientists will have to send more astronauts into space just to see what they can find up there, or send those other folks under the ocean in submarines to dig for oil. Mr. Cleary, he's my science teacher, he says the secret of the earth lies in the oceans and the seas. That sounds right to me. I'll bet scientists could find a way to get some of the heat from the sun and get everyone warm everyday of their life. I'll bet you could take all the heat of the sun away from it everyday and still have enough sun the next day. It would just heat itself up all over again during the night. Maybe they'll find something in those rocks they brought back from the moon too. Maybe they can get something out of them which will cure us of something or other." He was scratching his head.

"Scientists, you know, have to keep hunting. It's slow work, just like we saw over there today. You have to be patient, which is why I'm going to be a scientist. I've got the patience for it. You take a problem and you stay with it. Day after day, year after year if you have to. But you don't quit until you find the answer you're looking for; until you can find a way to cure somebody of something, even one person. Everybody's different, you know, so you need different scientists working for different folks. You understand?"

I nodded my head yes.

"We got to go back there someday real soon, Tom."

"We will."

"Can we?"

"Of course."

" 'Cause there's something else we have to know about."

"Which is?"

"I don't know what you call it, but they do these experiments on babies and children, real little kids, you know, just born, to see, like, how they're going to be when they're adults. What they got in them from their parents, I mean."

"Genetics," I said.

"Yeah. Genetics." He repeated the word with reverence in his voice. " 'Cause you know some people got things wrong with them when they're born, and only scientists can figure why they got them, and what we got to do to get rid of those wrong things."

"And lots of people have lots of not wrong things with them," I said, clumsily, "which geneticists also study."

"Geneticists. I think I'm going to be a geneticist," Keith

announced as we reached the corner of Taylor Street. He was look-ing up at the rooftops where he and his friends play. Children his age were running about on the street taking advantage of an unusually warm March day, but Keith took little notice of them. "I can see your college from right up there," he said, pointing to his favorite tower. Suddenly I saw the image of Galileo working alone in his private laboratory.

"That's the tower, eh, Keith?"

"Right up there, man. That's where my career in science is going to start, with your help."

"With *my* help?"

"You got to keep taking me to these places. I've got to get prepared. You got to help me get books and pass all my courses which are really going to get hard in a few years, like in high school. I'm counting on you to stay there at MIT and get me in there with you so I can become a scientist. Not a pretend one like I am now, but a real one. And if I get sick, you got to get me a doctor and get me the money, and get me all I need. . . ."

His friends calling to him to join them finally made it impossible for us to finish our talk. Within a few seconds he was running off, earnestly making me promise to take him to another college in a week or two, and reminding me of all the things I was going to have to do for him in the next ten or fifteen years. "I'll take it from there," he shouted as he disappeared into the burger shop on the corner of Taylor and South Plaine. "You got to take care of me, man, keep my head in one piece," was the last thing he said to me that afternoon.

Several minutes later I entered the house of Arthur and Estelle Downey, Keith's parents. As their apartment is on the top floor of one of the row houses, the stairs leading to the roof and towers are located in the hall just outside their front door. The apart-ment is unbearably hot in the summer and cold in the winter since the insulation in the walls and ceiling is inadequate and the basement furnace five floors below lacks the power to keep the few radiators warm. And since the heat is turned off at ten at night and not turned on again until seven the next morning, the hours of minimal comfort are precisely those hours when no one is home.

"So you took my boy to see all the famous scientists this

afternoon," Estelle Downey said, greeting me at the front door. As usual I had asked her permission to take Keith on our excursion, and she made certain to be home on our return. "Seems like that's all he's interested in these days is science and scientists. Books and television, and now he's got you, and MIT. Sure was a perfect day for it, wasn't it?"

"It was that," I answered my friend of five years, a mother of six children. I sat down in the living room. Estelle tidied up as she spoke to me.

"That boy and his love for science. Do you know how hard it is for me to keep my hatred of those folks away from his hearing?" She shook her head as she beat the dust out of the couch pillows. "Scientists," she muttered. "Rich folks is what they are, no different than all the rest. Sitting over there where Keith spies on them, playing with this and playing with that. Making up problems where problems don't really exist. Making things complicated when really what we need done is so simple a child could understand." Her tone was bitter. "Everyday I read in the papers about the money they get to do all those experiments with and whatever you call what they do over there. And not only over there at your place, but all across this land. In all the colleges! What I want to know is what good are they doing for this country? What good are they doing for black folks, and poor folks? What are they doing for folks who haven't learned to speak English yet? They just go on playing with this and that. What do they care that we have children dying over here in this part of town, and that boy of mine is sitting up there in his tower, ready to fall onto that street down there on his head, looking over across the river as if it were the promised land."

"Believe me, Tom, Arthur and I are very careful not to say a word in front of him, but don't you honestly think those folks over there know as much as anyone needs to know about things now? I mean, they got hospitals and doctors in this city, and they got colleges and all kinds of training schools. They even got a museum for science. Isn't that enough? What are they doing with all that money?" She was standing opposite me clutching a pillow to her breast. "What do they find to do with all that money? Making their experiments and all, and just who do they experiment on? Is that a question those scientist folks would like to answer?" She walked past me to the kitchen where she found a copy of the moring *Globe.* "On the second page, right here. You see that?" She folded the paper over and thrust it out to me. "Texas scientists this time, experimenting on babies. Newborn babies. Three hundred black babies, and making it out they were like little animals." She smacked the newspaper as she walked away from my chair.

"I read it this morning, Estelle."

"Well I hope it made you sick. They can't seem to find black children for the schools, they can't find them when they want to do something nice for them, but when they do their experiments, they find them all right. Three hundred of them! Now, you going to tell me that's fair?"

"Of course I can't," I said.

"You're damn right you can't. Not a scientist in this country right now seems interested in doing anything for black folks. You show me a scientist who's helped poor folks and I'll kiss her hand." She turned quickly to look at me. "You get what I mean?"

I nodded. "You want me to stop taking him to MIT?"

"Of course I don't. What would that solve? We're happy with any little break we can get. But I *would* like you to tell me why it is we can't get medicine and doctors and dentists, and the right food? I *would* like you to tell me how come they got buildings over there, Keith says, for their experimenting and we don't have the money to keep the heat on in this building we live in. That's what I'd like someone like you to tell me. I'd like someone to tell me where they get the money to train all these scientists and have to cut out the funds on every single project that just might help poor folks in this community." She paced back and forth in the small room, peering out the window down onto the street each time she passed it.

"My God, if what Keith says is even halfway the truth, and that boy doesn't lie to me or to his father, then they got better conditions for the dogs they do their experiments on than we've got for our *children!* You know, they actually feed them better, clean them better, take care of them better, and you want to know something, if those dogs and mice get sick, they cure them. It's not like here. What have we got here, scientists to help us? The hell we do. Child gets sick here, he's worse off than one of those dogs. Now, you're supposed to be some kind of a doctor or philosopher and I'm just an uneducated woman coming up here from a tiny farm in Georgia when I was eight years old. So we ought to be able to agree that I don't know and you do. So suppose you tell me how this society can keep going when your dogs on that side of the river are in safer hands than our children are on this side of the river? You go ahead and answer that question for me. That's the way you got to look at science, you know," she pointed at me, still pacing across the far end of the small living room. "It fits into the society somehow. Science can't be something so special that it doesn't affect the rest of us. Dogs and children are part of the society, but the way I look at it, the children right here on Taylor Street aren't getting as good a treatment as the animals." At last she turned to me and stood still.

"I blame the scientists," she said, seeing I wasn't about to speak. "I blame them all. They're specially educated, every one of them. Every single one of them has degrees, like a doctor or whatever. Just like you have." I said nothing. "They know what's happening in America. They know the children here are dying from the lead they eat in that paint. Not just black children either. And they know children are dying because no one takes their tonsils out in time or their appendix out. Scientists, doctors let that happen. Scientists don't make cheap medicine for us. They don't help us get the vitamins we need, or the medicine I need to give that boy of mine you like so much that he won't have one of his epileptic seizures at any minute. How many times have we gone without it? How many times have *you* carried him up those stairs with his head wagging back and forth and his eyeballs rolled back and his tongue half way down his throat? You tell me how many rich children have to go through that? You tell me how many children of scientists go hungry every night, or stay in this hot city every month in the summer. How many rich folks can't get the operations we need all the time and can't get? Scientists get the money to make their experiments and design all that research of theirs, but we don't get to see a dime of it. We just wait in those emergency rooms begging some twenty-one-year-old policeman to cure our children. Maybe that's what it's all about. You all got your wise old scientists, and we all got our policemen, most of them not that much older than my oldest boy. Policemen and doctors, 'cause we don't have anybody else. While your dogs do better. I sure as hell can tell without seeing them that they eat better." She was pacing back and forth again, stopping momentarily to glance out the window at the children and the traffic five stories below, then looking about the room trying to decide if she wanted to continue her cleaning.

"Where's my Keith now? Eating something? Eating grease and junk like that? Are they going to find some illness in him tomorrow, five years from now? They going to guarantee me that scientists are going to help him? They going to guarantee me that even if they know *how* to help him they can get their research over to this part of the city? Maybe he'll cut himself sometime and bleed, out there somewhere where no one can find him? Scientists! How they going to help that? With bombs? With rockets to the moon?" She spun around and looked straight at me. "I want them people down here, not walking around on the moon. I want them in their rich folks laboratories working on sickle cells. That's right," she shook her head. "Sickle cells. Let them work on *our* problems for the next ten years. Where's the money for that? Where's the money for curing *our* illnesses? Where's the research for black folks, instead of for white folks' dogs and white folks' mice, and all those little pigs they got too?

"And something else, which your face reminds me of too."

I sat up straight in the chair.

"What about *your* kind of scientists, Tom? Don't you belong to that certain kind of scientist too? Telling folks in the newspapers how we live and how our children are doing in school. Studying all these busing programs and deciding how well they're working or not working. Aren't you part of that?" Her voice had grown quiet. "Something tells me you are. Something tells me that *we* may get the sickle cells in here," she poked at the veins in her forearm, "but *you* folks are responsible for making the whole community filled with bad cells, evil cells. You scientist folks, it's up to you to cure the society. You're the ones who made it sick, and so far you all are just looking at it, doing your research, and going on about your business, not caring a damn about whether my child lives or any other child lives if you can use him in some experiment.

"Let me tell you, there's all kinds of experiments with infants, with little children, with children as old as Keith, that are going on in all these colleges. You can read about them in the paper there. And in everyone we're always being experimented on. We're always the dogs. You mess around on us, and then you leave us to die. But not a word comes from you, not a 'I'm sorry.' " She walked toward me and pushed her finger at the newspaper I still held in my lap. "Three hundred black folks gave their babies to those scientists. That means all over the country they're experimenting. We never get to say a word about it. Or if we ask, we can't be sure they're telling us everything that's going on. We're just your dogs waiting for you to play your rich games. But you never show your white faces around here. You never even say 'I'm sorry. I'm sorry for what's happening. I'm sorry that we got our white folks walking on the moon while you black folks are falling on your beds sick with hunger, and your stomachs rotting. I'm sorry that your boy is an *epileptic!*' " We were staring at one another, but with all the anger and feelings of betrayal, she still could sense what it was I naïvely hunted for.

"You don't have to worry, Tom. I'm not crying about this anymore. We've been your slaves so long, it's like nothing's changed. Only now," she said glancing at the paper, "they call it science. Everything in the name of science. But anyway you cut it, you're the masters, we're the dogs, and I just got to wait and see whether a seizure someday will take my boy away from me. And I suppose," she finished quietly, "away from you too."

SCIENCE AND SOCIETY: FUTURE

IX   The Unbalanced Revolution

DENNIS C. PIRAGES

ONE of the most unfortunate developments in human history took place as the Copernican Revolution began gathering momentum. This developing scientific revolution was nutured in a setting in which a rigid social structure, combined with religious constraints, directed development of inquiry outward into the physical world much to the neglect of analysis of social problems and human behavior. The Copernican Revolution marked the beginning of an intellectual assault on the mysteries of the physical and biological world, but taboos against the study of society were strong enough to foreclose the development of the social sciences for centuries. It was difficult enough for early scientists to debunk the myths that existed concerning the "natural order of things"; it was impossible for them to consider a science of human behavior.[1] These phenomena were in the province of philosophers and theologians and the emerging scientific method was not considered appropriate for the study of society.

The direction of the scientific revolution was molded by these early constraints and contemporary inquiry still suffers from its effects. Advances in the physical sciences have shaped the social world and diverted attention from the study of human needs, values, and institutions. Science and technology have leaped forward, creating social problems for which there are few apparent technological solutions. Too often science has acted as society's master rather than as a humane servant. Indeed, it is now questionable whether recent undirected technological advances might not have created less rather than more satisfactory lives for human beings.

Inquiry directed toward the physical world and physical technology has paid huge dividends and as a result there has been little incentive for science to focus on social areas in which theologians and philosophers originally set up camp. Industrial society has been able to avoid facing serious social questions by expanding production of material artifacts. The "problems in need of solution" in industrial society have largely been problems of production. Little attention has been paid to the overall quality of life or crises of human spirit.[2] For example, science has conquered many types of diseases and considerably lengthened the life-span. Agriculture now supports many times the number of people that could have been supported by the agriculture of the Middle Ages. New methods of communication have remade intellectual horizons for many around the globe. The potential for large masses of people living a quality life has never been so great as it is today. In short, technology has proved the forecasts of both Marx and Malthus to be in error, at least in the short run.

While physical science has been leaping forward and creating material abundance for the industrial segment of humanity, knowledge of the inner human, the roots of social behavior, and the generation of social values has not progressed much beyond the point it had reached when human beings still believed that the sun revolved around the earth. On much of the planet, people still breed as if they are in imminent danger of being extinguished, even though infant mortality has plummeted. Systems of stratification within nations and among nations, not dissimilar to those of feudal times, have remained in an era of supposed plenty. Industrial nations and supposedly intelligent people still act out their aggressions through organized warfare. Only the weapons have changed since human beings stopped living in caves. Physical science has created material prosperity which has permitted dangerous competitive principles to endure into an era when the complete destruction of civilization is a possibility and when cooperation is important to the continued survival of the species.

None of this has been the fault of any malevolent group of self-interested scientists. If anything, distinguished scientists have been at the forefront of the effort to open up new types of inquiry. But society has successfully frustrated those who would assault human "freedom and dignity" by studying human beings, their needs, and their motives. A psychology dedicated to understanding human behavior could hardly be nourished in a social milieu in which behavior supposedly resulted from a struggle between the dark forces of evil and the forces of good. It is a tribute to scientific persistence that psychological and social sciences finally did emerge in the early part of the twentieth century as legitimate forms of inquiry.

Physical science and the march of technology that has accompanied the scientific revolution have shaped a world view for industrial society that could be considered analogous to Thomas Kuhn's concept of dominant paradigms in the social system of science. Kuhn uses the word paradigm as a shorthand term for the model through which social groups, in his case scientists, interpret the world around them. Kuhn points out that scientists concentrate on "normal science," and he effectively raises the question of how new phenomena are ever discovered and investigated when they fall outside of the bounds of legitimate and "rational" inquiry as defined by the fairly rigid boundaries of a shared paradigm. Dominant paradigms or accepted views of research become entrenched in groups because members of such groups develop vested interests in established ways of thinking about or handling things. No scientist, for example, likes to face the prospect of sending a paper to a journal on a subject that isn't considered "appropriate" or containing revolutionary findings that go against many commonly accepted scientific wisdoms. The paper is not likely to get published quickly, and the scientist's reputation is as likely to be destroyed as enhanced by such an action.

Kuhn goes on to point out that dominant paradigms are occasionally shaken or even displaced, particularly when the number of unexplained findings and anomalies becomes large enough to disturb a significant portion of those adhering to the beliefs and methods of the old paradigm. A threshold is reached where contradictory evidence becomes so overwhelming that a "paradigm shift" is triggered, and a new view of the scientific world more parsimoniously explains scientific findings. Once a new paradigm emerges, however, pressures quickly develop to restrict free thinking, individuals develop new sets of allegiances, and a new paradigm dominates the group until anomalies accumulate once again.[3]

This approach to the nature of intellectual revolutions can be broadened to include social revolutions as well. A shared social paradigm is essential to any stable society. It consists of the values, norms, beliefs, and views of the social and physical environments held in common by its members. A social paradigm defines the bounds of appropriate social behavior, highlights social problems in need of solution, creates shared expectations that make social life possible, and makes some order out of an otherwise incomprehensible social universe. The shared paradigm is passed on from generation to generation through learning and socialization processes. A shared paradigm shapes social life and thought in much the way that shared scientific paradigms shape the scientific thought of those scientists working within it.

Social paradigms develop, shift, and are transformed in ways that are similar to the dynamics of change within shared scientific paradigms. When the information passed on by a social paradigm no longer gives valid guidance in dealing with the social world or the physical environment, pressures for paradigm shift are likely to follow. But vested interests in social life are every bit as important as vested interests in science, and therefore the social world can become severely disoriented before any major transformation takes place. Lewis Mumford claims that there has not been more than a handful of such drastic social paradigm shifts since primitive man. The scientific revolution represented one of those drastic transformations in social thought leading to a new picture of "the cosmos and the nature of man."[4] During the development of the industrial paradigm rational methods of scientific inquiry sent shock waves through the old belief system as one cherished myth after another fell before the sweep of scientific progress.

Science and technology have shaped the modern industrial view of the world. Willis Harmon has outlined several essential features of the shared industrial paradigm. They include:

1. Continued development and application of rational scientific method to all problems.

2. Science and technology wedded in rapidly accelerating material advances.

3. Industrialization through the division of labor and progress defined as economic growth.

4. Mankind seeking to control nature rather than living within nature's constraints.

5. Materialism and acceptance of a work ethic.

Harmon concludes that many critical problems faced by industrial society have their roots in the imbalances inherent in the scientific revolution and that they are "unsolvable in the present paradigm precisely because their origins lie in the success of that paradigm."[5]

But the orthodoxies of the industrial paradigm have become shopworn as many anomalies or imbalances have come to the attention of critical thinkers living within it. Particularly students and humanists, but also many scientists have challenged the unfettered advance of science within society.[6] This growing alienation is occurring at a juncture in history when science and technology will be most needed. Science, however, is now being asked to pay for society's past sins. The penance could be so severe as to threaten industrial civilization should an estrangement between science and society long persist. If an antiscientific ideology gains followers, society could well be forced back to the Dark Ages.

Science has created an incredibly complex society that is dependent upon more science and technology for its continued survival. It has led mankind away from the natural systems that can sustain human beings with current solar energy toward artificial systems that rely heavily upon science and technology to continue to develop substitutes for these natural systems and this naturally occurring energy. Little thought is given to the possibility that even the best that technology can develop cannot long be relied upon to ward off entropy in an overcrowded world.[7] The number of people on the planet is now many times that which can be supported by natural systems. Science has thus created a situation in which an economic or technological slowdown would occasion the sacrifice of hundreds of millions of lives. Science's staunch supporters argue that there is no alternative to greater doses of the same scientific medicine that has helped create present social problems. This is why the imbalances in the scientific revolution must be rectified. Science must be carefully managed to serve the future society or it will certainly be the first victim of citizen disenchantment during an impending epoch of social unrest and turmoil characteristic of periods of social paradigm shift.

It is an unfortunate human tendency to ignore possible long-term aversive consequences of presently pleasurable behavior. This accounts for part of the reason that mankind, within the present social paradigm, sees only material benefits from science and rarely looks to the long-term consequences of technology's gallop. The immediate results of science have been to prolong life, create undreamed of affluence for a good share of humanity, expand the power of human choice, and relegate many of the most odious production tasks to machines. But, materialism has been part of the spiritual price of this form of progress. Industrial humanity now expects material progress as measured by increasing production of material artifacts and increasing flows of resources through society. Little attention is paid to existing stocks or possibilities of sustaining such flows for long periods of time.[8] The negative side of science also lies in its failure to adequately anticipate the future. Population growth and higher levels of affluence are placing tremendous pressures on existing stocks of raw materials and are leading to significant environmental deterioration. Within the present social paradigm more technical fixes are suggested as a remedy for the problems science has already created.

Science has also been responsible for development of cherished institutions without which industrial man would feel most uncomfortable. Affluence has permitted the existence of a democratic political process. Democracy and political stability cannot easily exist with the wolf perpetually at the door. It is for this reason that political democracy has been rare in preindustrial societies.[9] Economic abundance has healed strife and made possible the peaceful resolution of disputes and development of democracy as part of the present dominant social paradigm in many countries.

But, on the negative side, science has created expectations of continued progress, growth, and affluence. If current predictions of troubled times ahead are correct, these expectations could well lead to the demise of democracy.[10] In a less affluent future demands may well outrun science's ability to produce, and present institutions will certainly come under considerable stress.

In retrospect it is easy to understand how society has been able to ignore the powerful hold that science has developed over it, as well as long-term aversive consequences of lack of social control of science. Scientific method has solved the most pressing problems of preindustrial society. For the first time in world history affluence has become the condition of the masses as well as the elite. Industrial mankind has unprecedented physical mobility, lives in comfort known only by the nobility in former times, and experiences a diet that is far superior to that which existed before industrialization began. The social and economic world has been remade by scientific discovery, and each success has meant greater social confidence in, and more freedom for, science.

In spite of all the obvious successes of science, most scientific advances have exacted a latent and hidden social price. For example, technology has prolonged the life span for many, but this has led to a rapid increase in the earth's population that threatens to destroy the present carrying capacity of the globe. Agricultural innovations have permitted the development of cities and nonagricultural occupations, but in the end this has resulted in crowded metropolitan areas and festering slums. Science has produced affluence in the industrial countries, but now these countries are dependent on the less-developed areas of the world for resources to support current levels of consumption. The United States, for example, has one-sixth of the world's population and uses more than one-third of the world's annual production of natural resources. The major industrial powers of the world, with the exception of the Soviet Union, are all net importers of materials essential for the continuation of affluence.

Thus, from a long-term perspective, unregulated science has offered a mixed performance. It has created a new social paradigm which it sustains with economic abundance. But this paradigm cannot long persist without tragic social consequences. It is a paradigm that encourages man to master nature and remake the planet in his own image. While this has worked during a pleasant industrial interlude, the present social paradigm cannot persist into a difficult future because of the imbalances inherent in it.

The Unbalanced Future

There are at least four types of imbalances that have accompanied the scientific revolution. Each of these represents a major problem on society's agenda for the future. Dealing with them adequately will require a revolutionary shift in present patterns of thinking.

The first and most important is an *imbalance between the power of science to shape social institutions and values and the powerlessness of society to regulate and plan the development of science.* Industrial society is technologically overdeveloped and socially and politically immature. In the words of Jerome Frank, galloping technology has become a new social disease.[11] Science is the leading institution of contemporary society, and its values and patterns of conduct have molded patterns of social behavior.[12]

One of the unfortunate by-products of galloping technology is that it legislates through uncritically accepted innovations. These innovations have changed the physical and social landscape. Each new development, usually accepted without open political decision or debate, blocks off sequences of alternatives that can never be reopened. These critical pieces of technological legislation set new rules for society, open up a different array of technical possibilities, but also close other sets of possible futures. The problem is that laissez-faire science and technology make these choices, not the people through their institutions that are designed for collective decision-making.

The automobile, for example, was an uncritically accepted technological innovation that did much more than offer a superior means of transportation. It occasioned the development of a huge complex of interstate highways, established a pattern of heavy fuel consumption and atmospheric pollution, was very much responsible for urban sprawl, and in general, reshaped the face of industrial society. Its development also effectively closed off possibilities for entirely different human transportation systems based on mass transit principles. In retrospect, it seems that comfortable mass transit would have been environmentally much sounder than private automobiles, but the decision was made, by social default, to foreclose this alternative. Science and technology will continue to legislate in this manner, acting as the master rather than the servant, unless society establishes methods of planning and regulation.

A related problem in the industrial paradigm lies in the *imbalance between available technologies in the physical and social sciences.* New technologies related to discoveries in the physical sciences have created an incredibly complex and interdependent society. There can be no rugged individualism in a crowded society where one person's actions have a very direct effect on the lives of others. Industrialization has meant greater population density, swelling of the size of urban areas, and development of more intricate and interdependent networks of production and supply. Adequate social insulating space is now at a premium, and social behavior and individual freedoms, appropriate in less crowded situations where an open frontier could reduce social pressures, cannot be tolerated in a closed system in which internal social and demographic pressures have no outlet.

While scientists have busied themselves with new physical technologies, old taboos combined with social realities have kept adequate technologies of behavior from being developed or employed. As B. F. Skinner has put it, "In trying to solve the terrifying problems that face us in the world today, we naturally turn to the things we do best. We play from strength and our strength is [physical] science and technology."[13] Skinner goes on to point out that to contain a population explosion, industrial mankind first solves the problem of producing new birth control devices and only later thinks about the norms, values, and patterns of behavior that often make birth control programs socially ineffective. In many similar situations contemporary science provides technological fixes while ignoring social and institutional aspects of problems. As a result contemporary society is held loosely together by a series of short-term technological fixes and promises of material abundance, but governs itself and its appetites no better than did the ancient Greeks. But abundance created by technological fixes will not last forever, and sooner or later social problems will have to be faced.

This imbalance stems partially from the problems of applying knowledge in the social sciences to social problems. Social technologies must be legislated, accepted by political bodies capable of implementing such decisions. Social technologies have no visible economic reward and by their nature are not attractive to private enterprise. In fact, they often involve denial of human desires and fly in the face of the predominant growth philosophies. Social science has rarely been given freedom to experiment. Social scientists run into resistance and prejudice when they make suggestions for change and have had little opportunity even to begin to develop a sophisticated behavioral technology. For example, the inefficacy of the penal system has been demonstrated time and again, but radical new penal philosophies have not yet been tested. Social science has no plots upon which to demonstrate the efficacy of its programs.[14]

The third facet of the industrial problem lies in the *imbalance in natural systems created by the environmental impact of science and technology.* In a materialistic society economic growth and development are out of control as science doesn't adequately anticipate the problem it creates. Each new technology that promises more material progress exacts an additional toll from an already assaulted ecosphere. The industrial revolution has been built around manipulation of nature, and with each new success humanity has become much more confident that nature can be completely subdued. But mankind has been continually moving further from natural ecosystems within which most evolution has taken place. The wheels of contemporary industry are turned by energy derived from fossil fuels, created and stored over millions of years. These energy resources are for practical purposes nonrenewable, and these essential building blocks of industrial society are rapidly disappearing. Thus, science has led to an increased utilization of a flow of resources, and little attention has been paid to permanent stocks. Apologists for technology argue that new innovations will produce the needed energy in time, but exponentially increasing consumption races to outstrip production. Despite its lofty achievements science has weakened mankind's future position in respect to nature.

The last imbalance resulting from various aspects of the scientific revolution has been an *exaggeration of the gap between rich and poor.* Basic questions of social justice have been put off by the increasing supply of material goods. This holds true both within industrial nations and among nation-states. The absolute magnitude of gaps in living conditions between the rich and the poor continues to increase. Growth is the cement that holds societies together in the face of such obvious class differences. To fail to submit to the will of science in the future is to risk uncovering these basic social issues of equity that have remained neatly buried over the last one hundred years.

In relations among nations, the uneven development of science and technology has created a two-tiered world in which the industrial countries keep the less-developed countries in perpetual subservience and sell the material artifacts created by new technologies to them. Control of technology by a few industrial societies threatens to keep the third world in an inferior position while these industrial countries skim off readily available natural resources to meet their own excessive needs. This new imperialism goes so far as piracy of human resources from developing countries, where they are needed, to developed countries where salaries are higher. As a result, the growth of technology remains unbalanced; innovation growing at exponential rates in industrial societies, and hardly at all in the less-developed areas of the world.

These four aspects of the unbalanced revolution will be much more critical in the future society. It is obvious that technology will play a more important role as the pace of change continues to accelerate and create problems for which technological solutions may not exist. Large sums will be earmarked from tax revenues for federally sponsored research and development as none of the many problems facing humanity have cheap solutions. This will open new opportunities for closer control of the scientific enterprise and cause more concern for its social spin-offs.

The imbalance between the physical and the social sciences will become increasingly apparent as growing populations and increasing demands create new social problems of distribution. As numbers of people increase and as expectations rise there will be a much narrower margin for error in the management of complex societies. Hence the need for much closer supervision of the scientific enterprise. Mankind is approaching an era in which everything must be done correctly in order to avoid serious consequences.

The imbalance between social and physical sciences must

also be righted because the problems of the future society will not fall clearly into the physical or the social domain. The two are rapidly becoming linked as it becomes apparent that most environmental problems have social roots. Predicted resource shortages, for example, can be dealt with through physical solutions, such as breeder reactors, or through social technologies designed to stem a universal revolution of rising expectations. Impending technological problems have very clear social roots, and new social problems will have links to the physical world. The drain on world resources and new pressures on the ecosphere are social in origin, and the high price of raw materials in the future society will have important social consequences. Social mobility might be constricted in a future society, and drastic changes in life-styles will undoubtedly be necessary in coming to terms with physical limits to growth.

Unless society determines to decelerate galloping technology and right present imbalances in the future, technology will continue to accelerate social life and change the social environment many times within each generation. New biological innovations will have significant social effects. Modern medicine promises new techniques to keep the aging alive for decades. This raises the social and moral questions of euthanasia and new institutions for housing the aged. Biologists will develop the capability of altering human genetics and changing sex and identity. Even now heart transplant operations are raising delicate philosophical issues in medicine. Cloning and the creation of life in test tubes will raise even more delicate issues for the future. It is clear that present laissez-faire science presents few constraints on the application of such developments.[15]

This generation is the first in history to possess the power to destroy life on this planet, by deliberate design as well as by accident and inattention. Mankind can consume itself to extinction just as readily as it can destroy itself with thermonuclear weapons. In this respect galloping technology is becoming more than a new social disease; it could well be classified as an epidemic. The antidote to stem the epidemic is careful scrutiny of scientific activity and rational scientific planning to meet future problems, something very much lacking in the present social paradigm.

Threading the technological needle of the future means checking to make certain that new innovations don't have unexpected side effects, both physical and social. Many of the by-products of the scientific revolution have had drastic consequences that were not recognized before they made a tremendous impact on the ecosphere. Dumping waste mercury into streams was not thought to be dangerous. But microorganisms converted it into methyl mercury which concentrated at many points in natural food chains. Many other developments have had similar lag effects. DDT was once welcomed as a beneficial pesticide. Now it is recognized that DDT concentrates in food chains just like methyl mercury and ultimately the effects on animals and human beings can be much the same. Suggested technological innovations must be carefully studied in the future lest they have similar adverse effects on ecosystems. A supersonic transport, for example, could well affect the ozone balance at high altitudes and permit radiation to do much damage to life on the earth's surface. By the time results would be apparent, however, irreparable damage could be done. Nuclear reactors promise to provide energy for the future, but also have the unfortunate problem that the resulting waste materials cannot now be safely transferred and stored given present levels of technology.

In summary, the technological society of the future will be much more complex, interdependent, and difficult to manage. It is for this reason that society must manage science, rather than be managed by it. Society needs science as never before to meet burgeoning problems, but at the same time disillusionment with technological society continues to grow. Latter day Luddites threaten to smash the scientific machine precisely at the point in history when a controlled scientific machine is most needed. It is preferable to harness the gallop of technology to the cart of social needs rather than to upset the cart in revenge for past misdeeds.

Righting the Balance

The first step in establishing social control of science and technology is to recognize the subtle ways in which science now makes political decisions. Political processes are defined as "authoritative allocation of values."[16] Obviously, science is just as political in this sense as is the legislative process. Science clearly allocates values as a result of innovations produced by research. In some cases these innovations result from initiative in the private sector. In others they are related to technologies developed through public funding for other purposes. At any rate, science is political and once this fact is accepted, ways can be found to make science responsive to collective direction.

Science not only legislates by default but also plays a critical role in the present political arena. Many political decisions involving social values are made solely on the basis of scientific expertise, or at least under the guise of scientific expertise. Power plant siting is a value issue, at least to residents of areas to be affected. But often the value issue is covered over by questions of technical feasibility and topography. In the past development of major weapons systems has been decided on issues of technical feasibility rather than on social choice issues. Consumer credit and the development of massive data banks in Washington, D.C., represent other legislation decided purely by technical capability. Since this capacity exists, this information becomes "needed" to keep tabs on millions of individuals. The trade-off in terms of personal freedom is largely ignored. Little thought is given to the whole range of values that are thus erased; the main question being "will it work?" Most recently the antiballistic missile decision and debate over the supersonic transport offered signs of hope that these value trade-offs will be made more explicit in the future, although even in these cases countless experts were lined up on both sides and the political battles were largely fought over technological feasibility. Only a small part of the debate centered on social desirability. Political dramas remain cloaked in scientific disguise, and political success is still often dependent on dragging the best scientific red herrings into the political arena.

Science also legislates by setting parameters within which social choices can be made. Feasibility studies and expert testimony are frequently used to bolster or reject value choices before debate begins. This works in two different ways. First, the testimony of "experts" narrows the universe of political discussion. Experts are drawn from a narrow stratum of society. Quite naturally their "solutions" and "suggestions" reflect work in progress, not work that could be in progress. Expert testimony also narrows choices in other ways. Benefits and values get reduced to technological terms. Those values that cannot be realized cheaply within the dominant paradigm are rejected as too costly. Little expertise is devoted to finding ways of translating socially valued goals into technological possibilities. Rather, existing technologies determine society's perception of possible goals. In other words, when technology offers a method of implementing a desired value, it can be readily done. When a new and expensive technology must be developed, it is less likely to be done.

There are three difficult issues to be settled in government-science relations. The first is to define an appropriate level of public versus private support of basic research. The second is to determine who will organize and who will control the direction of research. The third, and most important, is to allay the tension between independent science and demands of a democratic public. It is not clear that the public will continue to support expensive basic research as budgetary problems continue to mount.[17] New institutional forms must be developed and considerable collective thought must be given to an appropriate mix of priorities, carefully weighing which technological investments can be made in a society of increasing scarcity.

Once the impotence of the present management of science is clearly understood new institutions can be created to bring science back under social control. Most of these institutions will depend heavily upon what is called participatory technology. By this is meant "citizen participation in the public development, use, and regulation of technology." There are many forms that participatory technology can take, but all of them are dependent upon constructing a new social model within which new science-citizen relationships can be forged.[18]

Participatory technology is in one sense a contradiction in terms. Science, often by design, has remained apart from the citizen. The work and concerns of scientists have frequently not been understood by the public. Often their work has been done in secret. Scientists have been trained for years in the best institutions, speak an esoteric language, and operate on the frontiers of scientific research. They have felt no need to simplify their work and make it understandable. If anything, there has been a tendency to develop unnecessary jargon and terminology to keep scientific research mysterious and beyond the comprehension of the citizen. Obviously, the secrecy and the mystery surrounding research must be dropped if science is to be brought under social control. This idea will undoubtedly meet with intense opposition in scientific circles, but it must be done if science is to survive the developing challenge to its future.

In the political process participatory technology is already making itself felt in two different arenas. Many public interest groups have taken to the courts in an attempt to use litigation to force a public debate over value choices inherent in technological decisions. But the language of the law and the rules of the court are not always amenable to a thorough review of issues under discussion. Judicial decisions have given many citizen groups an opening to fight technology's challenges to the environment, but more legislation is needed to bring science under public control.

In another arena politicians have begun to show interest in long-term planning and formal types of technology assessment.[19] Technology assessment is a method for identifying and dealing with the social and scientific implications of applied research. Through technology assessment citizens can be informed of the impact and desirability of types of technological developments that are taking place. New forms of technology assessment should focus the process at the center of political decision-making and make covert research priorities more overt. In a sense this would be politicizing science, carrying on full-scale debates about the desirability of various patterns of technological development before an educated public. Technology assessment must become a routine part of the political process, and concerned citizens and scientists should be given every opportunity to examine the merits of science policies in an open forum.

Finally, every effort must be made to develop an adversary system or "forensic science" within the scientific community.[20] Scientists cannot remain a part of a mysterious monastic order. There are serious differences of opinion about the direction that science should take and about the allocation of future efforts within the scientific community. In the future, scientists must be encouraged to debate in public forums and speak out on matters of shaping priorities. The merits of nuclear reactors versus solar power should be thoroughly aired in front of politicians and the public before large social investments are made. The public should also be acquainted with the social costs and social benefits inherent in new mineral substitutions and fuel policies in society's attempt to come to terms with limits to growth. The environmental trade-offs involved in offshore drilling for oil should become public knowledge. In short, the public should be informed of the costs and benefits inherent in the scientific decisions that are presently being quietly made out of the public spotlight.

In combination, these new ideas and institutions could help to halt the laissez-faire gallop of technology and restore some balance between science's preferences and society's needs. Restoring the balance, however, means asserting the priority of social values over an undirected scientific establishment. Most important of all, social priorities must be drastically altered to turn the scientific revolution inward to the study of the inner man, his values, and institutions. This area of inquiry will be critical in the future society. There is a pressing need for assessment panels to identify critical *social problems* that can be solved through the input of new social research. These would complement technology assessment in the physical sciences. It took decades for American society to recognize racial discrimination as a serious social problem. In a similar vein we have now just come to recognize that equality of opportunity cannot really exist within the present type of social paradigm. New social indicators and new methods of monitoring society are clearly needed if society is to develop a social technology comparable to the physical science technology that now exists. Human expectations and behavior must be altered despite the best physical efforts to keep up with the pace of growing expectations and demands. Social and behavioral inquiry must be respected and rewarded, and old taboos against the study of human beings and society must be laid to rest.

In the end, all these suggested institutions pale in significance before the crucial question confronting twentieth-century man. This is the question of whether a democratically governed society can bring science back under its control. Laissez-faire scientific development has resulted in inquiry that is often divorced from human needs and human concerns. Today science is being challenged by many who see it to be out of control. But controlling science and redressing the balance requires a sophisticated citizenry willing to get involved in outlining future research priorities. Science has never been deliberately malevolent; at worst it has been self-seeking. In the absence of enlightened public direction, science and technology has been free to plot its own course. Ideally, a sophisticated public that understands science can be nurtured within the future society. An educated citizenry must manage technology in society's interest rather than continue to be managed and manipulated by an undirected scientific establishment.

NOTES

1. A more complete view of society and the role of hierarchy in the late Middle Ages is presented in J. Huizinga, *The Waning of the Middle Ages* (New York: Doubleday, 1954).
2. See Thomas Kuhn, *The Structure of Scientific Revolutions* (Chicago: The University of Chicago Press, 1962), chap. 4.
3. *Ibid.*, chap. 5., for a discussion of the priority of paradigms in science, and chap. 6, for a treatment of anomaly and the transformation of shared paradigms.
4. Lewis Mumford, *The Transformation of Man* (New York: Harper and Brothers, 1956), p. 231.
5. Willis Harmon, "Planning Amid Forces for Institutional Change," mimeographed (Menlo Park, California: Stanford Research Institute, 1971), p. 3.
6. See Robert Morrison, "Science and Social Attitudes," *Science*, July 11, 1969.
7. For further elaboration see Nicholas Georgescu-Roegen, *The Entropy Law and the Economic Process* (Cambridge: Harvard University Press, 1971).
8. See Herman E. Daly, "The Steady-State Economy: Toward a Political Economy of Biophysical Equilibrium and Moral Growth," in *Toward A Steady-State Economy,* ed. Herman Daly (San Francisco: W. H. Freeman, 1973).
9. Seymour Lipset, *Political Man* (New York: Doubleday, 1960), chaps. 1 and 2; see also Philips Cutright, "National Political Development: Its Measurement and Social Correlates," *American Sociological Review*, April, 1963.
10. See inter alia, Donella Meadows, Dennis Meadows, Jorgen Randers, and William Behrens III, *The Limits to Growth* (New York: Universe Books, 1972); Paul Ehrlich and Anne Ehrlich, *Population,*

Resources, Environment (San Francisco: W. H. Freeman, 1972); and "A Blueprint for Survival," *The Ecologist* (January, 1972).

11. Jerome Frank, "Galloping Technology, A New Social Disease," *The Journal of Social Issues*, no. 4 (1966).

12. See Joseph Haberer, *Politics and the Community of Science* (New York: Van Nostrand Rheinhold, 1969), chap. 1; also Bernard Barber, *Science and the Social Order* (New York: The Free Press, 1952), for a general overview of the relationship between science and society.

13. B. F. Skinner, *Beyond Freedom and Dignity* (New York: Alfred A. Knopf, 1971), chap. 1.

14. See Gunnar Myrdal, "How Scientific Are the Social Sciences," *Science and Public Affairs*, January, 1973.

15. For a discussion of biological impact on the future society see Leon Kass, "The New Biology: What Price Relieving Man's Estate," *Science*, November 19, 1971; and "Can Man Control His Biological Evolution," *Science and Public Affairs*, December, 1972.

16. Harold Lasswell, *Politics: Who Gets What, When, How* (New York: Meridian Books, 1958); and David Easton, *A Framework for Political Analysis* (Englewood Cliffs, N.J.: Prentice-Hall, 1965).

17. Carl Kaysen, "Government and Scientific Responsibility," *The Public Interest*, Summer, 1971.

18. See J. D. Carroll, "Participatory Technology," *Science*, February 19, 1971; and M. S. Baram, "Social Control of Science and Technology," *Science*, May 7, 1971.

19. See *Technology: Processes of Assessment and Choice* (Report of the National Academy of Sciences, Washington, D.C., July 1969), and *A Study of Technology Assessment* (Report of the Committee on Public Engineering Policy, Washington, D.C., July, 1969).

20. Dean Abrahamson and Donald Geesamon, "Forensic Science—A Proposal," *Science and Public Affairs*, March, 1973.

COMMENTARY

NOLAN PLINY JACOBSON

THE TOTAL impact of Professor Pirages's paper serves to remind us that the most important thing now occurring in modern science is not the widely publicized discoveries for which the

Nobel Prize is sometimes awarded, but the almost totally unpub-
licized efforts now being made to reconceive the nature of science.
Some of these efforts are being conducted at symposia such as this
one at Ann Arbor. Some are conducted in institutes set up for
continuing encounter with this task at major universities in Europe
and the United States. It is an entirely unavoidable endeavor since
science has not only moved out of its nineteenth-century self-
image as the progressive disclosure of objective knowledge of a
world governed by deterministic laws, but it has spread into every
major interest of man and it has become a multinational, multiracial, multicultural activity that can no longer remain one of the
stellar performances of Western civilization with its peculiar provincialisms and angles of perception.

Pirages's paper forces science in this direction by the sheer
power of the otherwise insoluble problems he presents. Science
must move on, out of traditional preoccupations, or else the
"unbalanced revolution" will bring on its own coup de grace. We
can get a focus on this thrust of Pirages's paper by examining what
he says about the "social paradigm" within which science always
works and moves and has its being. The controversies in recent
years over the concept of the "paradigm" often appear interminable and inconclusive.[1] The drift of Pirages's paper is to move
these debates off center, away from epistemological issues, and
into a new way of conceiving the "dominant paradigm" that
controls the way science thinks about itself, the kind of problems
it takes up for consideration, the constellation of questions it asks,
the angles of perceptions within which it works, and the presuppositions it accepts regarding the nature of life in the world.

The new paradigm in which science is already operating to
some extent is primarily social, rather than metaphysical or ideological in its character and scope. It is predominantly social because none of the problems Pirages presents in his paper is primarily a metaphysical or ideological problem. All of them require
a new social discipline or style of life if science is to work
effectively upon them. A new social paradigm will be needed,
Pirages is saying, if the vast powers of scientific research and
computerized technology are to be brought under control. Otherwise, the obsolete political and social institutions, the environmental pollution, and the exhaustion of natural resources can only
become even more destructive.

The new "dominant paradigm" is primarily social, more-over, because only a new social matrix of some kind can bring human history under some measure of control. Many prominent students of modernization agree with Gunnar Myrdal that "the course of history is largely outside our control . . . beyond our perception . . . [so that] we do not perceive its direction."[2] As Inkeles puts it, "we are experiencing a process of change affecting everything, yet controlled by no one."[3] In the age of stone weapons and the wheel man could exercise some degree of crude control and direct his activities toward some kind of desired outcome, at least some of the time. With the superindustrial age, man reaches an age in which he knows only one thing about the future; namely, that it will be different from the past. This can be spelled out, but it is probably obvious to all. Preconceived and predetermined goals join phlogiston today as casualties of scien-tific growth. In an age of innovation and invention it would be a gesture of primitive magic to foretell just what will make the future different from the past. The new social paradigm must predispose men and women to live in an unpredictable world.

The loss of control pervades all levels and areas of life. Local control systems such as family, church, and economic and political institutions no longer have the power to introject into each rising generation their own values and outlook on life. No parent can predict what his children will learn, nor how it will alter the way they have been reared, particularly since the children have often logged ten thousand hours of television by the time they enter first grade. No generation has ever come of age with its neural networks programmed with such an amalgam of unrecon-ciled meanings and values emerging from all the diverse cultures of mankind, and much that has been pressed against their transistor-ized senses will be available for instant replay. More and more, individuals are coming of age with practice in listening to the roar of this global communication; they have had their angles of perception stretched far beyond the limits of the sociological communities in which they were reared. Possibilities are present in their experience for correcting the one-sided provincialisms and distortions with which they were acculturated. Some observers conclude that such individuals are so spread out in their percep-tions across the grid of worldwide communication that the only culture and symbolic system appropriate to their experience will be one of their own individual choosing,[4] rather than one acquired automatically as in culture worlds of the past.

We can reinforce still further the necessity that the new social paradigm be predominantly social in its character and scope. An increasing number of men and women are becoming more fully exposed to events for which their own social and geographical location and culture world provide no adequate interpretation. They are forced to look beyond the local boundaries and open themselves to the opinions of people of radically different cultures. Such opinions and perspectives are carried constantly on modern media to the limits of the earth, broadening the merely local angles of perception with forms of awareness and new possibilities that no one can foresee. Individuals and groups are consequently multiplying who can no longer be controlled in what they see, know, value, and do by the sociological communities in which they come of age. Whether we like it or not, there is now no social institution, and none is conceivable, that can control the thought and feeling systems of even its most loyal members. The credibility of traditional institutions that spoke for God and country is vanishing. Even the nation-state, particularly in the most highly industrialized parts of the earth, has been stripped of its traditional power of control, the long war in Vietnam being the most recent and dramatic example. The supremacy of the nation as an autonomous authority is over.

We have no intention, of course, to separate the new social paradigm in which science will live from any and all metaphysical or ideological implications. To do this would only raise interminable arguments over the meaning of metaphysics and dull the clarity that can reasonably be expected as long as the new paradigm is conceived as a form of social existence. With each passing decade, moreover, modern science becomes ever more fully a multinational, multiracial, multicultural activity encompassing the globe. The dominant paradigm of such a global activity must itself be global in character and scope, unlimited by the special features of either Western or Oriental civilizations.

We can be quite specific on this point by distinguishing our position from that of Thomas Kuhn. Kuhn is correct in thinking of the paradigms of "science past" [see the present Symposium title] as constituted by fundamental forms of thought, metaphysical speculations, epistemological viewpoints, and concepts that "determine large areas of experience" in broad sweeps of reality.[5] He is also correct in arguing that such forms of thought have been highly resistant to change. But this must be understood in the light of the peculiar proneness of Western civilization to use the "theoretical component in experience," as Northrop calls it,[6] to control human thought and behavior. Many have written about this "Greek cognitive bias," and the way it functioned in the West as an instrument of social control as well as a vehicle of reliable knowledge. It is worth remembering that modern science in the West dawned upon Europeans who were exhausted from the interminable wars of the sixteenth and seventeenth centuries. It is part of the history of the time that science was seen as a new principle of authority for human living and a new way of controlling thought. And the tendency is still very much alive to think of science as a quest for certainty in an uncertain world. It is not science so much as the paradigm of this historic period to which we must look in trying to understand why the most reliable forms of thought became so resistant to change. Religious thought came under the same paradigmatic control, to such an extent that religion became a principle of social organization, a means for control of thought as well as an ultimate commitment and orientation to life.

This kind of thought-centered paradigm is what Pirages refers to as "the paradigm of the first Copernican Revolution." There are reasons for doubting, as Pirages argues, that this dominant paradigm can sustain itself much longer. I would go further and suggest that it has already passed from the most responsible centers of thought in the West. The whole effort to support life, and to organize its affairs, in the light of a conceptual system is rapidly dying out in the West where it once ruled supreme. Something has happened to man's capacity to believe. More and more, the balance shifts in favor of *methods* of inquiry, *ways* of believing, a *logic* of discovery, rather than the concepts to which such investigation gives rise.

The New Social Paradigm

A new paradigm with features suggested above is very rapidly emerging to dominance in our midst. This is a paradigm that science has itself played a role in creating. The knowledge explosion emerging from millions of centers of research throughout the globe, the computerized technology that can survive only through constant improvement and innovation, and the interpenetration and interdependence of radically different culture worlds of mankind all conspire together to create the new social paradigm for which Pirages is calling. The new paradigm is a new personal discipline and community of multiracial, multinational, multireligious, and multicultural interaction and communication from which fewer and fewer men and women can escape. It is scrambling and dispersing the greatest concentrations of entrenched power in history. It is currently breaking up all previous social paradigms and offering itself as the first discipline powerful enough to bring the gigantic powers of the superindustrial age into the service of widening and deepening and enriching life. This global interaction separates our own period from all preceding epochs in history. It is freeing us from the compulsive grip of the self-encapsulated, self-isolating, self-justifying, self-regarding culture worlds in which, Narcissus-like, men and women have lived since the rise of civilization in the upper reaches of the Fertile Crescent ten thousand years ago. The fruit of this cross-cultural communication is seen in the new perspectives emerging out of discussion of common problems such as hunger, disease, war, pollution, overpopulation, and many others, none of which can be solved by individual nations or multinational corporations acting independently of all the rest.

This interchange is rapidly becoming the vital center of life in the world, opening up beyond any limit the values men and women can experience and the power of control they can exercise together. It is bringing to dominance all over the earth a new interest in learning, a new concern for acquiring the forms of perception of other people, and a new willingness to reconsider and revise the assumptions and accumulated wisdom with which people interrogate their experience and interpret events. This process of interaction and communication is transforming man by transforming the conditions that nurture and control his growth. It is transforming him more fully into an inquiring creature who of necessity now can survive only by correcting the way he has been reared. He is a creature more capable of embodying in daily life the motto of the Royal Society of seventeenth-century England,

Nullius in Verba, "take nobody's word for it; see for yourself," and he sees science, with Karl Popper, as a concern for evidence that falsifies what is presently believed. Science, as Feynman puts it, is "the belief in the ignorance of experts." Flexibility in creating and changing conceptual models is what modern science is all about, whether it be in biology or astronomy or the behavioral sciences. Nothing would seem to incapacitate a scientist so much as basing his operations in socially recognized systems of belief which exist independently of convention. It is this that makes it so difficult for people to perceive the evidence that calls their beliefs into question.[7] It is this that incapacitates them from engaging fully and freely in rational criticism. A mere century ago few of Darwin's fellow scientists were able to agree with his evolutionary outlook, to accept which would have required them to rethink everything they considered true. A few short decades ago, on the other hand, the life-style of critical inquiry and self-correction was enacted in the work of Einstein who changed the thinking of all of us at the deepest level of our presuppositions regarding time and space, matter and energy, and the supposed simultaneity of events. The new life-style is based, not in the concepts of our most reliable knowledge, but in the process of inquiry and criticism in which concepts are corrected. Science has become a community of inquiry, rather than a fellowship of believers. Unlike any community in all the human past, this community daily acknowledges its need for correction. There is always a tremendous amount of inner controversy and infighting among the members, but all of it serves to deepen the community and make it more inclusive. This is because it is this community as a matrix of social behavior that embodies the self-corrective style of life and infuses it into the individual members.

   This new life-style of self-correction, presently exemplified in narrowly segmented areas of scientific research, is being communicated to a global community of cross-cultural interaction and communication which is dissolving or inundating the ancestral culture worlds of the planet, the most fully industrialized far more drastically than the others. The focal point of this global community, as suggested above, is the new social paradigm in which all the problems elaborated in Pirages's paper can be brought under control.

   The remarks that Pirages makes about participatory democracy point to a kind of creative interchange and self-corrective living in which men and women learn to cope with their most pressing problems. They learn to interact with their fellow creatures in ways that make everything that is known about these problems fully and freely accessible without suppression. The distinguishing mark of this new style and community of life can be characterized by five events that are always happening wherever this new social paradigm is rising to dominance against countervailing influences in contemporary life. These five events can be taken as the criteria of the self-corrective life, the vital center of the new global paradigm under discussion, the functional unity that is the major reality currently shaping history on our planet. Analyzed into its simpler elements, this functional unity is constituted as follows: first, in the emergence of new perspectives which come to individuals as novelties they could not previously imagine because their most creative imagination operates willy-nilly within the limits of what is already stored in their neural networks; second, in the merging of these new perspectives into the total apperceptive mass of what is already known and valued by men and women engaged in this communication; third, in the widening of the angles of perception, thus increasing the depth and breadth of what the self-corrective community is able to take into account; fourth, in the summoning of old stereotypes and culture-bound attitudes out of the psychological underground deep beneath conscious awareness, to face the contradictions and the one-sidedness of old learning increasingly apparent as new learning continues; and, fifth, in the phenomenon of forgetting and flushing away everything that has been falsified by new evidence and new encounter with other minds.

These five events constitute the essence of participatory democracy as it operates in our midst. They likewise point to a new spiritual ground and horizon toward which mankind is moving as the roar of global, cross-cultural communication enables individuals increasingly to accept responsibility for monitoring the total environment in which they live. In all likelihood it is true, as Pirages seems to be arguing, that the spectacular achievements of modern science can be carried into new and more fully human levels and horizons of development only if its own self-corrective spirit of inquiry can be embodied in a dominant social paradigm and cultural matrix of this kind.

NOTES

1. Lakatos, Imre and Musgrave, Alan, eds., *Criticism and the Growth of Knowledge* (Cambridge: at the University Press, 1970); see Nicholas Maxwell, "A Critique of Popper's Views on Scientific Method," *Philosophy of Science* 39, no. 2 (June, 1972): 131–52; also other issues of this journal of the Philosophy of Science Association.

2. Gunnar Myrdal, *Asian Drama: An Inquiry into the Poverty of Nations,* 3 vols. (New York: The Twentieth Century Fund, 1968), I: 702, 703.

3. Alex Inkeles, "The Modernization of Man," in *Modernization: The Dynamics of Growth,* ed. Myron Weiner (New York: Basic Books, 1966), chap. 10, p. 149.

4. Robert Bellah, "The Historical Background of Unbelief," in *The Culture of Unbelief,* ed. R. Caporale and A. Grumelli (Berkeley: University of California Press, 1971), p. 42.

5. Thomas S. Kuhn, *The Structure of Scientific Revolutions,* 2d ed. (Chicago: University of Chicago Press, 1970), pp. 17–18, 112, 120–21, 128.

6. F.S.C. Northrop, *The Meeting of East and West* (New York: The Macmillan Co., 1953).

7. Karl R. Popper, *Conjectures and Refutations* (New York: Basic Books, 1962), chap. 10.

X   The Ecology of Knowledge

JERZY A. WOJCIECHOWSKI

Introduction

FIVE centuries ago was born the man whose theory ushered mankind into the modern era. Five hundred years after the birth of Copernicus mankind has reached a new turning point. Half a millennium of revolutionary progress elevated humanity to unprecedented heights and at the same time faced it with the threat of extinction. It is therefore fitting that the celebration of the five hundred years' anniversary of the birth of Copernicus be an occasion for a reflection on the present state of knowledge, on the reasons of our contemporary predicament and on the future of the intellectual adventure.

The greatness of the Copernican theory consisted not so much in the fact that it was a better theory than the Ptolemaic one, but rather in this: it challenged deeply ingrained beliefs as well as universally accepted apparent evidence of the sense order. In other words, the value of the heliocentric theory was not limited to its strictly scientific content and to the contribution which it made to the science of astronomy. Even more important was its revolutionary effect on medieval *Weltanschauung* in general and on the theory of knowledge in particular. Today we are faced with the cumulative impact of science and technology which developed in the wake of the Copernican revolution, and we feel obliged to do something about it.

Copernicus stopped the sun. Men today ask whether it is not time to stop the development of knowledge as we have known it until now. The very fact that this question can be formulated is a measure of the power of man's reflective capacities. Moreover, it suggests that straightforward "classical" rationality may not be as satisfactory an attitude for man vis-à-vis reality as it has been assumed. The reflective capacities are man's supreme weapon in his constant struggle with himself. The great need to formulate the question of the value of further development of knowledge is a sign of the seriousness of the present situation of knowledge as well as of the progress in the depth and scope of the reflection about knowledge. Until recently, the principal responsibility of intellectuals was to think as much as possible and thereby to advance knowledge ever further and in all directions conceivable. Doubts and discussions centered around the difficulties encountered along the path of progress, but not about progress itself. Now it is the very idea of progress in general, and of the progress of science in particular that comes under close scrutiny. In this sense, one may say that the present-day difficulties usher us into the field of a meta-critique of knowledge. The meta-critique would differ from ordinary critique by the introduction of the question of the value of knowledge in general and of its branches. That is to say, knowledge would have to be assessed not only for its formal truth-bearing capacities, but also for its impact on and its value for the individual and for society.

Self-examination is an expression of intellectual maturity. Already Hegel realized that the heightened self-awareness, the "unhappy consciousness," is a painful but necessary factor in man's evolution. In this perspective it was therefore inevitable that knowledge be one day critically examined, not only from the strictly epistemological point of view, but also more fundamentally, as to its role in man's life and its overall value. Knowledge is a problem-creating factor. It can be a blessing or it can be a threat. Moreover, in itself, it becomes an ever bigger and more difficult object to understand. This situation may be formalized and expressed by means of the following law:

> *The need to think is proportional to the amount of knowledge available.*

It means that the greater our knowledge, the more we are forced to think about it and the more knowledge becomes problematic.

This strange situation is due not so much to the very growth of knowledge both in size and complexity, but rather it results from the fact that it is the nature of the intellect to transform into a problem whatever it considers as an object of reflection, even if the object is thought itself.

When knowledge becomes an object of knowledge, it presents difficulties which we encounter in the study of other objects plus some special difficulties not encountered in other studies. Man produces knowledge but this fact does not make knowledge fully intelligible. The act of knowing in general, and even that of thinking, is not fully comprehensible. Moreover, man does not exercise a total command over this act. This is even more true about the product of the acts of knowledge, i.e., the body of knowledge, or in other words, the intellectual constructs. Once formed, they lead a life of their own engaging in a dialectical relationship with other constructs and with minds. Man can invent new constructs, become disinterested in the older ones, but the net result is always more constructs—not fewer.

The difficulty of the critical approach in the field of knowledge is directly proportional to the importance of ideas scrutinized. The difficulty which mankind now faces is so great because it has to assess the most successful period in its history as well as the causes, the means, and the consequences of this success. In other words, it must take stock of the intellectual attitudes and beliefs which led to the birth of modern science and which sustained its development. It must also consider the nature of the scientific enterprise, the role of breakthroughs in the development of science, and the question of eventual harnessing of science. This of course is an incredible order. Somebody may object that it is unduly critical of science. This however is not so. It is a tribute to science and a sign of the normalcy of its evolution that the very success of the scientific adventure caused the need of evaluation of the scientific undertaking. One thing is certain: the progress of knowledge makes life easier for our muscles but not for our brains.

To Plan or not to Plan

The central question which contemporary man must ask and answer is whether man should be in the service of knowledge or whether knowledge should be in the service of man. Underlying it is another question, namely, whether and to what extent man can master his acts and their consequences. The situation is complicated by the fact that in order to make this ability meaningful and useful it is necessary to have an adequate knowledge of an adequate hierarchy of values. If knowledge is to serve man then he has not only to be able to master knowledge and its results, but also he must know what is best for him. In other words, one has to use knowledge to master knowledge. This sounds like a vicious circle. Whether there is a way out of this vicious circle remains to be seen.

We live in an era of overcrowded earth and of heightened awareness of the consequences of human activity. In this situation it becomes necessary to engage in global planning and to look at knowledge as a natural resource. Although this idea may sound revolting, and certainly is potentially a dangerous one, the time has come to ask whether it is possible to cultivate knowledge and direct its development so as to make it more useful for individual man, for a given society, and for mankind as a whole. The idea of economic planning is a familiar one. Instead, the idea of planning of and for knowledge and its development on a national and international scale with regard to the requirements of an overall, well-balanced culture has not yet received enough attention. And yet if mankind is to survive the present impass and if it is intent on improving the quality of life, it must develop this idea. The more thought that is given to it, the lesser is the chance of inadequate planning being imposed by some authoritative, undemocratic regime, or by a democratic, but inadequately enlightened government.

Until recently the progress of knowledge has been mainly the result of solitary efforts of individuals. Their work was being done on the sidelines of the life of society. Nor was the society very much concerned, intellectually or financially, with the scholarly pursuits of these few individuals. Now the situation has changed drastically. That it has changed is the result of various factors all of which are not yet fully comprehended. One of them however is rather obvious and of prime importance for our discussion. It is the internal logic of the development of science and the demands which it places on scientists. The progress of knowledge in general and the progress of science in particular exceed more and more the capacities of individuals. Thus research becomes increasingly a joint venture, a teamwork of a large number of researchers funded by ever greater outlays of capital. Hence the business of knowledge becomes increasingly a social fact and a public concern. The concern results in a desire of planning. Attempts therefore are made to impose order on the knowledge industry by assigning some priorities so as to contain it within the limits of national budgets.

Aristotle remarked that to think is to order, and he defined order as unity in the multiplicity. This abstract, metaphysical definition can shed much light on the problem of planning. The greater the multiplicity that the intellect has to face, the greater the need to find in, or to impose on, this multiplicity a unity, i.e., to order it. This situation can be expressed by means of the following law:

> *The necessity of planning is therefore, generally speaking, proportional to the number of units involved.*

Consequently, the need of planning the progress of knowledge is proportional to the number of people involved in the knowledge industry. This is why the planning of advanced education and research becomes a national priority in the leading countries. With the growth of population and the generalization of education the necessity of planning will only increase and never decrease. Plan we must; so plan we will. The problem before us is not one of choosing between planning and not planning, but how to plan best.

All planning is difficult. The more future oriented it is, the more difficult it becomes. Knowledge planning, to be of any value, must be a long-range planning. Knowledge planners must be consummate futurologists, possessed of great imaginative powers and capable of transcending presently existing situations and established hierarchies of value. Thus far, whatever planning there has been was based largely on the present model of consumer society dedicated to economic growth, exploitation of natural resources, and the satisfaction of many urgent and some imagined material needs. Nor was there much concern shown for balanced growth of various branches of knowledge necessary for the harmonious development of culture and the shaping of the society of the future. Thus, priority was usually accorded to investment which could quickly produce technology and applied research with the view of solving immediate, concrete problems. Humanities, fine arts, satisfaction of intellectual curiosity, i.e., pure knowledge, and culture in general were relegated to the second rank.

Obviously, the long-range planning ought to be subordinated to the needs and ideals of the society of the future, which will have to be quite different from the present one. It will have to be a society freed from illusions about the value of continued economic growth which rapidly exhausts the earth—a society respectful of ecological principles and dedicated to the development of the total man. In this type of society, the emphasis will be shifted from economic expansion to the intellectual, cultural endeavors. The satisfaction of all genuine, but only genuine material needs, will free man for the pursuit of personal fulfillment. The demands which man will then place on knowledge and education will differ from the present-day ones.

Tomorrow begins today. The planning for the society of the future must be undertaken now. To plan is to evaluate. Before any plans for the future can be made, it is necessary to find out the causes of the present predicament. This analysis should aid us in better evaluating the necessary reforms as well as the future development of knowledge and culture.

What Went Wrong?

One of the reasons of the tremendous development of science has been its independence from outside interference. Until now science has been developing as if it were an autonomous, self-centered, self-sufficient activity, possessing its aim and its justification within itself. It has been largely independent from, and unrelated to, other areas of human endeavors—a kingdom unto itself. But science was not all harmless contemplation. The birth of modern science coincides with the revolution in the views about the nature, the justification, and the aims of knowledge in general. The Greeks bequeathed to the Western world the ideal of contemplative knowledge which they considered as the most noble activity of all. Together with it went the idea of cosmic order, of which man was a subordinate part, and of natural law. Nature was not only the purveyor of all necessities of life, but also the great ruler. In antiquity, man looked at nature in fear and awe. He considered it to be eternal and transcendental, fit for contemplation and study, and he was unable to visualize nature as a matter to be conquered and used. Consequently, the development of theoretical knowledge far outdistanced that of technology.

Modern thinkers changed all that. Francis Bacon, the great prophet of the modern age, substituted for the contemplative, the practical ideal of knowledge. The aim of science was to be the knowledge of the laws of nature. It was to enable man to master nature and put its wealth at man's disposal. Two centuries later Auguste Comte expressed this ideal of knowledge in his famous dictum: "to know in order to foresee so as to be able to do." The consequence of this was the degradation of nature from the position of lawgiver to that of primary resource, and the abandonment of the idea of the subordination of man to the cosmic order. The movement of "man's lib" freed man from subjugation to nature and from old beliefs. Man the maker and man the master were free and rampant. The well-known result was the ecological predicament. There is, however, another result, perhaps less obvious though not less important, and of greater relevance to our discussion. It is the subordination of man to the idea of progress and to the process of progress with all that it entails in terms of conditions and consequences.

The liberation of man from the bonds of nature and the development of the demiurgic ideal went hand in hand with a radical change in the field of philosophy. Aristotelian realism with its idea of subordination of the knower to the object of knowledge and with its insistence on sense perception as the source of information about external reality has been rejected together with the scholastic *Weltanschauung*. Modern philosophy has developed on the Cartesian assumption, namely, the principle of immanence, that knowledge is formed within the limits of the intellect. Descartes conceived the intellect as a self-contained, knowledgewise self-sufficient spiritual substance, knowing within itself the perfectly intelligible ideas. This quasi-angelic notion of the intellect resulted in the thesis of radical separation of the mind from the body and in an "inward look" in the theories of knowledge. Intellectual knowledge became detached and independent from sense knowledge. Moreover, intellection ceased to be subordinated to the extramental object. Truth, therefore, was not to be sought in the conformity with sense data and the external world, but in the clarity of ideas, i.e., in the subjective certitude.

Whether in the continental form, the rationalistic-idealistic tradition, or in the British, empiricistic version of modern philosophy, knowledge was viewed in the subjectivistic perspective. The external world was thought to be either unknowable, as for Hume or Kant, or created by and conforming to the Intellect as in Hegel's philosophy. In either case, the knower neither had to conform to nor had to be a subordinate part of the external world and of the cosmic order. Consequently, the act of thinking was viewed as a fully autonomous, man-centered activity standing in no relation to anything outside of itself, and, not a part of a greater whole. With Hegel, the object had to conform to the intellect, its maker. Freed from subordination to nature, man embarked on a man-centered development taking himself for the supreme value and the universal yardstick for evaluating everything outside himself.

Recently, it became obvious that if humanity wants to continue to exist it has to come to terms with nature and respect its order. The ecological predicament obliges us not only to rethink our high-handed treatment of nature and our physical behavior. The revision should go deeper and involve the rethinking of the basic philosophical assumptions and of the *Weltanschauung* which goes with them. Of central interest to us is the problem of the nature and role of knowledge in general and that of science in particular. What is needed is a new vision of human knowledge and activity integrated into the context of human existence, and that of human existence in its ecological context.

Knowledge as an Existential Act

After over three centuries of subjectivistic approaches to knowledge, it becomes imperative to elaborate a new understanding of knowledge. As just mentioned, the new vision must be essentially a holistic view integrating knowledge into and explaining it as a component of human life and of the order of reality of which man is a part. Let us call this approach "the new realism." Its main characteristic should be the consideration of knowledge not only as a set of abstract notions, but as a concrete reality, an existential activity producing tangible results within and without the knower. In this perspective knowledge appears as an activity influencing the objective order of reality and in turn influenced by it, an integral part of a greater whole. This is therefore an ecological rather than logical or methodological point of view. It does not try to negate the other approaches, it simply intends to say something else. The new realism, and its preoccupation with and concern for the total man and his relation to the external world, is also a new humanism.

What the ecological point of view rejects is the Cartesian idealism. The monumental mistake of Descartes consisted in assuming that we think *inter limites intellectus* as if intellect was a self-sustaining, self-sufficient knowing entity. As it usually happens, in the case of great insights, this brilliant error had a good part of the truth in it. We do think within certain limits. But the limits are those of the total individual. We think *inter limites totius personae*, i.e., within ourselves. This affirmation defines the epistemological position from which the present paper is written. It has far-reaching consequences which lead to a very different understanding of knowledge from that of Descartes and his followers. It is the present writer's contention that this view of knowledge will allow an integrative explanation of cognition necessary for the discussion of the problem of ecology of knowledge.

The subject of the act of knowledge is neither the intellect and/or the senses, nor consciousness, but man, the total concrete individual anchored in space and in time, circumscribed by the spatio-temporal parameters of his biological existence. Such is the person; such is the thought. It is to the complex, concrete subject which is man that we have to relate the acts of knowledge and try to understand them accordingly. Knowledge is an organic process involving the total person and not just his intellect and his senses. It is the product of the existential act of the whole man, of the multilevel, complex immersion of the individual in the ambient world. It expresses directly or indirectly the sum total of exchanges which take place between man and the outside reality, and it serves as a mean in maintaining and furthering this commerce which is the essence of life. To consider knowledge in itself, abstracting from and forgetting about its relation to and its role in the life of the knower, is to take *pars pro toto*.

The intellect is the organ of thought but it does not operate apart from the rest of the person. Quite the contrary. It is an integral part of the human organism, not only present in it but dependent on it for its operations. It acts in relation to, and with the cooperation of, the whole body. Shall we therefore conclude that knowledge is necessarily not only man-centered but also unavoidably subjective and independent from the outside reality? Not at all! If knowledge is the product of the given individual, the individual in turn is shaped by the constant intercourse between the self and the ambient world. The latter is composed of other individuals, i.e., society and the physical reality. Both these elements, each one in a different way, model the individual. Man is not an island unto himself. This is especially true on the level of knowledge. Society and the physical world influence the thought of the individual in his very makeup in addition to providing the content of sense data. Physically and intellectually man lives in a dialectical relationship with the ambient world. The ambient world may be defined as the portion of reality composed of the sum total of elements with which we are in relation either through knowledge and/or physically, i.e., the things that we are aware of and others which act upon us or on which we act outside of our awareness. So defined, the ambient world is particular to each individual, and differing more or less from that of another man. This fact is as important existentially as it is for the understanding of the nature of knowledge. Sense knowledge and intellection are constitutive elements of the whole which is composed of man and his ambient world. They are at the same time a product of and a cause of this dynamic whole; they are man made and objectively caused, man-centered and related to the outside reality.

Knowledge as Participation in the Ambient World

Knowledge affects man in his very being. It shapes his behavior by conditioning his attitude to the external world and his participation in it. This is a very complex fact which is of prime importance for our discussion. Although man is always involved with the external world, whatever the level of his knowledge, nevertheless, the more he knows the richer, the more complex is his involvement. Let us express this situation by means of a law:

All other things being equal, the involvement of the individual with external reality is proportional to the amount of knowledge he possesses.

The progress of knowledge changes not only the amount but also the quality of man's involvement with external reality. The radius of activity increases enlarging the ambient world. As an indication of progress suffice it to compare the radius of activity of the caveman with that of the astronaut, the reach of the naked eye with that of the radio telescope. The human environment increases constantly not only in size, but also in complexity. Man's relationship with the external world is a two-way commerce. The more he acts on the environment, the more it acts on him. This may sound surprising because the avowed aim of technology is to give man mastery over the environment and shield him from at least a part of its influences. The fact is, the subordination of nature to man's will results from greater knowledge and better understanding of nature. Consequently, the relative independence, due to the mastery of the physical environment, presupposes a greater intellectual involvement, i.e., bigger and more complex contact with and subordination to the external world in the order of knowledge.

As to the change of quality of man's relations with the outside world, it will not suffice to say that it results from their growing complexity. The total human environment is composed essentially of two elements: natural and man-made, i.e., nature and culture. The important fact here is the building of culture. The roomier and the richer the cultural construct, the more it becomes a habitat for man. Culture not only facilitates life but moreover enables a fuller human development. The more man dwells in the cultural construct, the more he becomes influenced by its architecture. It is important to note that the human space engendered by culture is not the physical space of the material universe. The former is a much more complex and varied milieu than the latter. Because of the sphere of culture the relations between man and environment become more complicated. The relation man-nature is replaced by a triple relation, namely:

(*a*) man-nature; (*b*) man-his products (intellectual and material), i.e., culture; (*c*) man-his products-nature.

Culture is a historical and social construct. It always has a past and a territory. It transcends the individual both in time and space besides having a richer content than any of the individual contributions. Culture is the common property of a group, and the common bond uniting individuals and introducing a degree of homogeneity in their physical and mental behavior. Being the common denominator, culture facilitates communications, broadens the scope of exchanges between individuals, puts at the disposal of the individual the common patrimony of society and thus facilitates personal development. It is the influence of culture on knowledge that merits our attention. There is, in the first place, the *Weltanschauung,* with its point of view, basic ideas, hierarchy of values, and intellectual tradition habitually shared by members of the group. There is, further, the style of thinking, a predilection for certain types of intellectual preoccupations and some degree of consensus. These are all well-known aspects of the intellectual life of a community. What interests us here is the mechanism of the development of these communal aspects of the life of the intellect, and the laws which underlie it.

In this respect, knowledge in general, i.e., the knowledge construct, is the result not only of the intellectual activity and personal qualities of individuals, but also of the human group as a group. There exists, of course, a hierarchy of groups from the family up to the culture group which may be multinational as in the case of Western culture. Each group facilitates communications proper to it and each group-level has a definite potential for communications. The group which is of greatest importance to us is the one at the top of the group hierarchy, namely, the culture group. It has the highest potential for communications and for the development of knowledge. The personal and the social aspects are both constitutive elements of knowledge. In this respect, the most fundamental fact is the demographic one, namely, the very multiplicity of individuals in the group. The existence of a plurality of knowers is an essential condition of the existence and development of the knowledge construct. Because of the plurality there exists exchange of information, confrontation of judgments, and

verification. The result of this is intersubjectivity and objectivization of elements of knowledge. The following law may be formulated expressing the above-described situation.

> *There exists an interdependence between the size of the human*
> *group, the amount of communications within the group, the degree*
> *of objectivity of the knowledge construct, and the progress of*
> *knowledge.*

The interdependence is one of two-way causality and dialectical relationship. Each factor influences other factors and is in
turn influenced by them. Generally speaking, the larger the community, the more communications are necessary, the greater the
chance of divergence of opinions and possibility of confrontations,
the greater and more complex the sphere of intersubjectivity, the
greater the need for, and possibility of, forming objective, verifiable judgments, the more rapid the development of knowledge.
The greater the knowledge, the greater the means of sustaining
greater numbers of people, the larger the community, and the cycle
can again recommence. Of course, it would be a gross error to
consider all these relations as automatic, and univocally applicable
to all human communities. The fact remains, however, that the
individual knower is part of a community, of several communities
to be exact. For instance, an educated individual is a member of a
culture group, of a nation, of a social class, of an intellectual
group, which is first of all that of the educated people in his
nation. If he is a scholar, he is, moreover, a member of the group of
persons of similar education involved in the study of the same
field of knowledge. The latter group, by the way, may be international in scope. The inclusion in each group is expressed by the
interrelations and the interdependence which exist between the
individual and the group.

Because of the role of the group as a group in the phenomenon of knowledge, there is such a thing as the consensus of
opinion. Thomas S. Kuhn made us aware of the role of the
consensus, which he calls paradigm, in the development of science.
May we point out that the consensus plays an important role in all
branches of knowledge and of human activity, not only in science.
The progress of knowledge results in an ever greater number of
branches of knowledge with the proportional increase in the

groups of specialists and in the number of paradigms. One result of it is the complexifying of the society.

Another consequence of the multilevel group structure is the increase in the need of communications. The more there are different groups, i.e., the more complex the society, the more and the more varied communications are needed. The situation can be expressed in the form of a law.

> *The need for communication is proportional to the size of the society, the number of groups within the society and the amount of knowledge available.*

It is not, therefore, only the very size of the society that counts. The society may be very large indeed, like that of the Chinese, for instance, and yet the amount of communications generated by it and the actually felt need for communications may be smaller than in a numerically inferior society such as the American society. Until this century, the problem of communication has not been studied sufficiently. The above law explains the present-day preoccupation with this problem in its various aspects. Linguistics and cybernetics find in it their justification. In its light it becomes understandable that modern philosophers, in contradistinction to their ancient and medieval predecessors, became more and more preoccupied with the problem of intersubjectivity. Although, technically speaking, the philosophical problem of intersubjectivity is the result of Cartesian subjectivism, nevertheless, its study corresponded to an increasing need to explore this domain. Besides, it could perhaps be argued that Descartes's philosophy itself was an unconscious answer to this need, that it too has been conditioned by the social and noëtic development taking place since the Renaissance. The above law makes it plausible to suspect that there may be a correlation between apparently diverse aspects and fields of human activity on the one hand, and the size and physical development of mankind on the other.

If there is some correlation of this kind, then it becomes apparent that not only the act of knowledge of the individual is directly related to the concrete, existential situation of this individual, but also the development of knowledge in general as to its direction and quality is conditioned by the demographic situation of the society. The conditioning is reflected in the areas of prevailing interests, kinds of problems, and solutions suggested over more or less extended periods of time. Of course it would be futile to try to push these correlations too far. Moreover, it has to be remembered that all these relations must be considered as dialectical, i.e., two-way causalities occurring in an ever-changing situation. This means among other things that simple analogies cannot be used as explanatory models in the discussion of the development of knowledge. If, however, there is some such relation as described above, it does offer an important insight into one aspect of the problem of knowledge. Consequently, it should be useful for the discussion of the ecology of knowledge and for knowledge planning.

The Aims of Knowledge

Whatever relation there is between knowledge, size, and structure of society, the development of knowledge is the direct consequence of conscious acts of knowers. These acts are always goal oriented, i.e., finalistic, and this quite independently from the eventual existence of laws governing the preference for a given field or type of knowledge in given situations. Traditionally, knowledge has been classified into theoretical and practical, i.e., pure and applied. Until the nineteenth century, this distinction, based on the distinction of intended aims, seemed not only clear and obvious, but sufficient as well. Theoretical knowledge was good for knowing, and practical knowledge, good for doing things, but not vice versa. As long as the latter knowledge remained on the level of crafts and the former had few applications, there was little reason to doubt the adequacy of the distinction. Difference in aims was clearly reflected in the differences of modes of these two kinds of knowledge. Today, as in the past, the aim of an individual knower may be pure knowledge or practical cognition. However, knowledge resulting from these two distinct motivations may be almost indistinguishable as to its form and content. More and more, to know becomes almost automatically to know how to do, and vice versa. In this situation one has to ask whether or not in the future it will be possible to reverse this trend. Will it be possible to develop knowledge for knowledge's sake without making it a tool for the mastery of something or other?

The old distinction between theoretical and practical

knowledge went hand in hand with the belief in the clear-cut distinction between the observer and the object perceived. This belief was obviously based on the commonsense interpretation of visual perception, and, in the perspective of everyday experience, was perfectly valid. According to this belief the act of observation did not affect the object and left it unperturbed. The thing could therefore be perceived as it is in itself, and known adequately and objectively. Today we know that all observation is transformation. One cannot obtain a datum of observation without producing some change. The change is in fact a complex one: in the object, in the subject, and in the relation between the subject and the object. Moreover, we realize that all knowledge of nature is production. On the sense level it is the production of a sense datum; on the intellectual level it is the production of conceptual constructs, i.e., explanations. Although both the percept and the conceptual construct remain as such in the knower and do not seem to manifest themselves by some direct external effect, nevertheless, it would be obviously wrong to conclude that no external effect results from them. It would be difficult to overestimate the importance of this fact for our analysis and for the understanding of knowledge in general.

Knowledge is the expression of man's union with and of his involvement in the ambient world. Moreover, knowledge influences this relation. To a large extent it determines man's attitude toward the external world and his participation in it. This well-known fact may be expressed in the form of a law:

Man's behavior is proportional to his knowledge.

Thus quite apart from, and prior to, the communication of knowledge which is a natural and universally present consequence of the act of knowledge, the act of knowledge always produces a concrete effect first of all in the knower himself. Let us signify this fact by means of a law:

Knowledge always has a consequence which goes beyond the act of knowledge,

In other words, knowledge is not a kingdom unto itself; it cannot be contained in itself. It is a part of and participates in a bigger, more complex, and more fundamental reality than itself. This bigger reality is, first of all, the knower and his complex act of existence. The inability to fully comprehend the intrinsic involvement of knowledge in its existential context is a legacy of Greek thought from which, by and large, philosophy has not yet fully emancipated itself. Until this happens, philosophers will continue to indulge in rather fruitless considerations of man and his knowledge abstracting from his immersion in, and dynamic relations with, the environment, and will not grasp the full scope and meaning of knowledge.

Because knowledge is not self-contained, the classical distinction between theoretical and practical knowledge cannot be maintained in its traditional form. Whatever the intention of the knower, the act of knowledge has a concrete result and is therefore in fact, practical knowledge, in a broad sense of this word. If this is the case, one is allowed to wonder whether the very purposes of knowledge are adequately expressed by the simple alternative of knowledge for knowledge's sake and knowledge for producing things. It seems that a third eventuality has to be considered, namely, knowledge for the sake of the total development of the individual. This development must be viewed as an existential process occurring in a concrete subject present to and actively participating in the world. Because of the radical involvement of the individual in and with his environment, the third purpose of knowledge has multiple dimensions. Besides the personal aspect, it has material and social implications. Although for the purpose of analysis each one of them may be considered separately, in actual fact, they are inseparable facets of the same existential act.

Importantly enough, the third purpose of knowledge is an all-embracing one. It includes both traditionally recognized aims. The total development of the individual requires the disinterested quest for knowledge as well as the search for ever more efficient and varied technologies. It does not mean, of course, that every man has to strive in both directions, or that this is practically possible. In the future as in the past there will be men more intent on pursuing theoretical knowledge and others more interested in practical aims. However, directly or indirectly through society, they will be participating in, or at least profiting from the global development of knowledge. This rather obvious statement expresses an important fact of the nature of knowledge which philosophers do not seem to have taken sufficiently into account, namely the problem of the subject of knowledge. The subject of the act of knowing is different from that of knowledge in the sense of a body of knowledge. In the former case the subject is an individual, in the latter a group of individuals, i.e., a society. Knowledge which a society requires, especially a progressive one, for its very existence, and even more for its development, transcends in size and scope the actual knowledge and the capacities for knowledge of any individual as well as his need for it. The body of knowledge of which society is the subject and the cause, is necessarily composed of theoretical and practical elements. It shapes the context within which the individual evolves. Whether the individual's intellectual interests are theoretical or practical, he lives and develops in and through an environment determined by the total body of knowledge, its theoretical and practical aspects.

The more knowledge develops, the more it transcends the capacities of the individual, the more it becomes a social fact and the more determining influence will it have on the individual. It means that proportionally the individual will contribute less and less to the body of knowledge and gain from it more and more. The technological dimension of the life of society will be an increasingly more important and more massive subject for study and reflection. Man's involvement in and with society will continue to grow in size and complexity. Even the social protester or the potential hermit will have to take, therefore, ever more elaborate steps to achieve his aim. He will be obliged to think more about practical problems than his analogue in the bygone era. Not only the development of society, the demographic situation, but also the continually progressing merger of science and technology will make it imperative in the future to view the development of personality in a more global and balanced perspective than before. Both the theoretical and practical aspects and aims of knowledge will have to be included in the realistic ideal of knowledge.

For Plato the most noble occupation was the contemplation of Ideas. The sage, in order to live up to this intellectual ideal had to dissociate his thought from the body and things material. Our civilization followed much more the demiurgic ideal than the Platonic one. Technological progress freed our muscles and satisfied most of our basic physical needs. For the first time in history large numbers of people live in a condition of plenty. And yet the material abundance did not do away with a feeling of want. The hunger of the well-fed indicates that the ideal of practical knowledge is not a satisfactory one. Does it mean that we should revert to the Platonic ideal? Not at all! It cannot be conceived as an aim for people in general. Unavoidably, it must remain, even theoretically, exclusive, accessible only to a chosen few, therefore socially unjust. It will not do for our times. The large spectrum of needs of the society and of the whole mankind have to be catered to and satisfied. The individual and his development cannot be conceived in abstraction from or in opposition to the social fact. If advanced society offers more to the individual than the primitive one, the individual is supposed in turn to give more to the society.

Neither the Platonic ideal of contemplative withdrawal, nor the pragmatic aim of mastery of the environment and production offers acceptable alternatives. Instead the ideal of personal development conceived as occurring in intimate union with the ambient world and combining the contemplative and pragmatical endeavors, presents a plausible alternative for a realistic explanation of knowledge and its role in human life. Seen in this perspective, knowledge appears as an essential, creative factor of the concrete existential situation of man. It has an individualistic and a social dimension and relevance. It is therefore a personal good and a commonweal, a subject of interest and of concern for both the individual and the society. The more there is knowledge, the more concern they should manifest for it.

Knowledge and Evolution

Knowledge affects not only the knower but also external reality with which the knower interacts. The more there is knowledge, and the more there are knowers the greater the impact of knowledge on the world. Knowledge therefore is a factor of evolution and ever more important one. Man not only partakes in the general evolution of the world but moreover introduces into the world an additional current of evolution of his own making. The question is whether this man-made evolution is in some intrinsic relation to the general evolution of nature, or whether it is autonomous and simply superimposed on the latter. In the first alternative, the principles of man-made evolution would have to be sought outside mankind, in nature, i.e., in the general laws of evolution. Men themselves then would have little if any influence on the direction or form of that evolution of which they are supposedly the authors. The ecological predicament would have to be viewed as an element of the natural, i.e., biological evolutionary process, a step toward a more perfect situation or mode of existence. Although such an eventuality may seem too revolting or improbable, it should not be dismissed out of hand. More cogent arguments would have to be invoked to reject it. Even if it could not be accepted in its totality, i.e., as an element of a purely biological evolutionary process, it is rather obvious that the ecological predicament is an all-important factor in the evolution of mankind.

In the second eventuality, that of absence of intrinsic relation between the biological and man-made evolution, the principal problem would be that of coexistence of these two evolutionary processes. Biological evolution, at least according to Darwin, is a non-goal-oriented process of descent with modifications. In broad terms, the mechanism producing modifications is meant to be supplied by the trial and error method and natural selection. Whether this explanatory model is adequate for explaining biological evolution is not a problem which we intend to discuss here in any great detail. What interests us is the usefulness of this model for the explanation of a man-made evolution. The crucial question in this respect is that of goal orientation. Whether finality exercises itself in nature and to what extent are debatable questions. It is clear, however, that thinking, considered as a series of particular acts, is a goal-oriented activity. Whether the fact of the finalistic nature of particular acts of knowledge can be generalized is another matter. It does not follow necessarily from the finalistic nature of particular acts of knowledge that the development of knowledge in its totality is also goal oriented. One could perhaps argue that the evolution of knowledge similarly to biological evolution is or at least has been a non-goal-oriented process resulting from the trial and error pattern and from the rule of the survival of the fittest.

If this were the case, it would be only natural to ask

whether the development of knowledge will remain such indefinitely, or whether the progress of knowledge has reached a stage where man can take the development in hand and transform it into a goal-oriented process. Of course, the transformation of the development of knowledge from a non-goal-oriented process into a goal-oriented evolution would be a turning point in the history of mankind. As it has been pointed out earlier in the paper, the need to plan becomes imperative. Whether the development of knowledge has been until now an integral part of biological evolution or not, the very self-interest of mankind requires that a degree of harmony between the two evolutions be established. The simple fact is that in order to exist and to progress man needs both: knowledge and nature. Unfortunately, as the ecological predicament indicates, the relation between knowledge and nature is not such that the interaction between these two factors is automatically harmonious and beneficial for mankind. There seems to be a strange contradiction involved in the relation between man and nature. Let us express the situation in the form of a law:

The more man knows the more mindful he must be about nature.

Intellection distinguishes man from nature. The more knowledge he has the more he lifts himself above nature, and the more the difference between them increases—accordingly decreases the rule of nature over man and augments the possibility of discord. Thus, the more knowledge there is, the greater becomes man's capacity to influence nature. The greater this capacity, the greater is the chance of disharmony between man and nature and the greater the need to establish such harmony. It seems, therefore, that henceforth the evolution of knowledge will have to become an ever more controlled process. It will have to change from a spontaneous series of unrelated or little related acts into a sequence of deliberate, organized and coherent activity.

Although knowledge creates the greater possibility of disrupting the balance of nature, it makes also possible the control of the evolution of knowledge so as to bring it more in line with the requirement of harmony in nature. The better nature is understood the more it can be respected. The role of the knowledge of nature in the direction of the evolution of knowledge may be expressed in the form of a law:

The possibility of direction of the evolution of knowledge is proportional to the knowledge of nature.

The relevance of the knowledge of nature for the direction of knowledge in general is an important fact which thus far does not seem to have been properly understood. As long as theoretical knowledge remained exclusively theoretical, i.e., without concrete consequences, the very problem of the direction of knowledge did not manifest itself nor could the importance of the knowledge of nature for the direction of knowledge be properly grasped. However, the more knowledge becomes power to do, the more it is necessary to direct the development of knowledge so as to enhance the chances of a harmonious coexistence between man and nature. This can be achieved to the extent to which nature is known. Of course the knowledge of nature in itself cannot suffice for the direction of the evolution of knowledge. Other kinds of knowledge are required as well, namely, an adequate understanding of man and of the nature of knowledge itself. On the other hand, one cannot expect to achieve any worthwhile results in the direction of knowledge without an adequate knowledge of nature. The present ecological predicament is an indication and a warning for us in this respect. Man and nature are the two poles of the dynamic relation which we call life. In the perspective of life neither of these two poles should be considered apart. Both man and knowledge have to be viewed realistically, in the light of their involvement with the ambient world and along the lines indicated earlier in this paper. The direction of knowledge cannot, therefore, be accomplished on the basis of any particular branch of knowledge. The more global the knowledge and the more comprehensive the point of view, the greater is the possibility of an adequate grasp of the problem and the better the chance of success in the steering of knowledge.

Harmonious coexistence with nature is a necessity. It must therefore be an aim for the direction of knowledge, but it cannot be the only aim or even the most important one. The laws of ecology must be respected in order to allow man to survive and to develop. This is a necessary condition of human evolution but not a sufficient one. Man's primary duty is not the well-being of nature but concern for his own development. Nature and the ambient world are component elements of the human condition, the objective pole of the dialogue which man leads throughout his life. They are not, however, the essence of man's existence, or the justification of his strivings. Every person is concerned, first of all, with himself, and secondly with society. He looks at other elements of the ambient world as being conditions and means of and for his own existence and that of the society. Because of knowledge, man can make these distinctions and assign priorities. Value judgments become therefore an important factor of evolution. It is through them that man can consciously try to influence evolution and direct it toward an aim. The capacity to distinguish, to evaluate, and to choose is what sets apart man from nature and what allows him to become the creative element of the evolution of man.

To do this job properly, man has to know not only the facts, but also the values. Quantitative knowledge of the type found in physical sciences is needed along with value oriented cognition of ethical and aesthetical kind. Measurement and taste, moral sensitivity, and metaphysical insight are necessary tools for guiding man in the direction of the human evolution. Value judgments and aesthetic and moral perspectives are properly human dimensions absent from nature, and meaningless outside man. They cannot be inscribed into nature, or identified with it. Rather they can take nature into their province inasmuch as nature can be seen through them, and subsumed under them. In this respect these human dimensions are an integrating factor allowing us to position nature vis-à-vis man, and vice versa. Man can view nature in relation to himself and assess himself in relation to nature. This is necessary for man to become fully conscious of himself and to evaluate himself. Thus, the evaluative knowledge of nature is an important element in the evolution of man. Nature is not only the supplier of the necessities of life but also an object of knowledge and a mirror for man. In all three capacities nature serves him and his evolution.

Even if man distinguishes himself from nature, he is not outside nature. His biological self is part and parcel of the order of nature, subject to its general laws. There is no reason to believe that the laws of biological evolution do not apply to him. If this is the case, the human individual is inscribed in the evolutionary sequence, a particular instance of a universal process. Inasmuch as knowledge is an act of the total, concrete individual, it too, by its organic foundations, partakes of biological evolution and is influenced by it. Of course, biological evolution is a very slow process, and its effects on knowledge are therefore difficult to perceive and may seem negligible when compared with the rate of knowledge's own evolution.

Being subject to two evolutions, humanity is in a state of constant change. Instead of looking at change and progress as a special instance, a rather unusual event in the history of mankind, it would be more correct and more revealing to consider change and progress as a normal, i.e., permanent, state of humanity. In this perspective, immobility and fixity appear as exceptional, abnormal states although they may still be commonplace. Consequently, the following general rule may be established:

> *It is impossible for mankind to persist indefinitely in any given state, i.e., on any particular level of development.*

In other words mankind has to evolve, it has to transcend the state in which it finds itself at any given moment. Of the two streams of evolution, biological and cognitional, the former may be supposed to progress at a more or less even, very slow pace while the latter is an ever more accelerating process, producing constantly greater impact on the human species. The take-off point in the history of the evolution of knowledge is a very recent one, namely, the birth of theoretical knowledge in ancient Greece around the seventh century B.C. The next turning point was the birth of modern science, the first major triumph of which was the heliocentric theory of Copernicus.

The impossibility of persisting in an unchanging state, due to biological evolution, existed already at the moment of the emergence of man. But in the beginnings and for a long time after, as long as rational activity was little developed, the necessity of progressing was mainly due to biological evolution. Accordingly, this process, as applied to mankind, was very slow. The higher the level of mental activity, the greater is the necessity to change and the faster is the rate at which a given level of development has to be transcended. The situation may be expressed by means of the following law:

> *The rate of obsolescence of forms of existence, i.e., of the necessity to transcend these forms, is proportional to the level of intellectual activity.*

This is as true of explanatory models in science as of political and social structures with the underlying customs and moral norms. Strange as it may sound, the more man progresses, the more he is forced to progress. The problem of the subordination of man to the idea of progress and to the process of progress has been mentioned earlier. The law stated above explains the roots of this problem, but offers no solution. It merely makes us aware of the fact that the problem becomes unavoidably an ever bigger one. The need to consider this problem appears as a new factor in the human evolution. It results in the complexifying of the situation of knowledge and opens a higher level of the problem of knowledge, the meta problem of ecology of knowledge.

Ecology and Man

Man's specific habitat is the sphere of reason and of the products of rational activity, namely, culture. The proper human attitude and condition of his behavior is that of awareness. The development of the individual and the progress of mankind consist in and are measured by the extension of the sphere of reason, the realm of culture, and of the scope and depth of his awareness and responsibility for his behavior. Self-consciousness is intimately linked to social awareness in a dialectical relationship. The development of one induces the growth of the other and vice versa. Man defines himself in relation to and in terms of what he considers the supreme Reality, the environment in general and his human environment in particular. The evolution of humanity is concomitant with the broadening of social consciousness from that of a nuclear family through identification with a band, a tribe, a nation, to a global identity with the whole of mankind. The growth of social awareness involves the heightening of the feeling of responsibility for the behavior of individuals and groups with regard to the human race. This in turn leads to the idea of, creative, active coexistence of varied elements of world population and to the accommodation of different components of human behavior so as to assure the most propitious development of man.

As it has been pointed out earlier, the progress of knowledge increases the possibility of disharmony between man and nature. The question is whether or not a satisfactory balance can be reestablished automatically, i.e., by the sheer interplay of forces. In other words, the problem before us is one of self-regulation: is the system, man-nature, a self-regulating one or not? In the first instance, eventually no special interference by man is necessary, a laissez-faire attitude will suffice. In the alternative, special steps have to be taken to assure the ecological equilibrium between man and nature and within man himself. In previous chapters it has been argued that planning of and for the development of knowledge is necessary. Before engaging in the discussion of the ecology of knowledge it is only fair to attempt a final justification of the whole idea.

The basic difference in the discussion of the problem of ecology outside man and in relation to man is this: in the case of nature there is no question of consciousness. When we discuss the ecological balance between two species, say mountain lions and deer, we notice the concomitant variations of mountain lion and deer populations within a given area. The obvious interdependence between the population fluctuations of the two species is a manifestation of a self-regulating system the result of which is ecological balance. We look at this balance as a good and as an aim of, and a justification for, the relationship between mountain lions and deer. We evaluate it statistically, i.e., quantitatively, unconcerned about any feelings or desires of those animals. We treat them in this process as things not individuals, totally subordinated to the idea of a universal order, both causally and in terms of value. In other words, we contrast universals against individual beings. This way of evaluating goes unopposed because those concerned cannot challenge it. Obviously, in the case of man the situation is and should be different.

It is possible and often fruitful to apply statistical methods to the study of man and of human situations. But these methods have their limits. They are adequate in the examination of quantitative aspects of reality, but inapplicable or misleading in others. Important as numbers are, they are not the whole story and they cannot tell us the whole story about man. In the discussion of human ecology the individual cannot with impunity be reduced to numbers. His feelings, wishes, desires, and his very personality must be considered and respected. Abstract notions or universals must not suppress persons. Biological equilibrium occurs at the expense of individual members of a species, or even at the expense of entire species. As long as the books of nature are balanced, the human onlooker is satisfied. It is rather obvious that this approach will not do in relation to the problem of the ecology of man. What matters in the first place in the realm of living beings outside man is the species to which the individual organism is subordinated. Although each organism has in its behavior and development its own finality, which is the organism's well-being, nevertheless, we assume that this finality is totally inscribed in and subordinated to the overall finality, if any, of nature. This is why we feel justified in our approach to the problem of biological ecology.

The human individual cannot be adequately reduced to his species. Nor can the finality of a person be fully subsumed under that of nature. Consequently, man is not a part of the system of nature in the same way as is an animal or a plant. To the extent to which man is a rational and conscious being, he is not a univocal part of nature. As much as he transcends nature, he is outside nature's self-regulating mechanism. The question therefore arises whether or not men form a self-regulating system of their own. This question is more complex than it seems at first glance. An affirmative answer to it leaves us in the dark as to the mechanism of self-regulation. Does it occur automatically, under the influences of some biological causation, or is it the result of conscious direction on the part of man? If men do not form a self-regulating system, the problem is even more confusing.

If man was not a conscious being, one could accept the hypothesis of automatic regulation along the lines of biological ecology. But man is not a brute, the individual counts, and his well-being matters for himself and for other humans. The more man becomes conscious and knowledgeable, the less he is willing to leave it to nature and accept its treatment of individuals as mere elements of a greater and more important whole. The more he knows, the more he is able and willing to take the direction of his life in his own hands. Man's rational nature lifts him above nonrational nature and forces him to plan. Thus the problem of human ecology in general, and of ecology of knowledge in particular, is an unavoidable consequence of man's nature and of the evolution of mankind. It is also a turning point in the history of humanity.

The problem of ecology is one of achieving a dynamic relation of equilibrium among different elements and at the same time assuring them the best conditions of development. Two questions have to be elucidated concerning the above statement. In the first place, it is necessary to find out what kind of elements are involved. They are the subject of the ecological relation and, therefore, the component parts of it. From what has been said earlier in this chapter, it is rather obvious that the subjects of the ecological relation are different in nonhuman ecology from those in the ecology of man. In the first case the essential units involved are species and genera, i.e., classes of beings, while in the latter case they are individuals. Human ecology is about individuals and groups of individuals. Even if the whole human species is involved it should be viewed as a society of persons and not as an integral whole composed of subordinated parts.

In the second place, it is important to distinguish between the changing elements of the ecological relation and the unchanging or the much more slowly changing ones. Again one has to differentiate between human and nonhuman ecology. In the latter case the relatively permanent parts of the ecosystem are the classes of beings, be it species or genera. Instead, the individual organisms are the changing, dispensable factors. In the human ecology, because of the importance of individuals, the ecosystem must be viewed differently. As it has been pointed out earlier, the human evolution is a much faster process than the biological one. Consequently, the significant time intervals in the former process are proportionally shorter. In biological evolution the significant time periods are determined by the emergence of new morphological forms. The time of biological evolution is that of the sequence of morphological changes. It is measured in millions of years.

The distinction between the more permanent and the changing factors in the human ecology has to be found within the human species itself, or, to be more precise, in the history of mankind. The first obvious distinction is that found between man's biological nature and the modes of his existence, i.e., culture. Another one is a distinction within culture itself, namely, between the patterns of culture and behavior of individuals. The time of human evolution is therefore a complex one. On the one level, it is the measure of the changes in the patterns of culture; on the other, it expresses the pace of creative activity within a given culture. The more vigorous is this activity, the faster flows the time of human evolution.

One of the important consequences of evolution in general is the acceleration of time from cosmic time, through geological and biological, to the human time. The second consequence is the change of attitude from passive attitude of inanimate matter, through active but nonconscious participation of living creatures, to the conscious participation and, at least, partial mastery of time by man. The progress of humanity can be measured by the evolution of man's attitude toward time. The growing awareness of time results in more efficient use of it and in the development of the technique of planning. It is through planning that the future becomes increasingly a dimension of man's life, and the center of his interests.

The ecological predicament is an important moment in the history of man's attitude toward time. Ecology obliges man to look ahead and to evaluate long-range effects of man's activity. In other words, ecological thinking is intimately related to long-range planning. It is essentially a future oriented attitude. Time is of its essence. It views life in space and in time as a four-dimensional continuum. It involves man's thinking with time more thoroughly than it has ever done before. This situation may be formalized by means of the following law:

> *The greater the ecological predicament, the greater is the need to plan; and the more future oriented becomes man's attitude.*

The ecological problem is proportional to the size of the population and the amount of human activity. This means that the larger the world population and the sum total of man's activity, the more man must become future oriented and time conscious. Ecological thinking not only involves time but it also effects a change in the relation to time. The future becomes more intimately bound to the present. Instead of being seen as the result of a simple and unavoidable succession of moments, the future appears increasingly as an effect of the present situation and especially as the locus of the consequences of acts posited in the present moment. Accordingly, man feels increasingly more responsible for the future and envisions the future as a projection of himself. The ecological point of view adds a strong moral dimension to the thinking about the future. The perspective of the future forces man to consider himself as a part of a bigger whole. It is in the future that the consequences of his attitude toward society and toward nature as well as the consequences of the total behavior of society will appear. Thus, the concern with the future leads to the growth of man's awareness of himself as a part of a universal order. One of the consequences of the ecological predicament will no doubt be a renewed concern for ethics and a broadening of the scope of ethical problems. Besides the traditional questions of personal and social ethics, the ethical problems of mankind as a whole will have to be considered. Moreover, the relation of the human species to nature will have to be rediscussed, as well as man's attitude to time.

Greater awareness of time is a progress, but it does complicate man's life. The growth of awareness goes hand in hand with the acceleration of human time. The reason for it lies in the relation between time and activity, especially rational creative activity. Time is the number of motion, i.e., the measured measure of change. The greater the creative activity, the more change is produced and the faster the pace of change. Consequently, the faster the human evolution, the faster flows its time. Let us express this situation in the form of a law:

The rate of the flow of human time is proportional to the pace of the human evolution and to the degree of the awareness of time.

There is, however, a price which man has to pay for the heightened awareness of time; namely, the faster flows the time, the faster, and more creative, is human activity and the greater, and potentially more dangerous, are its consequences. The more important the consequences are, the greater is the need to evaluate them, and the less time there is to do it. The time to ponder and the time to act seem to be at odds. If mankind is to have a future, they will have to become adjusted to each other.

Ecology and Culture

Human evolution is, as was pointed out earlier, primarily the result of the development of knowledge. In the history of knowledge, the two crucial events were: the emergence of rational, theoretical thought and the invention of the scientific method. The process leading from pretheoretical knowledge to theoretical knowledge and the modern science is a history of progress of objectivization of knowledge. The objectivization is achieved at the expense of the totality of the *Weltanschauung*. The price of greater rationality of knowledge is paid in terms of increased selectivity of the methods used in the study of reality and in proportional limitation of the subject matter. At the root of the Western tradition of thinking lies the distinction between the subject and the object in the act of knowledge. This distinction leads to the opposition of the subjective and objective elements of knowledge and to progressive expurgation of the subjective components from the scientific method and world view.

An unforeseen consequence of the objectivization of knowledge was dehumanization of cognition with the resulting imbalance between science and nature and the alienation of man. The progress of science not only has augmented the possibility of conflict between man and nature, but also that between man and knowledge. Theoretical knowledge in general and science in particular do not express the total man. Science is supposed to be principally a method, an intellectual activity consisting in the application of this method, and a body of knowledge resulting from it. In fact, science is bound with a definite *Weltanschauung* and involves a hierarchy of values which is the backbone of the scientific world view. Its principal values are those of objectivity, verifiability, and communicability. These values are achieved through the systematic use of measurement and the identification of the subject matter with measurable properties, i.e., with number-measures. The consequence of this is the quantification of knowledge and the exclusion, as much as possible, of nonquantitative aspects of reality from the scientific language and explanatory constructs. Unavoidably, the scientific *Weltanschauung* inherits from science the onesidedness of its approach. It cannot, therefore, fulfill satisfactorily the role of the intellectual environment necessary for the harmonious development of man.

Man is the most complex of all creatures. His behavior is complex and the effects of his activities are proportionally complex. This relation can be expressed as follows:

> *The complexity of the effects of the behavior is proportional to that of the behavior which in turn is proportional to the complexity of the nature of the agent.*

Man's complexity has direct effect on the nature of the human environment. Environment depends on the subject just as the subject depends on the environment. There exists between the living being and the environment a bilateral relation of mutual interdependence. This relation consists in a dialectical interchange between the two poles of the relation, each one in turn becoming cause and effect. If man is what he eats, his food is increasingly the product of his labor and ingenuity. He eats what he is. The more man acts on the environment and shapes it according to his will, the more he interacts with it. The complexity of the environment is proportional to the complexity of the subject of which it is the environment. Human environment is the most complex of all.

The complexity of the human environment is not always adequately understood. Its complexity is a source of unintelligibility, an obstacle to understanding. This is why there is a constant temptation to satisfy the desire for intelligibility by explaining culture in an oversimplified way. It goes hand in hand with the inclination to form an oversimplified image of reality. The theoretical consequences of this tendency may be highly satisfactory to the intellect but are at the same time detrimental to the whole man. Man lives by his knowledge. The more knowledge there is, the more it influences man's development. Its role vis-à-vis that of nature increases. The process of development becomes less and less spontaneous and natural and more and more rational and directed. Consequently, it becomes increasingly more important that the image of reality in general and the human sphere in particular be as complete and adequate as possible. If knowledge forms a partial picture of man's environment, his behavior will be thereby affected, and his development compromised.

It is essential that man forms an adequate picture of culture and evolves a satisfactory culture. To be adequate, culture must express the total man in his individual and social dimension. All human endeavors and aspirations must find their expression in culture, together with fields of interests and spheres of activity. Culture is essentially a social construct, a framework for and a manifestation of the individual as an individual, as a social being, and of the society as a collective whole. Culture's primary aim is to facilitate human development so as to further the humanization of man. This is at the same time the raison d'être of culture. Its aim can be achieved only if culture allows for a complete expansion of human capacities, and for exchanges with all aspects of reality and not only with those which satisfy the conditions of measurability laid down by science. Culture, therefore, has to contain not only the objective, verifiable, and communicable elements of knowledge, but also the subjective ones. The advantage of possessing knowledge of an impersonal type, such as science, excludes neither the desire nor the need of personal, subjective knowledge. Consequently, the human environment must not only be the result of the desire for and the product of objective knowledge, but it must also express the need of the whole set of human values, both objective and subjective.

The man-made environment of culture is superimposed on the natural, physical, and biological environment. These two kinds of environment do not form a homogeneous whole. One of the important differences between culture and nature is the rate of their evolution, i.e., the difference in the pace of their respective times which they engender. This difference has far-reaching consequences for man. Natural evolution is very slow. Nature, therefore, can be taken as a static factor in human life. This is why, by and large, it is a stabilizing influence in the life of men. As long as a culture is largely subordinated to nature (let us call it nature-bound—as in the case of "primitive" cultures), the culture remains stable, i.e., static. The "emancipation" of man from the bonds of nature is a process reflected in the development of culture, the result of which is the transformation of culture from being nature-bound to being man-centered. Culture becomes dynamic, creative, and therefore changing. From past-oriented, it changes into present and future-oriented. Together with the "liberation" of man and of culture, the disharmony increases between culture and nature and the possibility of opposition between man and culture.

Contrary to nature, culture is a factor of human life of variable magnitude. Being the product of man's creativity, it is an ever-growing component of the human environment and exercises an increasingly important influence on man. Whatever its relation to nature, it has a radically different principle from nature. It is not only man-made, but also man-oriented. It has for its proximate principle not the universal laws of nature, but man's needs, desires, and his rational activity. Man is the measure of culture. But man is an individual person as well as a member of society. Moreover, he is a member of the human race. Culture, as the product of man, developed for the use by man, reflects his twofold character. It is shaped by two different factors. One factor is the set of universal, i.e., specific, characteristics of human nature present in all the individuals of the human species. The other factor is the individuality of the human agent. The tension between these two factors is reflected in the culture.

Individuality allows for the richness of variations of personal characteristics. It is also the source of the indeterminateness and endless variety of possibility of creative acts proper to man. Creativity plays a decisive role in the process of culture. It accounts for the flowering of culture, the richness of cultural forms, as well as the unpredictability of cultural developments and discords with the culture. Those facts have to be taken into account in the discussion of ecology of knowledge. Any planning of knowledge, in order to be conducive to a harmonious development of man, ought to respect the universal, specific factors in and of culture as well as the individualistic ones. Moreover, the development of knowledge must satisfy the need of completeness of the cultural environment, so that this environment is an adequate expression of the whole man. Only then will be served the ideal of personal development occurring in intimate union with the ambient world and combining the contemplative and pragmatical endeavors.

Ecology of Knowledge

The intellectual predicament of our times is essentially a crisis of traditional rationality. This rationality was the product of science and of the scientific *Weltanschauung,* i.e., of scientism. That it is in a state of crisis is a result of the consistency of the scientific point of view and, odd as it may sound, a result of the vigor and the success of scientific method. It would, however, be wrong to blame science for our present-day problems. Science, in its own field, is valid and adequate. Moreover, scientific method is a self-correcting process, subordinated to its subject matter as to the supreme measure of its value. If the traditional rationality is in a state of crisis, it is because its premises were too narrow to serve as basis for the construction of a human environment satisfying the higher needs of man. It would be easy to say that the solution lies in the broadening of premises. Even if this statement is basically true, it does not mean that it can be accepted as a satisfactory solution of the problem.

Specialization is a necessary condition of the progress of knowledge. If science is to remain progressive and successful it has to retain its restrictive view of reality. In other words, science is and will remain a branch of human knowledge, no matter how important this branch is now or will become in the future. The scientific point of view expresses one of the attitudes of man toward reality, but it does not express nor will it ever express the totality of man's relations with the world or with himself. This is a fundamental fact which lies at the basis of present-day difficulties. It is in the light of this fact that the problem of ecology of knowledge must be discussed. Man is a complex, multifaceted, multilayered being. Knowledge is one of his modes of behavior, even if it is the most perfect one. But through knowledge man forms an image of reality. This image expresses his understanding of the world and of himself. His behavior then is guided by this image, and his life is affected by the *Weltanschauung*. This is why it is so important that the image be an adequate one. The difficulty of the situation lies in this: although knowledge is not the totality of his behavior or of his attitudes toward reality, it has to express this totality in one way or other.

The problem facing intellectuals today is to devise ways and means of cultivating and developing rational behavior in such a way as to ensure the balanced and harmonious growth of culture. The problem is at the same time personal and social. It has national and international dimensions; it is a crucial problem fraught with danger. And yet it has to be faced squarely because it is unavoidable. Of course it remains to be proven that one can cultivate culture and reap a worthy harvest. To assure the success of such a venture it is necessary to adopt the most complete view of man possible as the point of departure. This creates the greatest difficulty and is the main bone of contention. Once the basic premises are formulated, the conclusions should not be too difficult to draw. Whether they can be implemented is another problem which can be discussed only theoretically on paper.

Ecology is a complex idea. It is a holistic point of view and an order-seeking approach. It is a method for dealing with dynamic wholes composed of many diverse parts from the perspective of time. It teaches us to look for and to evaluate the long-range effects of the behavior of the parts on the whole and on themselves. In contradistinction to the classical scientific point of view, ecology does not try to be value-free cognition. On the contrary, values and valuation are central to the ecological perspective. In biological ecology the principal value is the survival of the species or of a broader class of creatures. As it has been explained earlier, the subjects of human ecology are not classes of beings but individuals as individuals or as groups of individuals, i.e., societies. The principal value in human ecology is then the good of the individual as such and as a member of society. This good cannot be simply identified with the survival value. It must be viewed in the perspective of individual and social development and associated with the realization of human potencies. From the point of view of human ecology, the history of mankind appears as a good-oriented, value-motivated evolutionary process, and so does the life of the individual man.

A basic problem for the ecology of knowledge is the evaluation of long-range effects of changes in the intellectual environment. These changes are the result of intellectual activity. Consequently, the ecology of knowledge must deal with this activity as affecting the intellectual environment, and attempt to foresee its consequences. That this is a serious problem and a major task is due to the fact that man can choose to concentrate on one or on a few branches of knowledge and develop them more than others. The result of it is an unbalanced culture. In fact, this is the usual situation; it would be difficult to find a culture resulting from a harmonious development of all the principal fields of intellectual endeavors. For instance, contemporary Western culture in general and its North American version in particular, is heavily weighted in favor of practical knowledge, economic expansion, a dynamic, conquering attitude toward reality, short-range planning, and immediate success.

If we aim at the development of the best possible, i.e., most complete and richest human environment, then it becomes necessary to study the whole field of rational activity with the view of elucidating the nature of the various branches of knowledge. Their role and their place in the total knowledge should be analyzed and evaluated as well as the relations which exist or should exist between these various branches. Without attempting to discuss this problem in detail, let us mention that it is an extremely complex one. Moreover, its complexity increases proportionally to the constant proliferation of branches of knowledge and the resulting complexifying of the structure of knowledge. Traditionally, until recent times, philosophers considered it their duty to formulate classifications of knowledge as part of their philosophical systems. Nowadays the techniques of classification have been developed to such a degree that they constitute a specialized branch of knowledge. They were, however, elaborated mainly for technical purposes or for the need of librarians. Their primary aim is facility of retrieval, not a philosophical insight or an overall synthesis. Consequently, they cannot, of themselves, provide us with a philosophically satisfactory classification of knowledge. Present-day philosophers seem to be little, if at all, preoccupied with this problem. Not only do we lack an adequate classification of contemporary knowledge, but, moreover, it is not at all clear whether such classification can be elaborated. If this were really so, then the consequences would be very grievous indeed. The much glorified progress of knowledge would be a new version of the building of the Tower of Babel. This is a problem which must figure very prominently in the study of the ecology of knowledge.

Beyond the problem of the nature and the role of the branches of knowledge lies the question of the relation between knowledge in general and the material and human world. In other words, this is the problem of the role of knowledge and its value in human life. The very fact that we are convinced about the value of our knowledge does not dispense us from a careful consideration of this problem, nor from a study of the overall consequences of knowledge for the individual and for society. At the present moment there does not seem to exist a satisfactory study of the global effect of cognition. A further problem is the question of the relation between past, present, and future knowledge. This question increases in importance proportionally to the rate of development of knowledge and to its consequence, namely, the devaluation of the past. The more and the faster knowledge develops, the more different it becomes from past knowledge, and the less dependent on past knowledge it seems to be. The past, and together with it tradition, loses increasingly its value as the teacher and storehouse of experience. From past-oriented, knowledge and civilization become future-oriented. The relation between the intellectual present and the intellectual past can be expressed as follows:

The value of the past is inversely proportional to the creative, intellectual capacities of the present.

The more we can advance knowledge, the less the past is needed for that purpose, and the less it counts. The generation gap is an illustration of this situation. Another, even more serious problem resulting from the ever more rapid progress of knowledge is the obsolescence of language. The progress of knowledge results not only in the increase of the amount of knowledge but also in the progressive transformation of our understanding of facts known previously. Language can follow the quantitative increase of knowledge through the increase of the vocabulary. But the change of understanding of those things and problems for which words already exist cannot be reflected by a change of words. This is why language is at the same time a help and a hindrance in furthering knowledge. The faster the progress of knowledge, the more serious is the difficulty created by the persistence of words.

Underlying all these problems there is the question of the relation between knowledge and its justification; that is the problem of the aim, purpose, or reason of knowledge. This question is at the end of the list of questions which we have mentioned, but it is also at its beginning. It is the problem of the final cause. The final cause is first in the order of intention but last in the order of execution. It is therefore fitting that it should be considered in the preamble to the discussion of the ecology of knowledge and at the end. The question of the final cause is a patently obscure and controversial issue. Science has learned to avoid it carefully. In fact the very development of modern science depended, among others, on the elimination of the question of finality from the field of its investigations. One of the most important tasks of the ecological approach is to reintroduce this question into the mainstream of our thought.

Ecology of Knowledge and Creativity

The ecological approach to knowledge raises a host of questions, some of which are new and suggest novel aspects of the problem of knowledge, while others put into question beliefs which until now seemed to be well grounded. The present intellectual and biological predicament obliges us to ask ourselves whether or not our traditional attitude toward knowledge is a correct one, and whether it should be changed. In the light of what has been said so far, it is plausible to suggest that our attitude toward knowledge has been the result of insufficient knowledge! Surprising as it may sound at the first instance, this statement is simply the logical consequence of the fact stressed throughout this paper, namely, that knowledge, as an object of knowledge, is obscure and difficult to know. We do not have an exhaustive knowledge of knowledge. Since our knowledge of knowledge is only a partial one, our attitude toward knowledge resulting from this state of affairs reflects this incompleteness. There is, therefore, room for improvement. It seems to me that the ecological point of view contributes to the attitude toward knowledge an important element and is therefore a step forward, without, of course, pretending to be the final one.

As usual, progress is not an unmixed blessing. The ecological point of view leads quite naturally to the idea of planning. In the perspective of planning, the crucial problem is that of the possibility of harmonious coexistence of planning and creativity. Experience of totalitarian regimes teaches us that all too easily planning becomes synonymous with suppression of liberty resulting unavoidably in the limitation of creativity. Such a situation should be, of course, the opposite of the aim of the ecological approach. However, the requirements of biological ecology limit the freedom of our use of the material environment. The question before us is whether the ecology of knowledge will put a limit on our use of the intellectual environment. The problem in other words can be put thus: Is the ecological perspective not an intellectual straitjacket?

It is obvious that the requirements of ecological balance in the order of nature force us to limit the spontaneous, uncritical deployment of energy. They oblige us to alter the Baconian ideal of man-master-of-nature. Ecology reintroduces a new version of the ancient notion of natural law and teaches us to see man and nature as complementary parts of a bigger whole. These parts are so ordered to each other that they must play their roles in the drama of coexistence, the main rule being mutual respect. The difference between ecology and the original version of natural law consists in this: according to the latter, man was subordinated to the rule of nature which appeared as a formidable, transcendental force which man could do nothing else but obey. From the ecological point of view man is not a slave of nature but at least an equal partner if not more. He must respect nature not for the reason that he can do nothing about it but precisely for the opposite reason, namely, he can so easily alter and destroy it. The limits which ecology imposes on man's activity are not those resulting from impotency, but from too much brute power. They are introduced with the view, not of preserving a status quo, but of assuring suitable conditions for the progress of mankind. This means, among other things, that the aim of ecology is not to leave nature unchanged, a well-nigh impossible condition, but to arrange various factors so as to allow for a harmonious evolution respecting the laws and the balance of nature. The aim of ecology is not simply limitation but sublimation. The chief merit of the ecological approach lies in this: it makes us seek solutions to existing problems by enlarging the scope of our thinking about them, and by inducing us to transcend previous beliefs and positions so as to form a more complete and more balanced view of the complex whole which is our presence in this world.

Thus, it is plausible to believe that ecology of knowledge need not be an intellectual straitjacket. Being a holistic approach, ecology necessarily implies general notions and an overall synthesis. General notions and a synthesis may either be the result of the search for a common denominator with the ensuing leveling effect, or they may be produced by an approach respecting individual and specific differences and seeking unification through higher causes. In the latter case a hierarchy of values is necessarily introduced. It is this alternative that is proper to ecological thinking. Its principal value consists in this: that it safeguards the possibility of choice and offers a degree of freedom for the mind. In this way it ensures the basic condition for creativity and for its effect, the progress of mankind.

The above characteristic of ecological approach is of crucial importance especially in view of the progress of knowledge and the conditions which the progress imposes on us. The progress of knowledge, as it has been pointed out earlier, ceased to be the product of solitary efforts of a small number of individuals. Nowadays, the knowledge industry not only involves many individuals, but, and more importantly, it requires team effort, i.e., cooperation. Science especially sets aims before us which cannot be achieved by an individual. Cooperation implies planning. The relation between them can be expressed as follows:

> *The amount of planning required is proportional to the amount of cooperation needed.*

The larger the project is, the more thorough planning is necessary, i.e., the more long-range and global it must be. But planning is a mixed blessing, because it implies centralization of decision-making powers and the limitation of the initiative and the role of team members. Usually the following relation is true:

> *The role and importance of and the possibility of deploying initiative by a team member are inversely proportional to the size of the project and the amount of planning needed.*

However, planning not only presupposes joining of forces, but also implies a certain common denominator. The more you plan, the more similarity you assume, and the more difficult it becomes to make a place for the unknown, the unforeseeable, the creative. No matter how future-oriented the planning may be, you plan on the basis of the knowledge you have, and thus you predetermine the results. It is not, therefore, astonishing that great ideas are always personal. The personal component of knowledge and its role in the progress of knowledge must never be lost in ecological thinking. Of course, this complicates the task, but is nevertheless a necessary condition of the value of this approach. In the situation of global planning, unfortunately, it is the role of the individual that is most easily lost from sight. The greatest intellectual danger of the future is that of leveling, homogenization, regimentation; in other words, the danger of subordination of the individual to the rule of the common denominator. The loss of the creative impetus experienced by the scientific community in America in recent years seems to be an illustration of this danger.

To assure the progress of knowledge and of mankind a modus vivendi must be found between the values and the needs of the individual and those of the social group. This may well be the most difficult task for human ecology and especially for the ecology of knowledge. No fast and ready solution for this problem is available. Living beings can be viewed as self-regulating systems, the realm of living beings outside man as a self-regulating system of a higher order. The question is whether knowledge as well is a self-regulating system. In order to make any headway in the discussion of this problem, it is necessary, first of all, to define the meaning of the word knowledge. The question may have different answers depending on whether by knowledge we mean personal knowledge or the body of knowledge in general. In the first case there is a well-defined subject, namely, the individual knower. In the second case, the subject lacks such specificity. This is an important fact because knowledge is meaningful only in relation to a knower. One can, of course, say that the subject of knowledge in general is mankind. But obviously knowledge is not evenly distributed within the human species. Nor is this collective subject univocally responsible for the growth of knowledge. And yet, notwithstanding the difficulty of defining the subject of knowledge in general, it is the total body of knowledge per se that interests us here.

The principal question is whether or not the activity of knowledge in its global dimension forms an ecological system. If it does, then it means that it is a self-regulating system, i.e., it has a built-in mechanism assuring balance between the individualistic and group or social components. If it is not an ecosystem, it lacks this mechanism. In the latter case, the ecological approach would have to supply this mechanism. Perhaps this is the most important reason for the existence of this approach, and the greatest challenge at the same time. The principal difficulty seems to be not the coordination of efforts but the preservation of the conditions for it and the stimulation of personal creativity. It is in the light of this problem that the role of an adequate hierarchy of values in ecological thinking acquires its full meaning. It should never be forgotten that the intellectual situation of humanity is a highly complex one, being the result of various factors. In order to be appraised more or less adequately, it should be viewed from the angle of the progress of knowledge as well as that of the perspective of coexistence of various forms and levels of knowledge for the good of the individual, of a given culture, and of mankind in general. These confusingly different elements can be arranged into a coherent and meaningful pattern only with the use of an adequate hierarchy of values.

Ecology is the science of totality. The ecological point of view is characterized by a holistic approach. In this perspective singular elements of the totality, even if they are persons, tend to be viewed as parts of and subordinated to the whole. This is why, in the discussion of the ecology of knowledge, it is important to take into account the fact that the ecological point of view cannot be considered as replacing the personalistic perspective. Man in himself is not just the sum total of relations to other men and to the biological and physical environment, or a composite of characteristics resulting from these relations. Man cannot be adequately reduced to anything other than himself. His finality is not that of the environment. His life is a process of self-realization. That this process occurs in a concrete situation determined by a set of circumstances, ranging from physical to intellectual, is a well-known fact. But the principal characteristic of rational existence is the ability to constitute oneself in relation to these circumstances, thus creating an autonomous being.

Our analysis would be incomplete without at least a short discussion of the nature and role of the creative act. Its principal characteristic besides that of being personal is its unpredictability. Its unpredictability is in direct relation to the novelty which it produces. What is important in the case of intellectual creation, especially in the field of theoretical knowledge, is the influence of this act on the established ideas, i.e., on the paradigm. Thus, the creative act appears as the most important safeguard against the leveling effect of the common denominator, unavoidable in all planning as well as in all paradigms. Consequently, the creative capacities of man appear not only as the most precious power that man possesses, but also as the supreme defense against depersonalization, besides being the principal cause of progress. The question is whether or not it is possible to plan for creativity. The answer to this question is unclear to say the least. Perhaps our ignorance in this respect is a condition of the freedom of thought which is a necessary state of mind for creative thinking. Nevertheless, it is possible to indicate some conditions which any eventual planning for creativity must respect. The planning must not try to tie down the mind to the prevailing paradigms and classifications. It must accept as a premise the fact that thinking is not a fully autonomous activity. It is not a self-contained and self-sufficient order of reality. It is not a world in itself or for itself, but an expression of and existing for something which is at the same time more fundamental and more complex, namely, the concrete human individual. This means that the thinking process in general, and its creative element in particular, is not, nor can ever be, produced merely by thinking. It is, as has been explained earlier, the result of the total immersion of man in reality and therefore subject not only to the laws of reason but also to the broader laws of nature and of natural evolution.

Like many other things in human life, knowledge is an ambivalent factor. It is an offensive and a defensive weapon. It serves to explore and contemplate the universe and to master it. It can be creative and it can be destructive. As Abraham H. Maslow has pointed out, cognizing is to a great extent an anxiety reducing function of man. It transforms the unknown, the threatening into the known, the familiar. When unchecked, the desire to control reality develops into a pathological need. The result of such pathology of knowledge is to cut off the individual from the very sources of creative development and to enclose him in the limits of his inflated and threatened ego. If man wants to progress, he must never allow the desire to control reality overwhelm his thinking powers. This is a very serious question, indeed, for the problem of ecology of knowledge. All too easily the concern for ecology may

become an expression of a pathological attitude. It may very well
be that the present situation of mankind is a condition inducing
strong pathological desires. Greatest care and caution must be
exercised in order not to succumb to the temptation to control
everything within and around ourselves. This is why planning of
and for knowledge must make place for the element over which it
has, at the best, only limited power, namely, the dynamism and
harmony of nature.

COMMENTARY

J. M. BOCHEŃSKI

PROFESSOR Wojciechowski's concern was not with scientific breakthroughs, such as the rise of the heliocentric theory, due
to Copernicus. Rather, what he had in mind were the truly
"major" changes, not so much in the content of the theories as in
the basic attitude of the scientists. There have been two such
changes or "breakthroughs" in recent history: the one effected by
Francis Bacon and the other, "ecological," taking place now.
Consequently, we have three major periods in that respect: the
"Aristotelian," the "Baconian," and the "ecological" periods. The
basic views which characterized each may be described as follows.

The Aristotelian view, which dominated the Western world
up to and including Copernicus, has three assumptions: (1) there
are essences, clear-cut contents in nature, and, consequently, a
sharp opposition between nature and man, between contemplation
and action, between man and society, and so on; (2) nature is, at
least primarily, the "lawgiver"; (3) the aim of science is knowledge
of natural laws—contemplation.

The Baconian revolution consists in rejecting the last two
assumptions. External nature is no longer seen as a lawgiver, but
rather as a stuff to be manipulated by reason. And the aim of
science is not knowledge alone, but rather power: it is essentially
practical knowledge. However, insofar as the first Aristotelian
tenet is concerned, Bacon is still Aristotelian: with the old master,

he opposes man to nature, contemplation to action; in other words, he upholds such clear-cut boundaries.

With the rise of the "ecological" period, it is this very view which is rejected. Professor Wojciechowski argues that the opposition between subject and object, man and nature, individual and society, contemplative and practical knowledge vanishes.

The question arises: What philosophy constitutes the general background for the said "ecological" view of science? The answer is, so it seems, unequivocal: Hegelianism. It was rather surprising to note that Wojciechowski makes references against Hegel since most of his remarks indicated an assumption of Hegelian views: *das Wahre ist das Ganze.* No clear-cut essences. No fixed "nature," *physis.* No basic difference between contemplation and action. No clear boundaries between man and nature, subject and object. That is all Hegel—and Hegel at his best.

As a point of fact, there are many features of modern science which converge toward supporting the Hegelian view of the world and of knowledge. Thus, one can understand why philosophy is so dependent on Hegel today. According to Whitehead, the best characterization of Western philosophy is to say that it consists in footnotes to Plato. By way of paraphrasing Whitehead, one could say that our philosophy is a series of (usually bad) footnotes to Hegel. From Dewey and Heidegger down to Kuhn, everything seems to be more or less inspired by him.

Is this to say that we are bound to conceive the world and science in this way? It does not seem so. Surely, it is no easy task to challenge the all-pervasive philosophy and to search for an alternative view. Yet perhaps the philosopher's task consists precisely in the search for something other than the opinions of the hoi polloi. Moreover, it appears that this Hegelianism entails some undesirable consequences. Here are some sketchy remarks on but two of them.

One of these undesirable consequences, albeit a rather secondary one, is the rejection of the difference between theoretical and practical knowledge. There is no particularly new reason to reject this difference. True, we now recognize (to a greater extent than did the ancients) that theoretical knowledge usually brings, in the long run, very practical results. Yet this is basically an old medieval insight: *intellectus speculativus extensione fit practicus.* The denial of this difference simply does not fit the facts: there is such a thing as the study of an object which cannot as yet be manipulated (as, for instance, the study of quasars) or which, even if it could be, is not studied for the sake of any action (as when the art historian studies a medieval cathedral). Furthermore, many people are interested in such purely contemplative knowledge—witness the amount of information about quasars and medieval cathedrals found even in the most popular almanacs.

In order to justify this negation, it is said that both sorts of knowledge in fact serve the development of man. But this supposes a fallacious inference: because an action A results in B, B is the aim of A. The same fallacy was committed by the Hedonists, who argued that because achieving the goal of an action brings pleasure, pleasure is the aim of every action—which is clearly not the case.

So why this denial? Because of the very Hegelian rejection of any differentiations of aspects in man: between his appetite for knowledge and for a hot dog, between his spiritual and bodily needs, between the cognitive and emotional functions—all of which are seen as mere "dialectical moments" of one undifferentiated Whole. But one wonders why such a view must be accepted instead of adopting a little more sophistication in the matter, i.e., a differentiated philosophy of man.

The other and more important of the undesirable consequences of Hegelianism denies the very existence of "determined nature," of a *physis.* Nature is rather seen as a flexible stuff given to be manipulated by man's technology. It has to be studied in order to be used. Used for what? Obviously, for the service of man.

However, what this doctrine fails to take into account is that man himself is a part of nature. From the Aristotelian point of view, man has a *physis*—as do all other natural entities. To achieve the relative fullness of that human nature in each of us is, according to the old Master, the aim of every human activity, science included. Incidentally, the assertion that thinkers of the Aristotelian period (Copernicus among them) were practicing science for science's sake is false. In their estimation, science was in the service of man; but they saw this service as consisting in an attempt to realize their ideal of man's nature.

But if we think in Hegelian categories, there is no such thing as human nature. Man himself becomes a stuff to be manipulated. Note that our remarks do not concern the *fact* that science gives us the technical possibility of transforming man (e.g., the production of monsters). The problem at hand is a *moral* one: we are transforming nature and ourselves to what end? There is no human nature and, consequently, no fixed ideal toward which we can tend. What is the aim of science? What should we do? What do we want?

Of course, many will say that our immediate aim is the liberation of mankind from the many evils it now suffers. This provides our science with its moral justification. But the power of science is now so great that it is perhaps no fantasy to claim that we shall reach that goal very soon. It is also so great that we can do almost everything with ourselves and with nature. But *what* we should do is not clear and cannot be clear from this point of view

With his usual penetration, Hegel himself saw the consequences. For him, nothing finite can be the aim of human activity, including science. The only possible aim that remains is the endless change, the infinite process of dialectics—God. In other words, science is to become more and more powerful; it is to effect more and more profound changes in everything, including man, merely for the sake of its own expansion and for the sake of change itself.

Consequently, the objection may be stated in this form: if one admits Hegel's premises, then he is logically bound to follow Hegel to the bitter end, to the affirmation that the finite does not exist really—that man, humanity, its history, are only "dialectical moments" of the Infinite, deprived of any moral dignity in themselves. But then we have no right to talk about man and his development. We do not know what we are talking about. If, on the contrary, we adhere to the premise (common to Aristotle and Bacon) that science exists for man's sake, then it is necessary to reject these Hegelian premises. While one can admit many things the ecology of science teaches us, it seems necessary that our view of science be formulated within the framework of a completely different philosophy—a philosophy which would allow for a concept, whatever it might be, of a human nature, of a *physis*.

XI Panel: Has Science Any Future?

HAS THE SCIENTIST ANY
FUTURE IN THE BRAVE NEW WORLD?

C. A. HOOKER

I

"What Is Truth?"

An epistemological justification for a political act.

Pontius Pilate, ca. A.D. 30.

". . . to drop one's work in order to test some of Velikovsky's claims . . . would appear a culpable waste of time, expense and effort."

A political justification for an epistemological act.

M. Polyani, 1967.

II

THE panel discussion title "Has Science Any Future?" may suggest an exercise in science fiction—what will science look like, what sort of scientific theories and accomplishments can we expect, in the future? To this we all know full well there are no answers that do not read as science fiction, or at any rate none which ought not to be thus read (thereby demonstrating the importance of that literary category).

In this discussion, however, I wish to address myself to another question: What of the future for *scientists?* There is no science without scientists. Science is inescapably an anthropomorphic social, moral, political . . . cultural . . . epistemological enterprise. Thus, there are some inevitable repercussions for the role of the scientist that follow from the successful prosecution of science. All of these impinge upon the epistemological function of science to some extent, all of them impinge upon the development of human society, morals, culture. (The terms social, moral, cultural, etc., are all sufficiently vague and generally overlapping to be pretty obscure when used together. For the sake of clarity I shall hereafter keep to the following: "culture" I shall take to designate all aspects of the humanly constructed world—science, technology, social institutions, morals and mores, arts (Kultur); and as this list suggests I will distinguish within a culture the components listed, with "social institutions" itself being a compound of legal, political, economic, religious institutions, and so forth.) The question is, can we say anything useful about the future in this respect? Of course this too must have the status of fiction, but the issues are too important to leave alone, and the need for prior reflection on possible cultural, but especially social, responses to future circumstances too great, not to venture some speculations.

In current usage the terms "science," "scientific," have multiple meanings or dimensions to their meaning, which range from "The objective pursuit of truth" through "A method for solving problems," and "The organized research and development of technology" to "The prevailing cultural character (a *Weltanschauung*)."

Science has become all of these things. But it is of some importance to observe that it has acquired this protean guise only recently. Prior to the late industrial revolution scientists were not essentially involved in the large-scale technological organization of society. It is essentially only after the 1870s—when universities (or their equivalents) developed science departments on a significant scale and began systematically producing chemists, engineers, and so forth and when, following the success of chemical research for the German dye industry, industry began to realize the potential— that the union of science and technology took shape. This union dominates the institutionalization of science today. (I am considering all the institutional aspects of science, in particular its societal recognition through government support, and not just the research and development departments of industry.) It may come as something of a shock to the typically educated twentieth-century mind to learn that technology and, if we define that term narrowly, crafts before it developed largely independently of anything that could be called scientific research; that consciously applied scientific theory has dominated technological development only in the last fifty years. Although prior to this time individual men combined activity in "pure science" and technology, the former was an intellectual discipline with something of the same divorced-from-practical-cases flavor as pure mathematics and philosophy. Where there was interaction, it was almost always the case that a prior technological invention challenged the production of a scientific explanation. Only recently has scientific theory become sophisticated enough and technology sufficiently well organized and sophisticated for "applied science" to represent a significant category of activity.

This new role for science has effectively wiped out the category of pure science, at least with respect to its intellectual-moral overtones. It is no longer the case that scientists' choices of even *what kinds* of theories to pursue are without cultural ramifications, let alone the specific theories they devise. In every case now, knowing that the chain extending from the pencil and blackboard to the attitudes, habits and practices of society, though long and tortuous, is complete, the scientist must inevitably assess the practical impact of his ideas. The very success of science has removed any vestiges remaining to the distinction between *scientist* and *ideologist*. Not this idea, which is rather obvious on reflection anyway, but some of its possible consequences in the actual world we inhabit is the real theme of my discussion.

First, let me pause to remark that it has taken philosophers of science a long time to awaken to this change, and sociologists, with their often harsh positivist dogmatism, are still seemingly unwilling to even discuss such "unempirical ideas" at all. (This latter is excellent grist for my mill, as we shall see.)

Early in this century, at the very time when the socio-cultural role of science was changing decisively, the image of science was dominated, ironically, by the Positivist school for whom science was a relentless logical machine that ate basic sensory reports and spewed out more and more of the truth. This view has its extreme difficulties even as an idealized rational reconstruction of science, but here it is its tacit but firm exclusion from the image of science of anything but what was postulated to be objectively true that concerns us. Science was not a human discipline, except accidentally. There is nothing wrong in seeking for a rational reconstruction of scientific decision making, and parts of the positivist legacy will surely prove useful. But when we press it on the issue of fostering insight into the actual course of science on this planet among human beings, we discover a tacit ideology which lies beneath it: the science of science should also be positivist. But this leads inevitably to a neglect of the specifically human in our understanding of science since most of it cannot be captured (yet, at any rate) by the combination of basic sensory report plus logical machinery; one ignores decisions in favor of inductions and deductions (if any are identifiable) and one ignores intuitions, metaphysics, and the remainder of the human condition (morals, religious beliefs, psychological motivations) virtually altogether. And indeed, if the foregoing were not indictment enough, the approach fails on its own terms since, even behavioristically construed, specific scientific theories (e.g., of psychology) are needed to give an account of how science is possible for human beings, thus completing an epistemological circle in what was designed to be a rigorously, logically (hence epistemologically), linear space (the space of scientific ascent to understanding).

In recent times such writers as Kuhn and Feyerabend, and before them Popper, have effectively opened the way toward a new sociology of scientific knowledge. (The "old" theory was developed by Mannheim, Weber, and others around the turn of the century and has since fallen into a partially deserved neglect.) They have done this by concentrating on scientists rather than science, on beliefs rather than theories, on decisions rather than calculations. By recalling attention to an obvious, though neglected, truth—that at the center of the scientific matrix stands the scientist, who intuits, hopes, guesses, gambles, needs money, love, recognition; who reasons, argues, persuades, alienates. . . ; who decides—they and the like-spirited have opened the way for a larger understanding of the scientific process. Of course the old epistemic issues remain: Can we draw a distinction between *scientific* acceptability and reasonable acceptance by (i.e., relative to) *scientist* A? What, if anything, is to count as scientific knowledge, as opposed to justified belief for A? What distinguishes Scientific Method from any other way of arriving at human decisions? How is science possible for human beings? And it is not certain that the new reconstruction will retain a place for an analogue of the old positivist image of science as something different from the sum total of the beliefs of people (others, as well as scientists), coldly aloof, superior to the poor human intellect, and ivory hard from the rigorous manner of its arriving. Nor is it certain that the epistemological circle, and a corresponding one in the account of rationality, will not cause as much discomfort here—though dialectic ascent and gestalt shift are processes that defy induction and it is no accident that they figure prominently in the new writings. One thing is certain, the intellectual life will be more exciting and wide ranging than it has been for many years. Now we return again to the deep philosophical questions concerning science but in an infinitely larger context. From a more practical historical perspective I believe that this shift, predictable in its time lag from relevant socio-cultural shifts, has arrived none too soon.

It is hardly possible to consider all of the socio-cultural forces now acting to change the social role of the scientist at the present time. Instead, I shall consider the conjunction of just two of them which I consider to be especially significant: the emergence of scientific management systems in the context of the convergence of environmental disruption and the emergence of a scientific establishment.

III

What is the historical significance of the environmental disruption in our time? Getting straight on this issue is difficult because of the myths which surround the subject. I shall summarize here an argument presented in detail elsewhere. It has two premises and a two-part conclusion.

Premise I. Modern science (in very nearly every field, but especially ecology) forces on us the conclusion that life is an integrated (coherent) single system; each part is intimately linked to every other part.

This view is now a commonplace within ecology itself; ecology is possible only because living systems do in fact form integrated systems of this sort. Roughly speaking, ecological systems are energy flow systems, they represent the possible energy paths from incoming solar radiation to outgoing thermal radiation. The matter flows in ecological systems are driven by the energy flow. Matter moves so as to obey, roughly, two fundamental principles: (1) the movement of matter is cyclical; (2) the quantities of matter and energy moved are compatible with the preservation of life. Indeed, we can view evolution as the trial-and-error filling of possible energy pathways, with a species living off every sufficiently large energy transaction in the pathway.

This picture holds equally well of the physical analysis of the human industrial system which is also an energy-matter flow system. There are some striking differences, however, from the natural system. (1) The major sources of energy and matter at the present time are both terrestrial and distinctly finite. (2) Matter flow is not cyclical, rather it is linear, from resource exploitation through manufacture and use to dump. (3) Neither in quantity nor in type is the matter-energy flow so constricted as not to damage existing life forms. For these reasons the industrial system gives rise to resource shortages and environmental disruption, e.g., pollution.

But superimposed on the physical matter-energy industrial system is an economic system. The economic system is a money-flow decision system. Crudely speaking, the money flows in the opposite direction to the energy-matter flow; crudely, because this neglects the "nonenergy" transactions such as many services, investment, gambling, and so forth. In effect, the economic system acts as a feedback control system on the physical industrial system, assuming that economic motives are operative. Finally, superimposed on this are the legal, political, and remaining cultural systems of the society which act as controls on the economic system. (The cultural system I take to provide economic motives as well as the moral constraints on the possible means of realizing them and the pattern of normal decision making, or rationality, that prevails among the members of a society. I remind the reader that the legal and political systems are essentially subsystems of constraints within the cultural system, but I distinguish them because they are so distinguished, however artificially, by society.)

The whole edifice then presents the aspect of a complex

unity, a single system whose components wonderfully (and fearfully) interlock. No change at any level, no decision, is made but its repercussions are felt throughout the system. This is true equally of changes at the most abstract level—consider the consequences for practical policy of the differing cultural attitudes taken toward nature by the Western European Christian and some subgroups of the North American Indian—to those at the most concrete level (e.g., the decision to dump sewage in waterways rather than use it for fertilizer).

I have already pointed out that our industrial system has an inappropriate physical design for stable functioning on this planet. This instability is further enhanced by the explosive growth of population, forcing an even faster rate of industrial growth of the poorly designed sort (because it is cheaper) and elsewhere placing the natural ecology under great pressure in the desperate effort to provide food and living space. In consequence we are now witnessing an unprecedented convergence of environmental disruption in all countries of the globe.

Moreover, the foregoing has made it clear, I hope, that the effective control of these problems will require effective control of the entire human system: economic, legal, political . . . cultural. The system shows a high degree of coherence; you cannot simply step in at one level and alter things a little without altering a society at every level. In particular, you cannot hope for the global control that effective environmental management demands—control of size and distribution of population, of industrial type, physical distribution and practices (to control types of energy and material products use and the flow paths of these)—without demanding control of a society's economic and political processes and hence without gaining control of its cultural system, or by seizing dictatorial power (which may now yield this control ultimately anyway, or not succeed); the control of the legal system follows. Whether or not this authority to control is freely vouchsafed by the people in their own collective interest or lies in some degree along the continuum of selfishness and deceit toward absolute dictatorship, alters not the point that that control must effectively exist for environmental stability to be obtained. (Alas, God in his wisdom has not granted man predestination to a preordained harmony in this respect.) It is the facing of that historical fact that constitutes the first real environmental challenge to this and succeeding generations.

Let me rephrase what has happened historically in this way. Until relatively recently the environment for man was a natural system whose functioning he was largely powerless to alter. In these circumstances survival meant adapting oneself to this system; knowledge, learning its ways; planning, the ability to predict its course. Power consisted largely in the ability to design one's course of action so as to "have nature on your side" (whether it be a favorable wind in sea battle, or knowing when to plant crops) and perhaps by also very marginally manipulating the natural system. It is natural that the historical struggle to control nature should have centered on the obtaining of increasing manipulative leverage through the increasing injection of energy; this has been the essence of the technological process until very very recently. But so successful has this process been that man's environment is now increasingly a human artifact, not a natural system (e.g., cities, roads, weather control, irrigation). Artifacts, however, are *designed*—one does not simply adapt to them, one also adapts them to our needs. The ancient relation has been reversed. But coherent design requires control. Adaptability of the environment can be satisfactory only if it is thoroughly controlled. Indeed our own interventionist philosophy of control is now teaching us that very lesson; the more one intervenes in a coherent system the more chaotic and unpredictable it becomes (because of the complexity of the repercussions), unless sufficient information concerning the systemic function precedes each intervention and unless information increases much more rapidly than does our ability to intervene (because repercussions in general multiply much faster than the energy level of the perturbing intervention). Power requires control and control requires information. But much more than information, control of the global human artifact our world is fast becoming requires a global culture that is capable of utilizing the information to exercise total control, i.e., it requires a totally ordered culture, a society saturated with the decision process, a globally organized polity. It is the construction of a version of this society that is satisfactory to the majority of humankind that is the second great environmental challenge of the age.

Premise II. Because of the particular nature of the industrial revolution in Europe—it took a bourgeois capitalist form—our cultural, political, and legal systems are designed in a mold diametrically opposed to the achieving of the cooperative, coherent, community planning which environmental harmony and stability requires.

It is not possible to recapitulate here even in brief the history of the industrial revolution. Suffice it to say that it emerged first in England against a background of monopolistic control of the cloth trade with continental Europe and was developed in its initial stages by the great inventor—industrial entrepreneurs (e.g., Arkwright) who collectively developed such power that they brought about mass immigrations of population and the founding of large cities (e.g., Manchester). In these cities were found some of the most degrading working conditions yet known to man, but sufficient for a peasant disenfranchised from his land (sometimes forcibly).

In a relatively short time those in ownership of the industrial system wielded immense economic and, through it, social and political power. It was obviously of great importance to create circumstances in society which favored the continued evolution of industrial society. This goal was made considerably easier to attain by the effective disintegration of the medieval world view a century or more before (thanks to a train of cultural reactions, one branch of which has Copernicus in a prominent place; other branches concern the rise of the merchant class and the exploration of the "new world") and the rise of a rationalistic, science-oriented culture in the intervening century or so.

It was of crucial importance to the men who were recasting the structure of English society that they have freedom of action. This meant first and foremost freedom of economic action from either moral criticism by the church (in Medieval Catholic Europe economics was a subbranch of practical action, in turn governed by Christian morality, in turn determined by a Christian view of man and his society) or from too restrictive control by the government. These ends were achieved through the development of the view that economics was a value-free science, hence its laws as unalterable as those of Newtonian mechanics and moral criticism of them as irrelevant, attempts at government control as futile, as it would be of Newtonian mechanics; and the further development of the centrality of the notions of private property with the central right being the ability to do what one wished with it; and the free market which is its natural public correlative. The new order soon found learned apologists, e.g., the economist Adam Smith, the philosopher John Locke.

Today we inherit the legacy of those formative attitudes. Our dominant cultural orientation is individualistic, we define ourselves and our social position largely through the property we own, we have an image of democracy dominated by the notion of individual economic freedom. We persist in our view that economic structures are value free and even cling to the centrality of the notion of the free market as the central economic regulator. More subtlely we accept, without seeing the bias, the rejection, at least until very recently, of so-called externalities from economic decision making (all the unimportant things such as peace, quiet, convenience, good community relations, and so forth), and the assessment of the economic viability company by company (rather than in terms of the total impact of a product on the design of human life). We accept a legal tradition whose laws are dominated by private ownership rights and which is hopelessly weak on community rights (local or national) and an "engineering" view of the law and lawyers as merely blind (mechanical) servants of whoever is employing them. We accept the yawning gap between our industrial structure and our governmental structure, the resistance to rational communal industrial planning, and the ad hoc planning that results (or the secretive control of government by industry in its own interests).

Conclusion I. To bring about a society that is capable of rational environmental planning and long-term environmental stability will require a revolution at every level of our human system: cultural, political, legal, economic, and industrial.

Conclusion II. The kind of revolution required, if it is to be compatible with human happiness, must be one dominated by communal values, new forms of communal decision making and wide-sweeping global communal controls (the former are necessary if the latter are not to create a dictatorship); it must be one that accepts the description of our historical heritage as a critique of it; and one that aims at the environmental stability criteria tacit in the analysis under premise I.

To bring about this kind of a spiritual-cultural-institutional revolution constitutes the second real environmental challenge facing this and future generations.

IV

There is also at present a convergence of "scientific" management theories, i.e., theories that will permit us to predict and more finely control our various subsystems. Most of these developments are only in their early stages, yet there is already cause for hope, and fear.

First, there is general systems theory, or systems analysis. As the name implies, this approach attempts to exhibit the world as a collection of systems (perhaps interacting), each system being a collection of components connected by a set of laws determining their mutual states or, at any rate, assigning probabilities to state relations. General systems theory has some better known, successful applications in particular fields, some of which preceded the realization that they were instances of a general theory. Electrical circuit theory is one, machine analysis another. These days the approach finds application in the design of weapons systems, industrial production systems, and in the analysis of local and global ecological and meteorological systems. Recently it has come to be increasingly applied to social-cultural systems and human behavior within them. Two of its important social applications are to the theory of communications systems and to functioning hierarchies, for example, corporations. Cybernetics, the theory of intelligent systems, is another highly developed, supporting branch; in the case of human beings it needs to be supplemented with decision theory, compare below, because we do not yet have sufficiently detailed neurophysiology. As I conceive it, general systems theory includes the field of operations research, though many will be more familiar with this label; however, operations research, in some applications, may include a component of decision theory, compare below. General systems theory has even developed a philosophical dimension.

Another general theory of this sort is decision theory. This is the formal theory of decision making. It has origins in game theory and economic theory. It is an attempt to analyze rigorously the relations between decisions among alternatives and the "payoffs" (utilities) associated with those alternatives on the basis of various assumptions about how people will act, or ought to act (to be rational). One of its important results is the demonstration of a wide range of circumstances under which there is conflict between the interests of an individual *qua* individual and the interests of the community of which he is a member. A well-known and especially simple case of this has been dubbed the Prisoner's Dilemma. It turns out that many environmental situations, especially those involving communal resource allocations, satisfy the conditions of this kind of circumstance. The theory has been used with devastating effect to analyze cold war strategies and is now applied in many social contexts, such as trade union bargaining.

When we combine these two fields we obtain the means to treat legal, political, and perhaps even entire cultural systems as decision-flow systems subject to constraints (e.g., laws in the case of the legal system) as well as whatever principles express the principle of rational function in relation to the relevant utilities. This task is in its infancy, but it certainly holds out promise for a large-scale, social-action control theory given, especially for the cultural system, that we had enough psychological data on the participants to say what their preferences, rationality principles, and perceived constraints were.

Decision theory also lies at the roots of microeconomics where it has some of its most sophisticated applications so far. It is used to show that in a free market with an effectively infinite number of participants the equilibrium positions for mutual maximization of marginal utilities in transactions contracts to a single point. Ultimately, then, we are able to offer an in-principle analysis of the economic system using decision theory coupled to economic principles.

If we combine systems analysis with economic theory then we have a powerful tool for the analysis of the industrial system. Sophisticated theories of transportation systems, location of industry, optimum physical scales of industries and so on have already been produced. And this means too is increasingly turning toward social application; to the planning of population density distribution, purchasing patterns, and so on. On the other hand it matters just how the two are combined.

Systems designs are not themselves value free. The design assumptions always reflect basic, and too often tacit, value assumptions. If one carries out an operations research in a manufacturing industry to define work units, it is usual to assume that one is aiming at the specification of a task that is independent of any specific human characteristics, so that persons don't have to be individually fitted to jobs. A prize case is assembly line analysis. This assumption contains colossal subassumptions about the efficiency of such a view of industry, about the human desirability of such task definition, and about the relation of a worker to his work. In fact we know that it produces alienation, boredom, and breakdown, and so ultimately inefficiency. Vickers points out that cost-effectiveness techniques offered to the United States government build in colossal restrictions on the definition of policy goals. Rothenburg has recently severely criticized the widely acclaimed systems analysis of cities developed by Jay Forrester precisely for the value assumptions, as well as other assumptions, that the model tacitly makes.

Among the increasingly important applications of systems-oriented approaches are the developments in learning systems theory, advertising and market research theory, and the development of social indicators. All three disciplines are concerned with the systematic assessment of aspects of the states of persons and at least the former two are concerned with the alteration of the psychological states of human beings, respectively of skills and preferences. Learning systems such as learning laboratories and self-education program design are being increasingly employed in the educational system. They have found their profoundest expression in the "accountability-contract" approach to teachers and students now rising rapidly in importance in the United States. The activities of advertising research people are well known in a popular way; less well understood is the degree of sophistication and psycho-sociological depth marketing theory has achieved. Social indicators are a new phenomenon, arising largely out of a dissatisfaction with traditional constructs (e.g., Gross National Product) as a perceptive guide to the real quality of life enjoyed by a society. Clearly their proponents intend them to play an important role in the determination of future social planning. Once again, however, extreme caution is called for. Accountability criteria, with their emphasis on skills and performance, seem to have lost sight of all the really important educational objectives—creativity, confidence, reasonableness, criticalness, perceptiveness, responsiveness, a sense of perspective—that can only be brought about in intimate, sensitive student-teacher contact. They have concentrated instead on those "mechanical" skills which alone the economic "megamachine" recognizes and needs. Similarly, the tendency is for social indicators to also restrict themselves to "megamachine" categories and to ignore, as thoroughly as does G.N.P., all the truly important, but intangible, features of human life which constitute its quality—harmony, peace, richness of cultural experience, sense of belonging, and lack of personal stress. The morality of advertising theory neither needs nor deserves further comment.

However, decision theory is a powerful tool—as yet insufficiently exploited—for revealing the basic value assumptions behind both economic theory and actual economic behavior. (Basic value assumptions appear not only in the latter in an obvious way through the rank ordering of a person's preferences or utilities, but they also appear even through the former because of the way the decision function resolves community versus individual interests, for example.) Given that we are irrationally coy in our society about talking about the value-laden foundations of economic theory and practice this treatment provides a refreshing critique.

Systems analysis plus decision theory, supplemented by specific constraints and theories (e.g., laws: physical, psychological, legal; principles: economic, moral) bids fair (and a little foul) to provide us with a global theory of the human condition; all the levels of interacting systems can be taken into account. And we have seen that in many areas the development of these techniques into detailed theories and so into finely honed management techniques is well advanced. The remaining areas, and the global integration, will surely follow. Moreover, the reader will have observed that many of the management techniques now in hand are of just the sort which we saw the integrated global control of our environment to require; the remaining management theories and skills will surely follow The question is: By whose values are the designs to be made, who is to control these, and how?

V

Anyone who is to effectively control these management processes must understand the theories involved: he must belong to the scientific elite or he must be certain of the loyalty of those who do belong to this elite. In either case the theoretically real power lies, as Galbraith has stressed, with the scientific elite. Spaceship earth, like any other spaceship, has a survival potential no higher than that provided by the management skills of the scientific elite who design and operate it.

Moreover, this role is thrust upon the scientific elite just by the very success of science. And so it has always been. No one who has ever done anything that bids fair to enhance the survival value of human beings has ever been left alone by his society unless he bequeath his knowledge or skills to another so that he personally becomes irrelevant.

It is here that the recent emergence of a scientific establishment begins to assume its correct proportions, namely, as a historical event of the first magnitude.

This "event," you may recall, occurred sometime after 1870; before 1935 the scientific establishment had assumed recognizable proportions. World War II secured the political importance of scientists and the cold war became a largely technological battle. Scientists have probably always regarded themselves as an elite of sorts (back in the "good old days" when essentially only monks or private individuals carried on science as an intelligent man's hobby, scientists would have belonged to existing elites, defined in *other* terms anyway), but the term "scientific *establishment*" connotes more than this. It connotes the recognition, mutually by society and the collection of scientists, of the importance of the latter's role in shaping the society and determining its fate—a recognition measured in terms of state support of scientists.

Part of this phenomenon of recognition is a numerical increase in the ranks of the scientists, with many from the socially nonelite now viewing the role of scientist as highly socially desirable. This adds personal motive to the social process of creating a true *establishment;* the sort of technocratic elite that Chairman Mao evidently feared. Of course the creation of an establishment is not a mutual love affair between everyone—society is replete with diverse groups all competing for power and prestige—but it does betoken a genuine recognition, however grudging, of the relative importance of the role played. Finally, it does not follow that the major path of scientific influence will always be directly with the supporting government of the time. At least at the present time this does not appear to be true of any major power, despite the ring of science advisers and governmental support of science. What the situation will be when the hard necessity of socially global environmental planning is upon us is more difficult to say. There will be some increase in direct influence certainly. At present the influence is exerted more directly through innovation in industry, through approaches to public services—medicine, transportation, pollution control, urban design, education—and through the military. These changes have an influence on government and society generally only in the most fantastically complex, indirect ways, though this does not prevent the influence from being pervasive and profound.

VI

When we speak of an establishment we intend, by innuendo, to indicate the ability and the will to fight for survival, power, and prestige. This characteristic of establishments is well borne out by historical experience. The science establishment of our day seems no different.

The classical case of the reaction of the establishment to scientific challenge is that of Immanuel Velikovsky. In the late 1940s and the 1950s Velikovsky wrote a number of articles, some letters to important men of the scientific establishment, and a book outlining certain views of the solar system and the earth's history (geological and human-historical) to which he had been led by his own largely historical research. These views not only sharply contradicted the particular predictions of the established (=accepted) view, they ran entirely counter to the philosophical presuppositions of the established outlook. Whereas the established theory of the solar system is based on Newtonian mechanics, refined where necessary by relativistic physics, and hence is dominated by the gravitational force, and is uniformitarian (like a clockwork), Velikovsky proposed a *recent* (!) *catastrophic* history of the solar system in which *electromagnetic* and possibly molecular and nuclear forces played a very significant role. In similar fashion Velikovsky offered a very different account of the earth's geological history, offering unorthodox explanations which contained the same features of catastrophe and actions of unorthodox physical forces. But it is in the area of ancient history, where his investigations began, that his views lead to the most thorough-going changes. A major revision of dating, and of dating methods, is called for, myths are taken seriously as recountings of historical events, and so on. All of this makes Velikovsky's theory extremely controversial. Yet it is a theory which has a high degree of coherence from its own point of view. (Though broadly educated, Velikovsky is no physicist, geologist, or paleontologist; he arrived at his views in these areas by deducing the logical consequences of the historical core of his theory.) All of this should have made the theory an exciting one, one that would be eagerly put to the test. It is hardly every day that one meets with a radical challenge to many established theories, a challenge that is based on careful research and leads to straightforwardly testable consequences.

The scientific establishment, however, reacted to the challenge in quite another fashion; they reacted with rejection, hostility, suppression, hysteria, propaganda, and lies. At first it is perhaps hard to believe this (though it is excellently documented); the scientific community behaved like an ideological party, their conduct governed by a politician's morality. But after the shock has passed and the evidence remains, one begins to take stock of the lessons to be learned concerning the character of the contemporary scientific community.

The inescapable conclusion is that the scientific *community* behaved as a scientific *establishment,* i.e., it reacted *politically* to an *epistemological* challenge—this is the distinctive trait of any ideology. Velikovsky challenged the prevailing philosophy of the time thus attacking the dignity (prestige) of those scientists who espoused it, and of those who acquiesced unthinkingly in it; worse, as an outsider challenging central dogma, Velikovsky indirectly attacked the relevance of expertism, specialist training, and departmental empires. In short, Velikovsky was perceived as attacking the very roots of educational dogma and the basis of all power (economic and political) within academe and its extensions into the larger society. Finally, scientists were so committed to huge programs of expenditure in various fields (e.g., astronomy, moon exploration, accelerators, where millions of dollars are involved) that the possibility that their experiments were designed on false assumptions was virtually unthinkable because of the appalling consequences for them should that belief ever become widespread. These were Velikovsky's chief sins.

Let us pause to reflect a moment on the situation that our ancestor, whose natal day we here do celebrate, faced when he reproposed a heliocentric system of astronomy. There was a scientific establishment of that time, actually, precisely because virtually all of the science that was carried out was done by people either belonging to the first of the two great establishments, the Roman Catholic Church, or patronized by the other, the High Nobility. Moreover, there was a scientific orthodoxy of quite formidable proportions precisely because it flowed from a general religious view of the world—the newly emerged Aristotelian-Christian synthesis brought about most notably by St. Thomas Aquinas. We are familiar with the Cosmology of the view, a "womb" world, the earth at the center of a collection of concentric spheres, God's heaven beyond the last, star-carrying sphere. And this was not merely a physical hierarchy, for the descent from God's realm to the center was a moral descent and a descent in the order of creation—from God and his peculiar perfection to the perfection of the heavens and so down below the lunar sphere into the sphere of corruption at whose center stood the corrupt earth, the creatures living thereon, and even the inanimate objects, holding similar places in a strict order of creation. Therein, of course, lay the great power of the edifice, everything flowed downward from the religious center—from theology, morals; from morals, social order, public conduct, epistemological propriety (intellectual conduct); from public conduct, economics (as we would call it), and so on—the social hierarchy, even the cosmological structure itself, repeated the moral structure and in doing so glorified God.

To introduce into this harmonious and great edifice the heliocentric cosmology was to attack it at its *religious* heart, that of course was the real seat of Holy Roman Catholic Church opposition—earlier to the Pythagorean, at the time to Copernican, and later to Galilean cosmology, not to mention Giordano Bruno and a millenia of other lost souls. Gone is the great cosmological structure, gone the place for God "on the map," and gone with them the material foundation for the moral hierarchy, for speaking of God as a person, in something like a recognizable sense of that term, for viewing man as having a special, or perhaps even a significant, place in the universe. This is not to say that there were not formidable scientific difficulties in the road of making the heliocentric view plausible. There were. Indeed, we now must admit that these difficulties were much more immense and pervasive than had hitherto been imagined (the necessity of a new principle of planetary kinematics, a new theory of dynamics, and of optics, etc.), but that is not the point. The Church responded *ideologically* to suppress serious *public* consideration of such alternatives, notwithstanding its learned scientists who actively discussed the scientific issues in Latin. The Church responded as the cultural establishment it was.

In Copernicus we had a man trained from within the establishment but standing outside it in respect of the theory he offers. He was, so it seems, cowed by it; but then the cost of open-mouthed courage was high (compare Giordano Bruno). (I pass over the suggestion that it also seems he set unorthodoxy in train as a genuine, if misguided, attempt to return to a purer orthodoxy!) In Velikovsky we also have a man trained from within the establishment but standing now outside it in respect of the theories he propounds. Velikovsky is also more open-mouthed, in keeping with the times. They don't burn one for unorthodoxy loudly proclaimed today, though the asylum is a distinct possibility (compare Szasz), and positions, opportunities, and friends can still be lost. But now see the establishment! Such a reversal of roles! *Here is indeed an anti-Copernican revolution.* Ideology in the guise of epistemology.

(Aside. Before I turn to assessing the historical significance of this reversal let me say a few ameliorating words about the complexities of life. I am well aware, as I said, that not everyone in the Holy Roman Catholic Church was unsympathetic to inquiry. That would be a ludicrous position to hold given the intellectual changes that took place at the time—Nicholas of Cusa challenged the womb world as well as Giordano Bruno, various "schools" worried the problems of Aristotelian dynamics—and I am aware that Galileo, and to a lesser extent Copernicus, were propagandists [Galileo was both inveterate and superb at this], as evidently they had to be since their view was both insufficiently developed and founded at the time to argue on its own merits and it clashed ideologically with the established view, nor could it as a matter of practice receive the necessary foundation without the ideological clash being won. Likewise, I am aware that not every scientist in the contemporary establishment concurred in the treatment meted out to Velikovsky. Although very few protested publicly, I am sure that many were too preoccupied with their own research to hear of radical alternatives at all, and many more who did were doubtless too busy to respond, just as there must have been many monks in these same categories in the time of Copernicus and Galileo. It must also be admitted that the researches which Velikovsky's view calls on to be made are often costly and time consuming and that from within the contemporary circle of scientific ideas such expenditure might hardly seem justified; so too for the situation of Copernicus and Galileo among their contemporaries. I shall return to this shortly. And yet the historical outcome was clear enough: Copernicus and Galileo belong to the path of scientific progress, and the opposing schoolmen did not.)

The scientific community responded to Velikovsky as an establishment. But a feature of this situation I find more frightening than any of the overt acts of hostility (which can at least be identified and rejected as stemming from unworthy, unscientific motives) is the neglect, the effective rejection, sanctioned by a cold-blooded appraisal of the situation. The question is, how likely is it that Velikovsky's view is true, or will lead to fruitful insights? The answer to this question helps determine the extent to which it will be pursued. But there are other factors involved in this latter decision: How costly will it be to pursue the implications, experimentally or theoretically? What is the likelihood a funding agency can be persuaded to support the research? Only all of these factors taken together can begin to provide an adequate basis for the explanation of the decisions of scientists. But the answer to these questions determine the interest the scientific community shows in a theory, and indeed there is a feedback mechanism here which tends to generate an upward or downward spiral of interest. In Velikowsky's case the theory is "way-out" radical, many of the experiments costly, most perceived as unmotivated distractions from the business of the day, the likelihood of serious funding and payoff small. When a National Aeronautics and Space Administration (NASA) team invests millions planning an incredibly intricate scientific trip to the moon on the accepted assumption of a moon long dead, you do not easily rearrange it, especially not on an amateur's insistence that its surface was recently molten and highly magnetized. This is good sound business procedure—cost-benefit analysis would support it, I'm sure, if the prior probabilities could only be known—and plain common sense. Now the bigger the programs, the more dollars, careers, and societal consequences hanging on them, the more convincing this argument is going to become. *Science, through its very success and societal entrenchment and by applying only the accepted canons of reasonable procedure, becomes less and less tolerant of unorthodoxy.* It is apparently the rational thing to do!

May I repeat this. As the costs of a mistaken research program mount, the willingness to take epistemological risk decreases, properly so, on any reasonable cost-benefit analysis. Thus, by being ignored, neglected, not to mention rejected with more or less hostility, alternatives which are risky are effectively suppressed, and increasingly *less* risky alternatives become unacceptably costly as the costs mount. "Let it first prove itself, then we will follow." Scientists tend instead to study "safe" applications of already established theory. But the costs are largely determined by the institutionalization of science, i.e., by factors which are external to the standard picture of science. The massive institutionalization of scientific control in our society has already increased the costs of error enormously. The traditional philosopher of science, incidentally, might have appreciated this epistemological feedback from social conditions more easily if he had been less keen to reject the scientist and the process of creating science from his account.

And there is another effect of institutionalization: Scientific activity becomes inextricably bound up with the life of the society; scientists become willy-nilly politically important possessions; and it is correlatively demanded that they behave in like fashion. The course of scientific research is tied to other societal objectives, security, G.N.P., control; the various establishments are *made* to merge.

Possibly the most famous recent case of this phenomenon is the development of the atomic bomb and the later trial of Dr. Robert Oppenheimer. His trial is matched in history by that of an early successor of Copernicus—Galilei Galileo. Though there are dissimilarities, the role of establishments seems to be strikingly similar. (The two have been compared by de Santillana.)

We all know of other obvious instances where this merging has occurred, namely, the use of science in warfare. But it is also occurring in the increasing resort which business has to scientists to defend their practices and create new ones which they find attractive, and it is perhaps on the threshold of occurring in genetic manipulation and personnel selection and training. The future holds the prospect of a great increase in this area as management and management skills increase; scientific effort and the "national interest" become increasingly indistinguishable.

VII

Let us put the various fragments together. For example, let us imagine ourselves in a society in which science is much more highly socially entrenched, in fact a society in which all of those societally global scientific management techniques, which we earlier saw the control of ourselves vis-à-vis the environment to require, have been instituted. And not only the environment is producing this requirement, but it is also one of many factors pushing in this direction. Present criteria of efficiency and control in business and government, criteria borne out of systems management, and communicational structure are other examples. It is just likely to be the historically most important catalyzing factor, in my view. What we have imagined is a society with the most powerful establishment ever created—knowledge *is* power in 1984.

Why does the prospect strike us with terror (some of us anyway), why is it not rather the final arrival of the rational, intelligent utopia? Why do we read what might be the dawning of the golden age of science—of true knowledge and the enlightenment of wisdom, that great day whose glory has been proclaimed by visionaries throughout history—why do we read this instead as the possible beginning of all the science fiction worlds of refined horror? The answer lies in the characteristics of establishments in general and the impact of establishment on science in particular. We picture a scientific community all but perfectly intolerant of unorthodoxy, thoroughly compartmentalized, with 99 percent of the scientists either "worker bees" slaving away on details, or overseeing that work. Moreover, the very social responsibilities science will have then acquired will have added a strong moral weight, let alone considerations of security, to the already justified insistence on not deviating from orthodoxy. And by then we will have perfected those educational skills which see that it is so. And so the intellectual oppression, and hence the social oppression, the blindness, becomes complete.

I am not antiscience; this article is not antiscience. To the contrary, science has been one of mankind's great intellectual adventures. (It is overshadowed by the adventure in creating value systems, communication, culture, civilization and is fated to be permanently thus eclipsed, I think; compare below.) It may even be true that there are relatively more men of integrity and good-will in science than elsewhere. Certainly I believe that society will be relatively better cared for if the development of societal management systems is dominated by men of science than others with more obvious vested interests. Nonetheless it is still possible for men of goodwill to be conditioned by their institutions and to be overwhelmed by the social roles their expertise apparently demands they play; the combination of the academic pressure to specialized expertism and the social prestige of designer is immensely powerful. The drift to relatively inflexible specialization and social entrenchment is now well advanced. If this in turn leads to a controlled world equally rigidly compartmentalized, it will only have shown what other species have shown in the past—that, lacking the resources to respond effectively to our changing circumstances, we have fastened rigidly on one temporarily successful configuration and thereby backed ourselves into an evolutionary dead-end. There are other evolutionary traps of course. We might simply not respond to the challenge of global control and design—the scientist pleading his amorality and asociality in the search for "truth," the technologist his role as servant, the businessman his role as pure marketeer; we might create an unlivable environment, or a culture which ensures our demise through increasing mutual predation and conflict. But science need not be so, nor technology, nor economic activity. It could provide us with the true vehicle to fully express our truest human aspirations in the design of our global village.

The real question of the age, as I said earlier, is how to ensure that it is a golden age of science that we usher in and not a nightmare.

VIII

Let us begin by examining the responsibilities of scientists. I realize that in many circles talk of the responsibilities of scientists is anathema. Science is the objective pursuit of truth, so the arguments run, and as such has none but epistemic responsibilities. This defense seems possible only because of the switch from *scientist* in the first remark to *science* in the second, i.e., only by ignoring the process of creating human science. The scientist is both an epistemic and a moral agent; he has a relationship with his society which is not extrinsic to the structure of scientific work but is determined in fact largely by the institutionalization of scientific activity; these relations, I have stressed, become an important part of the value judgments every scientist constantly makes. Even *qua* epistemic agent the scientist is involved in moral values; respect for truth ultimately requires respect for other persons as agents in the critical process of inquiry—science is public knowledge.

The relevant question, then, is what are the responsibilities of those who create science? I deal first with moral responsibilities. Consider the following argument. Social freedoms are possibilities for choice among alternatives created by society. Freedom of speech is the creation of the genuine possibility, first of expressing oneself in a language, then of saying diverse things publicly without reprisal. To be free is to possess the freedom to express oneself, to have the ability to exercise a variety of specific freedoms in such a fashion that the choices reflect one's individual personality. The social freedoms are disanalogous to physical freedoms that are given, not created. The physical freedom to walk, skip or run is the physical possibility of choosing, without physical constraint, among these modes of perambulation. But the acquiring of these abilities may not be easy, it may require, for example, rigorous training or a specific education; more fundamentally, to have the capacity for a wide variety of freedoms requires a certain kind of personality. This in turn presupposes a rich variety of experiences, for example, loving care when young, imaginative and nonauthoritarian teachers, and so forth. To be free to jump six feet high, or join the next Olympics, or outrun one's father may all require training of various degrees of rigor without which any choice is absent. To be able to employ language at all to communicate requires that one have been raised in a human linguistic community. To be truly free to choose between a career in business administration or mathematics requires the possession of the requisite educational background and intellectual capacities. To be free to criticize contemporary society, or science, requires the ability to be imaginative in a fashion at least, partially unconstrained by contemporary conditions, and to be able to critically evaluate a variety of social or scientific alternatives. This in turn will come about only through educational and other experience of a certain kind. To be free to really love people, in the appropriate circumstances, requires the ability to be affectionate, thoughtful, responsive, to enjoy other persons, and so forth. So many people possess this ability in only meager proportion, and largely, I assume, because they have been emotionally deprived and/or their social experience has molded them into selfish, competitive material acquisitors. So then, society creates the conditions of freedom in its members through the experiences and the social structures which it offers them. An important part of the experiences is epistemic, i.e., the acquiring of specific skills, knowledge, and critical abilities.

If we agree that the increasing of human freedoms of these kinds is a good thing (it is the maximizing of human potential), then it follows that we all have a moral responsibility to act so as to do so. But there are special responsibilities acquired by those people who are in privileged positions with respect to those processes governing the acquisition of the abilities to exercise the various freedoms. A parent has a special responsibility to his child; the teacher to his students; the lawyer, politician, and businessman to society at large. Among these, scientists have special responsibilities. He (collective) it is to whom society bequeaths the right to critical knowledge and the skills to manage our development. On this view scientists have special responsibilities (1) for maintaining, in fact stimulating, the ability for critical thought; (2) for imaginative innovation; (3) for ensuring the proper flow of all relevant information to the public in the most lucid terms; and (4) for the human consequences of applied science, hence for the control of scientific research. Of course, he may not succeed in controlling the consequences of his work, but that does not relieve him of the moral responsibility. It simply becomes a factor that ought to be taken into account in advance and striven for afterward. Moreover, it should spur his responsibilities to publicly discuss the cultural implications of his work. The scientist has the responsibility of ensuring, or attempting to ensure, that the evolution of science increases our freedoms—intellectual and social—rather than decreases them.

Our freedoms will surely decrease if the nightmare world of rigid global control, discussed earlier, is realized. Part of the responsibility of scientists is to see that this does not happen.

The reader will have long appreciated that this view binds science inextricably into other human processes, cultural, political, legal, and so forth. It calls for a much more socially active scientific community than we have. It explicitly rejects the bifurcated view of the human scientist as antiseptic amoralist epistemic agent *qua* scientist and as layman *qua* human being. In my view the human community has a right to expect the scientific community to play a socially responsible role, precisely because they have entrusted them with so much that is of social importance: the education of its children, criticism of its myths, setting of health standards, global planning of its future, and so forth.

This attitude applies to all of the professions. Every profession at present seems to insist that a central part of its professional image is the unthinking, but expert, playing of its narrow professional role. Many lawyers would have us believe that their only task is the exercise of law, that they are slaves to the text, they would tacitly have us believe that the law is not at all the social expression of currently desirable contractual arrangements, that it is not dynamic, and that its development does not in turn both help determine our freedoms and the kinds of social contract we will accept. As the custodians of legal expertise, lawyers have the responsibility to develop the laws in directions that will promote new freedoms (e.g., the development of community rights in environmental contexts) and to be active in educating and persuading the public concerning these developments, as well as in the nature of, and exercise of, the law in general. There is a tendency for doctors to treat just the so-called medical dimensions of a problem under the guise of being "scientific"—patients become cases, and cases are forcibly seen as instances of formulae. If we were less keen on the divorce of economics from ethics, we might begin to see the colossal value judgments professional business management experts constantly make and the social ramifications of esoteric fields such as the theory of industry location.

Think what the consequences might be if the lawyers really brought to the public a critical appraisal of public legal freedoms; if the trained planners discussed publicly our cities and how planning decisions are really made and what some imaginative alternative might be; if doctors (and others) really pressed home modern insight into the contribution that ill-designed jobs, ill-designed community structures and facilities, poor environmental quality, food practices, etc., make to the incidence of unhealth, family stress, and other "social diseases."

Some will no doubt argue that it is better not to have these groups politically active since then their power would be too strong. But the fact is, they already are politically active and their powers are immense. And not just politically active, everywhere we see the cultural consequence of professional activity. The situation could hardly be worse than at present where professional people operate "silently," the public largely unaware of their activities and/or gulled into acceptance by the myth of professionalism. This is not a conspiracy theory, the consequences come just through being hooked into today's cultural web of social practices. Better to have things out in the open where value judgments and social consequences are clear and where there is some chance of achieving a measure of social responsibility and public education.

Moreover, those who object to the politicizing of all elites thereby show that they have not grasped the implications of the need for global control. In a world where virtually every decision on every aspect of life has nonnegligible consequences there is no room left for the nonpoliticized since every activity reflects a multitude of value judgments, namely, that its various consequences bring more benefits than costs. The human artifactual world achieves its design through the human value structure and all of its controlling operations are correspondingly value laden.

Incidentally, the traditional philosophy of science effectively prevents the development of this point of view because of its emphasis on science as an abstract, value-free product. In this it proves not only epistemologically inadequate and intellectually arid but morally culpable as well.

IX

If the recognition of the social-moral responsibility of scientists and a socially more active role for scientists is important to the creation of a disestablishment world, it is only a necessary condition. It is necessary too to have a nonauthoritarian methodology and educational system. But to achieve this it is first necessary to have a nonauthoritarian epistemology.

Once again we observe that the traditional philosophy of science is morally culpable, it has a thoroughly authoritarian approach. If fact, all foundational epistemologies fall into this category, for the "sacred heart" of knowledge, the privileged data at the foundation of knowledge, must either be given by authority and/or confer authority on those to whom they are given. Indeed, the traditional account of the history of science is famous for its attempt to foist the accretion model of authoritative growth, that follows from the epistemology, on the historical account.

Without elaborating, we see that what is required is a foundationless epistemology, and a thoroughly naturalized one. For this offers not only a truer account of man's epistemological position (man begins in ignorance and cultural poverty and only laboriously learns and constructs through trial and error) but, precisely for this reason, it is the only view which opens to man the full potential for his freedom. *Everything* is constantly open to criticism and revision. Our best views, even of our capacity to know (hence the naturalism), are no better off than our current science and this may be overthrown tomorrow. Our best views, even of human values and moral responsibilities (hence the naturalism), are no better off than our present understanding of our cultural world, and we may reconstruct that world, or overthrow our understanding, tomorrow.

All foundational epistemologies generate authoritarian methodologies as well: if you know the privileged set of data then the only sound methodology must be one which maximizes these and builds theory from them, builds only theories which conform to them and no theories which go unnecessarily beyond them. (Actually, the very word "unnecessarily" gives the game away here). It is methodology based on the assumption of an accretion view of the history of science. By contrast, the methodology to which the foundationless epistemology leads is of the "Let the hundred flowers bloom" type. The best conditions for the progress of knowledge are where every idea, every claim, is subject to criticism (genuine criticism, not merely the claim that it is unorthodox) and where a maximum of innovative, unorthodox ideas are encouraged; for it is the case that most fundamental ideas can only be severley tested by comparing them to opposing views of the matter. These are also the circumstances under which the intellectual life is most vigorous and free, not to mention interesting— overspecialization tends to stunt the soul and bore the mind.

Indeed, traditional epistemology and methodology is doubly deficient. Not only does it lead to an authoritarian structure in a methodological straitjacket, it tacitly assumes that we are still adaptors to the natural order; what it offers is quite out of place in the artifactual world. In the natural world where man clearly experienced his impotence to alter the scheme of things it was appropriate to speak of objective facts and objective laws, namely, those belonging to the natural system. As adaptors to the system the primary epistemological goal was to discover its functioning. One could, seemingly, afford to concentrate on just the system descriptions—the *science*—and neglect everything else. Methodology became the shortest route to this epistemic goal. But in the artifactual world one cannot speak of objective facts and laws in the same sense since the scheme of things was designed by us and may be changed tomorrow. Of course, one can describe how things are now, but even that description cannot be made penetratingly unless one divines the inner principles of design. But these latter are value systems and correlative theories of how to realize them. In the artifactual world knowledge of values dominates, and then come theories of how to realize those values. After that the facts and generalities of the actual design of the moment may make a humble entry. By excluding, or attempting to exclude, values from *science,* the traditional view leaves us singularly unprepared to deal philosophically with the artifactual world and helps to create the harsh and limiting demand by the social sciences to stick to "objective empirical data" that sadly crushes creative debate at present. Methodology, like epistemology, will be radically different in the artifactual world, for it is no longer a search for facts, except at an extremely low level within a given design, but a search for design alternatives and the correlative theories of realization, though the latter have certain facts of psychology relevant to them. Sifting the social facts before one's nose in the Positivist hope of insight helps not a whit if one is searching for a design alternative to that which created those facts! Instead one needs bold experimentation in design alternatives, one needs to change the "facts" and to see how the new situation squares with one's value system. We seem not yet to have an account of how to go about this rationally—no doubt because of the little attention this most important of problems has received vis-à-vis *science.*

But to achieve this we need an appropriate educational practice. Today's education tends to be authoritarian and boring, giving little place for individuality and imagination, for true criticism. The teacher is an authority figure by dress, location in class, and behavior—often it is an "ego-trip"; the emphasis is on getting things right (right often meaning agreement with teacher), not on the development of understanding, critical ability, creativity. Most important areas of life are never discussed—emotional development, human relations, morals, and government. The school life is divorced from that of the larger society; the artificial divisions of knowledge are imposed at a time when they will stultify innovative ability, and so on. We will have to rear children to be not only vigorously critical of their elders, but also to take the responsibilities that participative criticism brings with it; to be masters rather than victims of their social institutions; to be creative, innovative, and free in spirit and mind.

We find in these considerations, I believe, something of the real importance of the planning process as a supraindividual process. The issue of whether there are irreducible communal values is a central one for decision theory, indeed, for any theory of rational conduct, for it determines whether we can both understand another's actions and plan our own, entirely in terms of rationality principles that comprehend only considerations of the individual involved (his preferences and present condition, the consequences of various actions for him, and so forth). It would seem that recent work (e.g., that by Arrow) has answered this negatively. Now liberal thinkers claim that the reason one cannot seem to create satisfactory collective decision-making processes even for those parts of the design we are now forced to decide about (e.g., distribution of income, education, and so forth) is that there is no corresponding market. Ironically, what the methodology which I suggest would do is to create the only relevant market, a market in alternative ideas such as theory and design. But this very possibility can be realized only (*pace* the analysis of freedoms) assuming a certain kind of educational system and cultural institutions to back it; even this market freedom presupposes global planning! Indeed, this must be true if, as is true, all of us are the persons we are in large measure because of the design in which we were reared. In the past it has indeed been the case that the resulting design has been more nearly the accidental outcome of the actions of myriad individuals acting more or less incoherently relatively to one another. Against the essential naturalistic background of this phase there was a measure of truth to the individualistic model, but the resulting "design" was chaotic and uncontrollable. Keynes pointed this out forcefully for economics just before the advent of nuclear strategy made it obvious. When we turn to the future artifactual world that must be globally controlled to ensure survival, there will be no room for such accidents. It must be the specifically communal values that dominate the design if the design is to be controllable to the degree survival demands; the design process raises the planning process and its values to the fore.

This profound change also creates the circumstances in which at least a partial resolution of the conflict outlined in section VI can occur, i.e., the clash between the pressure to choose "safe" research projects and the epistemic necessity to explore radical alternatives. That conflict is especially sharp at the level of the individual scientist, for it is *his* credibility, prestige, and security that are at stake. But an entire society might more rationally allocate funds to the pursuit of radical alternatives. Of course, some degree of conflict remains; unless resources and human pliability are infinite not all radical alternatives can be pursued. This is a trade-off a society must make between small-scale and large-scale innovation, between short-term social betterment and long-term epistemic betterment. How it is made will be determined by the cultural temper of the society.

X

Also necessary to the achievement of the disestablishment society is an appropriate set of social institutions. These institutions must prevent the formation of closed elites that can become rigid establishments. The key to achieving this is communication. I have already remarked how increasing intervention requires increasing communication for stable management. We must add to this that the method sanctioned historically for balancing central authority, namely, the creation of many local political jurisdictions, is an ineffective anachronism in the artifactual world. Ask anyone attempting rational management in a federal system such as Australia, Canada, and the United States of America! What is needed to balance the inevitable growth of global management is the creation of a communicational network that will directly open up decision-making authority to all "levels" of society, indeed, to abolish levels. Global planning leads to bureaucratic centralization only where communication is so inefficient, willfully or otherwise, that only a few can be brought into the decision process. But we have the means now to create a truly integrated, "organic" society whose decision processes run through all levels. With, and only with, an appropriate communications system can the information necessary to responsible decision making be widely disseminated and a wide variety of people effectively participate in the decision process. Indeed, once we understand the full impact of the fundamental role of value systems in planning the artifactual world, we see that our communications system must cope with the conveying of different value systems, different "world views" (outlooks, appreciative systems), a task to which it has hitherto been miserably unequal. That is, the communicational institutions must complement the educational process and the critical methodology with the permanent public discussion of, and permanent participation of the public in, matters affecting the human condition. In this way the exercise of the public responsibilities of elites, in particular the scientific elite, will be made possible and to some extent assured.

In addition to the communicational structure it is necessary to have a correlative governing structure so that decisions are made, and implemented, at levels appropriate to the communities affected. In fact these structures will surely be the mirror images of the communicational structures; as such, they will have to be fluid, "organically organized," emerging as required and submerging again back into the community when not required, drawing on a wide variety of persons, proceeding through multiple-person, multiple-valued decision patterns. It is not hard to see that this government structure will be very different from the rigid, entrenched structure we know today. The question of what precise forms our social institutions should take is a complex one and I have no neat solutions to offer. One can, however, offer some suggestions about getting the process of change under way. Perhaps some generalization of the Maoist policy of all workers holding regular discussion sessions concerning public policy and of sending intellectuals to the countryside from time to time would be initially helpful. Certainly, immediate changes in the laws regarding secrecy in government and industry are called for. At the more local level, community ownership of community services (health, recreation, information, environment, planning) with community centers in which all these services were centered and in which community education and public discussion were carried on would help to develop the habit of participation, remove the fear and apathy at the unknown, and counterbalance concentrations of power. But these are not more than suggestions; the fuller discussion of how our institutions (political, legal, economic, and so forth) must be arranged to lead to the maximization of human freedoms in these respects I leave to others at this time.

XI

We can all project the expected scenario for the familiar model of the scientist. As global management theory and our technology become increasingly sophisticated, the boundaries between government and industry blur; effective control is lodged in the hands of an increasingly small elite; education becomes highly specialized and "skill" oriented, perhaps even down to the very first year. Most people, increasingly far removed from decisions which affect them, both through bitter experience and education accept apathetically, or with growing alienation, their roles as cogs in a well-managed machine. The anti-Copernican revolution is complete.

By contrast I have tried to briefly outline an alternative culture, one in which global management and technological sophistication are still possible, and hence where rational environmental control is still possible, but where the culture contains its own dynamic set of checks and balances to the concentration of power among scientific elites. This is the culture which is vigorously alive to its values, in which the value assumptions behind, and social implications of, every planning move are publicly debated, and where community structure and educational experience reinforce these values. Here is a genuine possibility for a golden age of science, where scientific knowledge is truly in the service of man to promote first freedom, and then well-being. This would be a fitting crown to that intellectual undermining of church establishment that we refer to today as the Copernican revolution.

REFERENCES

The bibliographical references made in the following notes are in no sense intended to be exhaustive. They are intended simply to serve to introduce the interested reader to some of the background material and issues which lie behind my comments in the text. In some instances the references are even to entire bibliographies. A variety of journals, including *Daedulus, Minerva, Operations Research, Policy Sciences, Science, The Bulletin of the Atomic Scientist, Technology and Culture, Science and Technology,* and *Science Forum,* often contain interesting material which bears on portions of this discussion, and the reader is invited to consult them to obtain a feeling for some of the current debates.

Section I

The remark by Pilate may be found in the Christian scriptures, and that by Polyani was made in an article defending the response of the contemporary scientific community to Velikovsky:

> Polyani, M. "The Growth of Science in Society." *Minerva,* vol. 5, 1967, 533–45.

See also Section VI below.

Section II

On the development of technology vis-à-vis science see:

> Allen, F. R. et al. *Technology and Social Change.* New York: Appleton-Century-Crofts, 1957.
> Bernal, J. D. *Science and Industry in the Nineteenth Century.* Bloomington, Ind.: Indiana University Press, 1970.
> Fryth, H. J., and Goldsmith, M. *Science, History and Technology.* London: Cassell, 1965.
> Marsak, L. N., ed. *The Rise of Science in Relation to Society.* New York: Macmillan, 1964.
> Sarton, G. *Introduction to the History of Science,* 5 vols. Baltimore: The Williams and Wilkins Co., 1926–47.
> ____. *The Study of the History of Science.* Cambridge, Mass.: Harvard University Press, 1936.

For bibliographical material on this and other aspects of science see:

> Ferguson, D. *Bibliography of the History of Technology,* 2 vols. Cambridge, Mass.: M.I.T. Press, 1968.
> Caldwell, L., ed. *Science, Technology and Public Policy.* Bloomington, Ind.: Indiana University Press, 1968.

For a discussion of the effect of the machine on civilization see:

> Mumford, L. *The Culture of Cities.* New York: Harcourt and Brace, 1938.
> ____. *Technics and Civilization.* New York: Harcourt and Brace, 1934.

For the role of the universities see:

Ashby, Sir E. *Technology and the Academics.* New York: Macmillan, 1958.

Brubacher, J. S. *Higher Education in Transition; A History of American Colleges and Universities 1636–1968.* New York: Harper and Rowe, 1968.

Mountford, Sir J. F. *British Universities.* London: Oxford University Press, 1966.

It is noteworthy that that blight on intellectual freedom and development, the university department, seems to have arisen at the stage when science was thrusting its way into the university against the opposition (enlightened of course!) of those teaching the entrenched subjects (theology, law, medicine). It evidently stood then, as it stands today, as an effective political tool to monopolize money and influence and eject all outsiders.

When I refer in the text to the scientist as ideologist I intend to mean that, since each of his acts brings with it social repercussions, it is inescapable that his actions will exhibit a social philosophy by implication. Thus the scientist and the more overtly socially active operative (e.g., politician) become indistinguishable. For discussion of the entrance of value judgments of various sorts into scientific decisions, see e.g., Churchman, Leach, and Rudner below, Rapoport and Rothenburg in Section IX, and the references in Sections V and VIII.

For discussions of the traditional approach to science see:

Achinstein, P., and Barker, S., eds. *The Legacy of Logical Positivism for the Philosophy of Science.* Baltimore: Johns Hopkins Press, 1969.

Braithwaite, B. *Scientific Explanation.* New York: Cambridge University Press, 1953.

Carnap, R. *The Logical Structure of the World.* London: Routledge and Kegan Paul, 1967. (First published as *Der Logische Aufbau Der Welt.*)

_____. "The Methodological Character of Theoretical Concepts." In *Minnesota Studies in the Philosophy of Science,* Vol. I., ed. H. Feigl, and M. Scriven. Minneapolis: University of Minnesota Press, 1956.

Hempel, C. *Aspects of Scientific Explanation.* New York: The Free Press, 1956.

Nagel, E. *The Structure of Science.* New York: Harcourt, Brace and World, 1961.

Representative writings of Feyerabend are:

Feyerabend, P. K. "Against Method." In *Minnesota Studies in the Philosophy of Science,* Vol. IV., ed. M. Radner and S. Winokur. Minneapolis: University of Minnesota Press, 1970.

_____. Knowledge Without Foundation. Mimeographed. Oberlin College, 1961.

_____. "Problems of Empiricism." In *Pittsburgh Studies in the Philosophy of Science,* Vol. II., ed. R. Colodny. Englewood Cliffs, N.J.: Prentice-Hall, 1965.

_____. "Problems of Empiricism II." In *Pittsburgh Studies in the Philosophy of Science,* Vol. IV., ed. R. Colodny. Pittsburgh: University of Pittsburgh Press, 1969.

For Kuhn's views see:

Kuhn, T. *The Structure of Scientific Revolutions.* Chicago: University of Chicago Press, 1962.

Lakatos, I., ed. *Criticism and the Growth of Knowledge.* Cambridge: Cambridge University Press, 1970.

Popper's early views will be found in:

Popper, K. R. *Conjectures and Refutations.* London: Routledge and Kegan Paul, 1963.

_____. *The Logic of Scientific Discovery.* London: Hutchinson, 1962.

_____. *The Open Society and its Enemies,* 2 vols. London: Routledge and Kegan Paul, 1966.

His views are critically discussed in:

Bunge, M., ed. *The Critical Approach to Science and Philosophy.* New York: The Free Press, 1964.

While his later epistemological views are heavily loaded in favor of Science and the exclusion of the scientist, see:

Popper, K. R. *Objective Knowledge.* Oxford: Oxford University Press, 1972.

My own views on these matters may be found in:

Hooker, C. A. "Critical Notice: Against Method, P. K. Feyerabend." *The Canadian Journal of Philosophy.* 1 (1972): 489–509.

_____. "Empiricism, Perception and Conceptual Change." *The Canadian Journal of Philosophy.*

_____. Systematic Realism.

See also:

Hayek, F. A. von. *The Counter-Revolution of Science.* New York: The Free Press, 1955.

And for a brief presentation of some themes similar to mine, see:

Skolimowski, H. "Science and the Modern Predicament." *New Scientist.* February 24, 1972.
_____. "Science in Crisis". *Cambridge Review* January 28, 1972.

See also Habermas's "Technology and Science as ideology" in:

Habermas, J. *Towards a Rational Society.* Boston: Beacon Press, 1969.

For the early sociology of knowledge see:

Mannheim, K. *Ideology and Utopia.* London: Routledge and Kegan Paul, 1960.
Weber, M. *The Methodology of the Social Sciences.* Edited and translated by E. A. Shils and H. A. Finch. Glencoe, Ill.: The Free Press, 1949.

For the early Marxist approach see, e.g., Bernal and Crowther in these notes; contrast Ben-David and Hahn (Section V), and for more recent developments see:

Curtis, J. E., and Petras, A. W., eds. *The Sociology of Knowledge; A Reader.* London: Duckworth, 1970.

Cf. also:

Barber, B. *Science and the Social Order.* New York: The Collier, 1970.
Barber, B., and Hirsch, W. *The Sociology of Science.* New York: The Free Press, 1962.
Elkana, Y. "The Problem of Knowledge." *Studium Generale* 24 (1971): 1426–39.

The new knowledge needed to successfully bring the new revolution to fruition is slowly being developed. This includes decision theory and general systems theory (see notes for Section IV) as well as the role of values in scientific judgments and epistemic acceptance criteria; see:

Churchman, C. W. *Prediction and Optimal Decision.* Englewood Cliffs, N.J. Prentice-Hall, 1961.
Levi, I. *Gambling with Truth.* New York: A. A. Knopf, 1967.

Leach, J. J. "Explanation and Value Neutrality." *British Journal for the Philosophy of Science* 19 (1968): 93–108.
Rudner, R. *Philosophy of Social Science.* Englewood Cliffs, N.J.: Prentice-Hall, 1966.

On ethics developing from the world view and practices of science see:

Bronowski, J. *Science and Human Values.* London: Penguin, 1964.
Disch, R. *The Ecological Conscience.* Englewood Cliffs, N.J.: Prentice-Hall, 1970.

See also the references of Section X.

Section III

The general conclusion arrived at here is discussed from an even more general point of view in:

Vickers, Sir G. *Value Systems and Social Process.* London: Tavistock, 1968.

See also the papers by Sir Geffrey Vickers and Eric Trist, entitled respectively, "Science Policy and Social Policy" and "Social Aspects of Science Policy" in the proceedings of the Senate Special Committee on Science Policy, Ottawa, Canada, 1968–69, as well as the remainder of the hearings and the report of that committee. Available from The Queens Printer, Ottawa.

Similar conclusions to these are today being arrived at from very different starting points; see:

Hunter, R. *The Enemies of Anarchy.* Toronto: McClelland and Stewart, 1970.
McLuhan, M. *Understanding Media: The Extensions of Man.* New York: Signet, 1964.
Starrs, C., and Stewart, G. "Gone Today and Here Tomorrow." A working paper prepared for the Committee on Government Productivity. Ontario, Canada. Queens Printer, 1972.

The need for such global planning and its present lack because of the divisive specialization of knowledge and skills (cf. my remarks on university departments in Section II) is increasingly often noted, cf.:

Galbraith, J. K. *The New Industrial State.* Boston: Houghton Mifflin 1967.

See also the works of Bateson, Boulding, and Odum cited elsewhere in these notes.

I have discussed the particular premises adumbrated in the text at greater length in:

> Hooker, C. A. "The Systematic Significance of Environmental Problems." Proceedings of the 1972 meeting of the Law Society of Upper Canada, Continuing Education.

Much the same conclusion is reached by:

> Heilbronner, R. "Growth and Survival." *Foreign Affairs* 51 (1972): 139–53.
> ____. *An Inquiry into the Human Prospect.* New York: Norton, 1974.

For material on the industrial revolution see:

> Bury, J. B. *The Idea of Progress.* Toronto: Oxford University Press, 1920.
> Ferkiss, V. *Technological Man.* New York: Mentor, 1969.
> Polyani, K. *The Great Transformation.* Boston: Beacon Press, 1944.
> Trevelyan, G. M. *English Social History.* London: Longmans, Green and Co., 1942.

Much of the "future" literature that is culture oriented is now attempting to come to grips with some of the massive cultural changes involved. See:

> Fuller, R. Buckminster. *Utopia or Oblivion.* New York: Bantam Books, 1969.
> Landers, R. L. *Man's Place in the Dybosphere.* Englewood Cliffs, N.J.: Prentice-Hall, 1966.
> Toffler, A. *Future Shock.* New York: Random House, 1970.
> Theobald, R. *Futures Conditional.* New York: Bobbs-Merrill, 1972.

Two interesting authors who also treated the recent growth of science and technology with the social role of science very much in their minds are Bernal and Crowther; see:

> Bernal, J. D. *Science in History.* 4 vols. London: Watts, 1969.
> Crowther, J. G. *Science in Modern Society.* London: Cresset Press, 1967.

These men are Marxist oriented and their values show through rather (too) strongly at times. Nonetheless they serve to quicken us to the social relations of science and it is significant that they should have nearly dominated this field between the European "world" wars.

On systematic developments of ecology and human society which display its integration see:

> Bateson, G. *Steps to an Ecology of Mind.* New York: Ballantine, 1972.
> Odum, H. *Environment, Power and Society.* New York: Wiley-Interscience, 1971.
> Watt, K. E. *Ecology and Resource Management: A Quantitative Approach.* New York: McGraw-Hill, 1968.

See also the references to systems analysis found in the notes to Section IV.

Section IV

On general systems analysis see:

> Bertalanffy, L. von. *General System Theory.* New York: George Braziller, 1968.
> ____. *Robots, Men and Minds.* New York: George Braziller, 1967.
> Laszlo, E. *The System's View of the World.* New York: George Braziller, 1972.
> ____. *System, Structure and Experience.* New York: Gordon and Breach, 1969.
> ____. *Introduction to Systems Philosophy.* New York: Gordon and Breach, 1972.
> Laszlo, E., ed. *The Relevance of General Systems Theory.* New York: George Braziller, 1972.

This last contains articles on biology, physiology, economics, psychology, psychiatry, communications, education, among others.

In addition the *Society for General Systems Research* has published, under the editorship of Anatol Rapoport and Ludwig von Bertalanffy, 16 volumes since 1956 in a series entitled *General Systems,* Washington, D.C. which contains a great deal of valuable material.

On the application of general systems theory and systems philosophy see:

> Ashby, W. R. *Design for a Brain.* London: Chapman and Hall, 1952.
> ____. *An Introduction to Cybernetics.* New York: Wiley, 1963.
> Boulding, K. E. *Beyond Economics.* Ann Arbor: University of Michigan Press, 1968.

Buckley, W. *Sociology and General Systems Theory.* Englewood Cliffs, N.J.: Prentice-Hall, 1967.

———. *Modern Systems Research for the Behavioral Scientist.* Chicago: Aldine, 1968.

Easton, D. *A Systems Analysis of Political Life.* New York: Wiley, 1965.

Elsasser, W. M. *Atom and Organism.* Princeton: Princeton University Press, 1966.

McRie, T. W., ed. *Management Information Systems.* London: Penguin, 1971.

Maslow, A. H., ed. *New Knowledge in Human Values.* New York: Harper and Row, 1966.

Pfeiffer, J. *New Look at Education; Systems Analysis in Our Schools and Colleges.* New York: Odyssey Press, 1968.

Pugh, D. S. *Organization Theory.* London: Penguin, 1971.

Sakman, H. *Computers, Systems Science, and Evolving Society.* New York: Wiley, 1967.

Simon, H. A. *The Sciences of the Artificial.* Cambridge, Mass.: M.I.T. Press, 1969.

Wiener, N. *The Human Use of Human Beings; Cybernetics and Society.* Boston: Houghton Mifflin, 1950.

See also the references to Bateson, Odum, and Watt in Section III. Most of the books referred to contain basic bibliographies of relevant works which the reader may explore at will.

For material on the development of modern decision theory and its applications see:

Arrow, K. *Social Choice and Individual Values.* New Haven: Yale University Press, 1963.

Black, B. *The Theory of Committees and Elections.* Cambridge, Mass.: Cambridge University Press, 1958.

Buchanan, J. M., and Tulloch, G. *The Calculus of Consent.* Ann Arbor: University of Michigan Press, 1962.

Davis, M. D. *Game Theory.* New York: Basic Books, 1970.

Friedrich, C. J. *Rational Decision.* New York: Atherton Press, 1964.

Kaufmann, A. *The Science of Decision-Making,* New York: Mc-Graw-Hill, 1968.

Kyburg, H. E., and Smokler, A. G., eds. *Studies in Subjective Probability.* New York: Wiley, 1964.

Leach, J. J., Pearce, G., and Butts, R., eds. *Science, Decision and Value.* Dordrecht: Reidel, 1973.

Luce, R. D., and Raiffa, H. *Games and Decisions: Introduction and Critical Survey.* New York: Wiley, 1967.

Michalos, A. *Foundations of Decision Making.* Amsterdam: Elsevier, 1973.

Miller, D. W., and Starr, M. K. *The Structure of Human Decisions.* Englewood Cliffs, N.J.: Prentice-Hall, 1967.

Morgenstern, O., and Neumann, J. von. *Theory of Games and Economic Behavior.* Princeton: Princeton University Press, 1947.

Olsen, M., Jr. *The Logic of Collective Action.* New York: Schocken, 1971.

Savage, L. J. *The Foundations of Statistics.* New York: Wiley, 1954.

Thrall, R. M., Clyde, H. C., and Davis, R. L., eds. *Decision Processes.* New York: Wiley, 1954.

The dangers of the system-decision theory approach, in its present crude form, should be constantly borne in mind; the two greatest dangers being oversimplification and the uncritical incorporation of value assumptions into the structure. On this score, see:

Rapoport, A. *Strategy and Conscience.* New York: Harper and Row, 1964.

A critique of:

Kahn, H. *On Thermonuclear War.* Princeton: Princeton University Press, 1961.

And:

Rothenburg, J. *Problems in the Modelling of Urban Development.* M.I.T. Working, Department of Economics, No. 81, October, 1971.

A critical review of:

Forrester, J. *Urban Dynamics.* Cambridge, Mass.: M.I.T. Press, 1969.

See also Galbraith (Section III).

On the development of the application of these techniques to policy formation see Vickers (Sections III, VIII), the journals *Policy Sciences, Operations Research,* and:

Bauer, R. A., and Gergen, K. G., eds. *The Study of Policy Formation.* New York: The Free Press, 1968.

Dror, Y. *Design for Policy Sciences.* New York: American Elsevier, 1970.

Kuenzlen, M. *Playing Urban Games.* Boston: The I Press (G. Braziller), 1972.

Lasswell, H. D. *A Pre-View of Policy Sciences.* New York: American Elsevier, 1971.

On the development of social indicators and their critique, see:

Bauer, R. A., ed. *Social Indicators.* Cambridge, Mass.: M.I.T. Press, 1966.

Michalos, A. "Quality of Life: A Comparative Social Report for Canada and the United States." Mimeographed. University of Guelph, 1972.

Stewart, G. "On Looking before Leaping." In *Social Indicators,* edited by N. A. M. Carter. Council on Social Development. Queens Printer, Ottawa, 1972.

On learning systems and accountability see:

Olsen A. V., and Richardson, J. *Accountability: Curricular Applications.* Scranton, Pa.: Intext Educational Publishers, 1972.

Roberson E. W., ed. *Educational Accountability through Evaluation.* Englewood Cliffs, N.J.: Education Technical Publishers, 1971.

For refreshing alternative attitudes and approaches to education in general see:

Holt, J. *Freedom and Beyond.* New York: Dutton, 1972.

_____. *How Children Fail.* New York: Pitman, 1964.

Illich, I. *De-Schooling Society.* New York: Harper & Row, 1970.

Kozell, J. *Free Schools.* New York: Houghton Mifflin, 1972.

Reimer, E. *School is Dead.* New York: Doubleday, 1971.

For the term "megamachine" and its critique see:

Mumford, L. *Pentagon of Power.* The Myth of the Machine, vol. 2. New York: Harcourt, Brace and World, 1967–70. (Vol. 1: *Technics and Human Development.*)

What these last areas show is that the development of systematic quantitative or semiquantitative measures of personal and social states are at far too crude a stage yet to rely upon to any great extent in human planning. This complements my earlier criticisms of the systems-decision theoretic approach. I would not have the reader leave this section, however, with the feeling that these approaches had no value whatsoever. To the contrary, in the long run they will be absolutely indispensable, it is just that we stand at the historical dawn of their use and must recognize the present dangers of adopting them too dogmatically. It is also the case that we must work strenuously to develop ways in which the intangible dimensions of the quality of human life can be brought more explicitly to the fore.

Section V

On the social institutionalization of science see:

Adams, J. B. "Megaloscience." *Science* 148 (1965): 1560–64.

Armytage, W. G. H. *The Rise of the Technocrats*. London: Routledge and Kegan Paul, 1965.

Ben-David, J. *The Scientist's Role in Society*. Englewood Cliffs, N.J.: Prentice-Hall, 1971.

Carter, C. F., and Williams, B. R. *Science in Industry: Policy for Progress*. London: Oxford University Press, 1959.

Cox, D. W. *The Scientist's Rise to Power: America's New Policy Makers*. Philadelphia: Chilton Books, 1964.

Creighton, H. B. *Political Control of Science in the U.S.S.R.* Oberlin: Oberlin College, 1950.

Daddario, E. Q. *Centralization of Federal Science Activities*. Report to the Sub-Committee on Science, Research and Development of the Committee on Science and Astronautics, U.S. House of Representatives, 91 Congress. Washington, D.C.: U.S. Government Printing Office, 1969.

_____. *Toward a Science Policy for the United States*. Report of the Sub-Committee on Science, Research and Development to the Committee on Science and Astronautics, U.S. House of Representatives, 91 Congress. Washington, D.C.: U.S. Government Printing Office, 1970.

de Solla Price, J. J. *Little Science, Big Science*. New York: Columbia University Press, 1963.

Gilpin, R., and Wright, C., eds. *Scientists and National Policy Making*. New York: Columbia University Press, 1964.

Goldsmith, M., and Mackay, A., eds. *Society and Science*. New York: Simon and Schuster, 1964.

Hahn, R. *The Anatomy of a Scientific Institution*. Los Angeles: University of California Press, 1971.

Lapp, R. E. *The New Priesthood: The Scientific Elite and the Uses of Power*. New York: Harper and Row, 1965.

Price, D. K. "The Scientific Establishment." *Proceedings of the American Philosophical Society*. 107 (1962).

____, *The Scientific Estate.* New York: Oxford University Press, 1965.

Weinberg, A. M. "Criteria for Scientific Choice." *Physics Today* 17 (1964).

Note also the references to Barber and Hirsch (Section II) and Crowther (Section III), the references in Section VIII and some of the earlier references of Section III. For other reflections of the social importance of scientists connected with the 1939–45 war see:

Bailyn, B., and Fleming, D., eds. *The Intellectual Migration.* Cambridge, Mass.: Harvard University Press, 1969.

Section VI

The key work by Velikovsky which initiated the controversy is:

Velikovsky, I. *Worlds in Collision.* New York: Dell, 1967.

See also now:

Vekilovsky, I. *Earth in Upheaval.* New York: Dell, 1968.

These books were first published in 1950 and 1955 respectively. The account of the controversy has been nicely captured in:

de Grazia, A. *The Velikovsky Affair.* New York: University Books, 1966.

But the reader should also consult recent special issues of *Pensée* dedicated to the Velikovsky theory and its acceptance. *Pensée*, (1972)—available from P.O. Box 414, Portland, Oregon, 97207, U.S.A. *Pensée* is to devote a substantial set of issues to further explorations of Velikovsky's work.

For the circumstances of Copernicus and Galileo the reader might consult:

Dijksterhuis, E. J. *The Mechanization of The World Picture.* Oxford: Oxford University Press, 1961.

Koestler, I. E. *The Sleep Walkers.* London: Penguin, 1964.

and innumerable other books on the history of science, but especially the references to Feyerabend (Section II) and the other articles in this conference proceedings.

The trials of Galileo and Oppenheimer are compared in:

de Santillana, G. *Reflections on Men and Ideas.* Cambridge, Mass.:
M.I.T. Press, 1968.

See his essay "Galileo and Oppenheimer"; cf. also Wharton (Section VIII),
and Bertolt Brecht's play *Galileo Galilei.* A similar and enlightening comparison is made between the Spanish Inquisition and the Contemporary mental
health movement in:

Szasz, T. S. *The Manufacturer of Madness.* London: Routledge and
Kegan Paul, 1971.

Compare also:

Szasz, T. S. *Ideology and Insanity.* New York: Doubleday, 1970.

For the "cold-blooded" approach to the choice of research see,
Polyani (Section I) and nearly any "sensible" discussion of the criteria of
scientific research, e.g., some of the articles referred to elsewhere in these
notes.

The Velikovsky affair is also paralleled by the prevailing attitude to
psychic research in most quarters, despite some apparently spectacular advances recently, especially in the U.S.S.R. (there is little reason to doubt
some of the reports); see:

Ostrander, S., and Schroeder, L. *Psychic Discoveries behind the Iron
Curtain.* New York: Bantam Books, 1971.

Some insight into this circumstance is given by the remarks of a group of
Soviet nuclear scientists who, having given one of Russia's best known
psychics a thorough test, produced a disparate explanation for the phenomena they had witnessed (and for which explanation they have evidently been
criticized) and then remarked that "if they admitted it (the psychic phenomena) they ought to leave physics and start studying parapsychology" (Ostrander and Schroeder, p. 83).

It is also paralleled in reverse by the Lysenko period in Soviet
biology; see:

Caspari, E. W., and Marshak, R. E. "The Rise and Fall of Lysenko."
Science 149 (1965): 275–78 (cf., Kaellis. 149: p. 1443).

Section VII

The works of science fiction are filled with examples of controlled
worlds, and as such they offer important explorations of what the design of
our world might be like in those circumstances. See:

Brunner, J. *The Sheep Look Up.* New York: Ballantine, 1972.

Lewin, R. C. *Triage.* New York: Warner, 1973.

Orwell, G. *Nineteen Eighty-four.* London, Seeker and Warburg, 1949.

Huxley, Aldous. *Brave New World.* London, Chattœ Windus, 1932.

Herbert, F. *The Eyes of Heisenberg.* London: Sphere Books, 1966.

Miller, W. M. *A Canticle for Liebowitz.* Philadelphia: J. B. Lippincott Co., 1959.

And many, many others. Science fiction writers seem to have concentrated on the negative outcomes—worlds controlled by evil—understandably, perhaps, because of the extreme need to avoid those consequences. But one gets the feeling that they too are locked into the preglobal planning assumptions, that their tacit recipe for a good outcome is a politically anarchistic one: do as little planning, allow as little institutional authority, as possible. If only they would help us by imaginatively exploring the ways in which a globally planned planet might reach a happier end!

On the methodology of science and its relation to social morality see the works of Feyerabend, Popper (Section II), Wolff (Section IX), and Vickers (Section VIII).

Section VIII

Philosophers have been writing about the nature of freedom for a long time, though often without achieving any useful insights. My own views have been conditioned by John Locke's remarks in chapter 22 of his *Essay:*

Locke, J. *An Essay Concerning Human Understanding.* 2 vols. London: Everyman, 1965.

Few of the traditional discussions seem to have touched the notions of social freedoms in any deep way largely because philosophers have been tacitly committed to the view that freedom, if it exists at all, is a purely intrinsic feature of individuals (a hangover, one suspects, from Christian theology formulated in the days when the pre-artifactual view of the world dominated). Recently, however, there have been some changes in this orientation; see:

Bieeman, A. K., and Gould, J. A., eds. *Philosophy for a New Generation.* Toronto: Collier Macmillan, 1970.

Burr, J. R., and Goldinger, M., eds. *Philosophy and Contemporary Issues.* New York: Macmillan, 1972.

For other important reading see Holt (Section IV) and also:

Vickers, Sir G. *Freedom in a Rocking Boat.* London: Pellican, 1972.

Correlatively, virtually none of those writing on the social responsibilities of the scientist offers any analysis of freedom or apparently sees any need to base their position on such a theory. An important exception to this is Skinner; see:

Skinner, B. F. *Beyond Freedom and Dignity.* New York: A. A. Knopf, 1971.

On the cultural roles of scientists, see:

Baier, K., and Rescher, N., eds. *Values and the Future: The Impact of Technological Change on American Values.* New York: The Free Press, 1969.

Beer, J. J., and Lewis, W. D. "Aspects of the Professionalization of Science." *Daedalus* 92, no. 4 (1963).

Bernal, J. D. *Social Function of Science.* Cambridge, Mass.: M.I.T. Press, 1967.

Boulding, K. E. *The Meaning of the 20th Century.* New York: Harper and Row, 1969.

Brown, M. *The Social Responsibility of Scientists.* New York: The Free Press, 1971.

Charlesworth, J. C., and Eggars, A. J., eds. *Harmonizing Technological Developments and Social Policy in America.* Philadelphia: The American Academy of Political and Social Science, 1970.

Crowther, J. G. *The Social Relations of Science.* London, Cresset Press, 1967.

Dupree, A. H. *Science in the Federal Government.* Cambridge, Mass.: Harvard University Press, 1957.

Ellul, J. *Technological Society.* New York: A. A. Knopf, 1967.

Feuer, L. *The Scientific Intellectual.* New York: Basic Books, 1963.

Greenberg, D. S. *The Politics of Pure Science.* New York: New American Library, 1968.

Handry, R. *The Measurement of Values.* St. Louis: W. H. Green Inc., 1970.

_____. *Value Theory and the Behavioral Sciences.* Springfield, Ill.: C. C. Thomas, 1969.

Hardin, G., ed. *Exploring New Ethics for Survival.* New York: Viking Press, 1972.

_____. *Science, Conflict and Society.* San Francisco: Freeman, 1969.

International Studies of Values in Politics. *Values and the Active Community.* New York: The Free Press, 1971.

Lakoff, S. A. *Knowledge and Power: Essays on Science and Government.* New York: Free Press, 1964.

Laszlo, E., and Wilbur, J. B., eds. *Human Values and Natural Science.* New York: Gordon and Breach, 1970.

Leys, W. A. R. *Ethics for Policy Decisions.* Englewood Cliffs, N.J.: Prentice-Hall, 1952.

Mesthene, E. *Technological Change: Its Impact on Man and Society.* Cambridge, Mass: Harvard University Press, 1971.

Nelson, W. R. *The Politics of Science.* New York: Oxford University Press, 1968.

Obler, D. C. and Estrin, H. A., eds. *The New Scientist.* New York: Doubleday, 1962.

Polyani, M. "The Republic of Science: Its Political and Economic Theory." *Minerva* 1 (1962): 54–73.

Ravetz, J. R. *Scientific Knowledge and Its Social Problems.* Oxford: The Clarendon Press, 1970.

Ridgeway, J. *The Closed Corporation.* New York: Random House, 1968.

Shils, E. *Criteria for Scientific Development; Public Policy and Goals.* Cambridge, Mass.: M.I.T. Press, 1968.

Scholler, D. *Science, Scientists and Public Policy.* New York: The Free Press, 1971.

Snow, C. P. "The Moral Un-Neutrality of Science." *Science* 133 (1961).

____. *Science and Government.* Cambridge, Mass.: M.I.T. Press, 1961.

____. *The Two Cultures: And a Second Look.* New York: Mentor, 1964.

Toulmin, S. "The Complexity of Scientific Choice: A Stock Taking." *Minerva* 2 (1964): 343–59.

Vig, N. *Science and Technology in British Politics.* Oxford: Pergamon, 1968.

Walsh, W. *Science and International Public Affairs.* International Relations Program. Maxwell School of Syracuse University, 1967.

Wharton, M., ed. *A Nation's Security: The Case of Dr. J. Robert Oppenheimer.* London: Secker and Warburg, 1955.

Ziman, J. *Public Knowledge.* Cambridge, Mass.: Cambridge University Press, 1968.

See also the works by Bernal, Crowther, Sarkman, etc., cited earlier.

Section IX

For a discussion that inspired the present one see:

Nelson, B. "Society: Today and Tomorrow, or What Can We Make of a World like This." Ontario Camping Association, Toronto, March 29, 1972. Mimeographed.

See also Vickers's work (Sections III, VIII). The work of Arrow is cited in Section IV; compare also:

Wolff, P. *The Poverty of Liberalism.* Boston: Beacon Press, 1968.

Some relevant works on education are cited in Section IV.

HAS SCIENCE ANY FUTURE?

ALASDAIR MACINTYRE

IF the title of this paper is a request for a prediction, then that request is not going to be met and for two reasons. The first is the notorious unpredictability of the future content of the sciences, an unpredictability in which we have a number of good reasons for believing. Consider just one. Suppose that some character in a piece of science fiction were to claim that he had invented a type of computer and a type of program which enabled such a computer to predict which as yet undiscovered theorems in some branch of mathematics would be proved within the next decade. It would follow, of course, that the program in question would embody a procedure which would enable us to select from the set of those well-formed formulas of a given calculus, which at a given point in time had been provided neither with proofs nor with disproofs, a subset which was provable, but not yet proved. That is, we should possess a decision procedure, although perhaps of a limited kind, for the calculus in question. But the fact is that we already have a rigorous proof that for any calculus rich enough to provide expression for arithmetic, let alone topology or the equations of the differential calculus, no such decision procedure can exist. It follows that we can have precisely the same degree of confidence that we repose in the most assured findings of logic and mathematics that the future of mathematics is unpredictable. But if the future of mathematics is unpredictable, then, given the place of mathematics in the development of the sciences, the future content of science must also surely be unpredictable. The notion of a predictable future for science must remain science fiction.

Yet fictions have an undeniably powerful role in our social lives. Every practice breeds distorted and fictional images of itself and such images may be as, or more, powerful in their effects than is the practice. The natural sciences are no exception to this. Nor are they exempt from the possibility that the distorted images of a practice may be among the key influences in determining the future of a practice. The image that we have to fear in this regard is that of science as primarily an instrument for producing natural and social effects rather than as a central form of rational inquiry. About this distorted image of science we should first note the part that it has played in making it easy for science and superstition to coexist in our society. This coexistence I take to be a crucially important social fact.

Consider California. The sociological function of California is to be an illuminating parody of the rest of the United States. California supports something over 20,000,000 people, of whom under 9,000 are professional physicists and 30,000 are professional astrologers. I select astrology as an example from among the irrational and antirational practices of the present age partly because of its intellectual pretensions; just because it is degenerate science, just because it mimics the claims of a science, the connection between the flourishing of astrology and the influence of distorted images of science may be easier to establish and to portray than a similar connection in the case of other irrational cults.

When astronomy and astrology enacted that divorce which is recorded in the statutes of the Savillian Chair of Astronomy at Oxford in 1611 (the first occasion when teachers of astronomy were forbidden to teach astrological doctrines), what was at issue was not only the truth or falsity of certain claims within science, but also the whole question of the place of science in human life. The methods embodied in astronomy involve a notion of the continuous questioning of received doctrine and an openness to conceptual and theoretical innovation, which together lead inexorably to the view that we cannot know what we are going to discover tomorrow and to the corollary that since we cannot know the character of tomorrow's theories, we cannot know how our lives will be affected by those theories. The same mode of discovery that established the determinate character of the planetary movements and the predictable future of planets and stars established simultaneously the unpredictable character of the human future. Contrast the relationship between the knowledge of the human future which astrology claims and its character as a closed body of doctrine. Astrologers "knew" all the key predictions of human fate in the time of Dr. John Dee or indeed of Ptolemy; were their claims true, the unpredictable would not be a key influence on human action, and it follows that the as yet undiscovered sciences would have to remain marginal to the determination of the human future.

The natural sciences then present us with an unknown future, astrology with a known. The regularities of a social world informed by the practices of science will be quite other than those of a social world informed by antisciences such as astrology. To note this is to raise sharply two questions: the first is, if science and this type of superstition are so mutually inimical, how can they coexist in our society? And second, can we expect this coexistence to continue? Must not one type of practice undermine the other?

First as to the question of coexistence: we ought to note how this possibility was never foreseen by our scientific ancestors in the climates of eighteenth-century materialism or nineteenth-century positivism. They foresaw the growth of science as resulting in an intellectual enlightenment which would destroy the roots of all irrationality and superstition. The process of secularization, so it was prophesied, would drive equally from the educated mind orthodox religion and sectarian cult. In fact the erosion of orthodox theism has been accompanied by a quite new flourishing of irrational sectarian cults and astrology has a considerable educated public. At the very least the relationship of theism, science, and superstition is not what a Condorcet or a Comte believed it to be.

To this it may be retorted, and rightly, that it is not the sciences themselves but those distorted images of science of which I spoke earlier which coexist so easily with superstition. But this retort may miss the point that the prophets of the enlightenment and of positivism supposed the growth of science and of education in science would necessarily be accompanied by a correct understanding of science, and in this lay one of their central errors. What did they miss? What are the roots of misunderstanding of which they failed to take account? To answer these questions, consider the different ways in which in different cultures individual agents envisage the relationship of past, present, and future. One important difference between so-called traditional cultures and so-called modern or modernizing cultures lies in the degree of predictability of the future. In modernizing cultures, as in other cultures subject to rapid and dislocating social change, the individual agent may find himself forced continuously to make plans which outrun all reasonable predictions. In such situations the impulse to search for hidden forces which control the future, to exhibit the apparent uncertainties of social life as merely apparent and superficial, is naturally at home. It is not surprising that in the period immediately following its intellectual discrediting in seventeenth-century England, a period of radical change in both social structure and political power, astrology flourished as never before. Those who are forced to calculate in a way that outruns the resources of rational prediction have a choice between socially paralyzing skepticism and looking for resources elsewhere.

As in seventeenth-century England, perhaps so also in twentieth-century California. If so, then we can perhaps also understand the link between the flourishing of astrology and the distorted images of science that are abroad. For the image of science as primarily an instrument producing natural and social effects, as primarily a means to power, consorts easily with an image of the future as an expanded version of the present, already latent in that present, and an image of ourselves as moving along predictable pathways to a predetermined and already predicted future. In such a cultural habitat we should expect a widespread flourishing of false sciences of prediction; and of course this is just what we do find in the present age. Futurology and Hermann

Kahn, the Club de Rome study and Dennis L. Meadows give the same intellectual form to their claims as do the astrologers. So does psychoanalysis in the versions of some of its practitioners. If these then are the characteristic forms of superstition of the present age, can the natural sciences hope to survive, menaced as they are by systematic social misrepresentation?

I said at the very beginning of this discussion that I was not going to predict the future of science for two reasons, the first of which was the unpredictability of the future content of science. I now come to my second reason, which is our inability to extrapolate successfully from social trends. The social trends with which I am particularly concerned are not only those embodied in the growth of irrationality and superstition, but also those embodied in changes and continuities in the social organization of science. Such trends are usually best represented by exponential curves, and we do not know how to specify functions for such curves in a way that would enable us to extrapolate from the past and the present to the future. So trends in scientific publication and in the uses of scientific manpower remain unpredictable. But it remains salutory to remember that it is not so very long since it was calculated that the rate of increase in scientific publication was such that, had it continued unchecked, the mass of such publication would by the year 2000 have been greater than that of the earth. You will be glad to know that this trend has not continued, and perhaps those trends in the division of labor in science which result in more and more scientists doing crossword puzzle science—the science represented by the subject matter of the vast majority of doctoral dissertations in chemistry, for example—will also not continue. But so long as they do, a fragmented specialization will intervene between the science that most scientists do and their own understanding of science. Scientists for whom science is only a set of discrete, technical tasks can be invaluable to science at every stage; but if the vast majority of trained scientists are this kind of person, then the tendency for science to be misunderstood will be generated within science. The kind of education that we actually give to natural scientists encourages the true thought that of any given individual it is unlikely that he will become a Copernicus or a Lavoisier or an Einstein without that other true thought that among those individuals who were unlikely to become a Copernicus, a Lavoisier, or an Einstein, were Copernicus, Lavoisier, and Einstein. The problem is not now that, confronted with the facts of scientific discovery, "every man alone thinks he hath got to be a Phoenix," but that far too few students are induced to think this. This in turn is the reason we have not yet grasped the point that you will not in general be able to educate scientists who understand science if you only teach them science. It is a familiar truth that those who have been educated in the philosophy of science and in the history of science are apt to make egregious mistakes if they have not also been thoroughly trained in at least one of the sciences. But science without the philosophy and the history of science lacks precisely that self-consciousness which any science must possess if it is going to have the resources to survive the allied threats of the rise of superstition and of the contemporary forces of the division of scientific labor. But the philosophy and the history of science cannot flourish unless philosophy and history flourish, both because science requires educated scientists and because science requires an educated public of nonscientists.

This is the point in the argument and it may be the first point in the argument with which everybody is liable to agree. But the ease with which we agree to propositions of this kind is itself suspicious, especially when it is contrasted with the difficulties we encounter in agreeing on any concrete proposals for educational change. Those forces which threaten the future of science threaten intellectual life in general and they do so at the climax of a period in which the impulse to professionalization and the varying provision of government and private money have had far more impact on what and how we teach than have the content of the disciplines themselves.

But my chief aim in this paper is not to substitute the accents of Cassandra for those of Condorcet. To do so would be to substitute just one more version of a superstitious belief in a predictable future for those that are now prevalent. I have tried to identify the threats to science with a view to specifying tasks rather than making predictions. There are, in particular, two tasks the importance of which I want to underline. The first is to establish that those discussions which bring to the fore the question of the relationship of the rationality of the sciences to other forces of rationality are of crucial social importance. How do we draw a line between degenerate sciences, such as astrology and phrenology, and corrupted sciences, such as German physics or Lysenko biology, on the one hand and genuine sciences on the other? That discussion of such questions which links the names of Polyani, Popper, Kuh, Lakatos, and Feyerabend is one example of the kind of intellectual enterprise which ought to have a higher priority for financial support than either cancer research or inquiries in neurophysiology. For to prolong physical existence in an intellectually corrupt world is a doubtful blessing and to go on giving overriding priority to doing science itself may in the end be destructive of science. Foundations please note.

The second task is closely linked with the first. It is that of not allowing superstition to flourish by default. The academic community ought to be scandalized and to feel itself disgraced by the high sale of astrological books in university towns. Such symptoms of educational failure are never taken seriously at the moment. But if the exposure of astrology and of the needs on which it feeds is urgent, so also is that of its academic counterparts among the social scientists and the psychoanalysts. Here again it is a matter not merely of social arguments but of curricular reform and of financial priorities. Foundations again please note.

The cry that is likely to go up will certainly have the authentic note of pain which is always produced by a threatened injury to vested interests; and those who cry out are all too likely to understand the right to intellectual freedom as the right to propagate superstitious nonsense, for it is true that nobody has as yet succeeded in drawing an intellectual or a social line that will safeguard the first without allowing the second. We therefore ought not to legislate against astrologers, but we ought systematically to teach our children that pernicious nonsense is what it is and not another thing.

The performance of both of these tasks will not necessarily be sufficient to ensure the future of science. But to fail in them is to render that future far more dubious. The question of whether science has a future or not is a question of what those of us who are guided by its norms are prepared to do now.

INTIMATIONS OF UNITY

MICHAEL W. OVENDEN

AT the outset, I have two problems. The first is to understand why I have been asked to join this panel. I still do not know why. However, I console myself with the thought that, since the "science" with which "The History and Philosophy of Science" deals is (for better or worse) the product of *scientists,* the views of a scientist (not a "galloping technologist"!) of such matters are not wholly irrelevant. So, should you find my presentation trivial or banal, I ask that you accept it as an empirical datum, an anthropological case history if you like, and make of it what you will. And should you find some of my remarks emanating strangely from scientific lips, I can but assure you that I speak in all sincerity. As Dean Swift said apropos of the writing of *Gulliver's Travels,* "It is my chief . . . end . . . to vex, not to divert."

My second problem is to acquire a meaning to the question "Has science any future?" I still do not know the meaning of the question—but I do know the *answer;* it is "yes and no." Yes—if by "science" you mean that curious activity of homo sapiens whereby he and she seek order in the apparent chaos of their experiences; this activity will surely continue as long as homo sapiens endures. No—if by "science" you mean the current techniques and philosophies which homo sapiens uses in his search. In this sense, science, current science, has no future; it never has had and it never will have! We may single out certain figures and so intellectually emaciate them that they may stand as symbols of significant changes in the scientific viewpoint; but the development of science has been and will be, in truth, a continual flux in which the fulfillment of an understanding comes only with its supersession.

With the completion of these preliminary remarks, I now proceed to my more formal presentation, which I have entitled "Intimations of Unity."

We have but to look around us to discern that there is a deep malaise in the relationship between science and society. No doubt the attitude of the nonscientist to science has been always an ambivalent amalgam of faith and mistrust. But the omnipresent retreat, lemming-like, from "rationality" betokens a profound lack of confidence in science as a panacea for the problems that science, with its twin technology, has helped to produce. This lack of confidence is well justified! The failure of what we currently conceive to be the rational, scientific approach to problems of ecology and the social sciences is proving little less than disastrous. It is easy to discern the symptoms of the malaise—less easy, perhaps, to diagnose the root cause. Yet it is vital that the correct diagnosis be made, and the necessary remedial action be taken, if science (and indeed Peoplekind) is to have a future.

I have found a prevalent tendency for the nonscientist to blame the failure of science in a social context on the sheer technical ineptitude of scientists—an example of the aforementioned ambivalent attitude. This is too facile an answer. The real problem lies much deeper and is of much greater significance. It is, I submit, the failure of what I shall call the "analytical" method; the belief that the way to understand things is to break them down into the smallest possible pieces (molecules, atoms, subatomic particles) and understand these smaller units. Indeed, it used to be that the smallest known subatomic particles were called "fundamental" particles, until, with the increasing power of particle accelerators and the discovery of an increasing multitude of particles of less and less stability, the joke began to wear thin. I would like to quote from an article published recently by the much-respected Vancouver architect Arthur Erikson:

> Part of our Western outlook, which we must grow out of, stems from the scientific attitude and its method of isolating the parts of a phenomenon in order to analyze them. This has become a universal thought process having the unfortunate results of isolating phenomena from their context.
>
> But there is evidence that this is changing. There is a slow but increasing awareness of the interrelatedness of things. We are becoming less prone to accept an immediate solution without questioning its larger implications.
>
> And perhaps the single greatest influence on this change has been the issue of the environment and ecology. These are fashionable issues at the moment, shortlived maybe in their present intensity,

but out of the environmental-ecological issue has come an increasing consciousness of the interrelationships of systems.[1]

A not wholly unfair example of the analytical method would be to take a wristwatch, smash it into its component molecules with a sledgehammer, and then exclaim with interest, after much thought, that out of these molecules it would be possible to build a watch! How could such an absurd idea of the way to understand the world have arisen? It came to fruition with Cartesian philosophy and Newtonian mechanics; with the purely practical necessity of treating *isolated* systems in order to limit the number of parameters with which one had to deal. It carried with it the corollary that, in order for an isolated system to be a meaningful concept, it was necessary to suppose that objects were located in space and time.

It is interesting to trace the development of the geometrization of space and time in the Western consciousness in fields outside science. For example, in musical notation, the ninth-century neumes used for recording chants had no time scale (the words themselves determining the duration of the "notes") and no fixed pitch scale. The staff began to be developed in the late tenth century, but it was only from the twelfth century and onward that geometrical representations of time values began to be developed. It is no coincidence that this development was proceeding with vigor not long before Nicholas of Oresme began to treat concepts such as velocity and acceleration geometrically; no coincidence that Newton was not until the time when the present notation had become more or less stabilized.

It is also of interest to note that the idea of *determinism* is inextricably bound up with the concept of an isolated system. For it is *only* if a system is isolated that it is determinate, in Newtonian mechanics. In nonisolated systems, there are unmodeled "forces" whose effects, over a long haul, will significantly distort the system in unpredictable ways. Lest it be thought that science would be impossible if systems were not determinate, I would remind you that there is an essential indeterminateness associated with quantum mechanics, and that in quantum mechanics isolated systems do not exist.

Furthermore, the idea of *experimentation* is also closely bound up with the possibility of treating systems as isolated. In a typical experiment, we suppose that the effects of the rest of the universe on the subsystem of our interest may be mimicked by suitable boundary conditions. When this is not possible, we must view with severe reservation the process of arriving at insight by experimentation. We meet such a difficulty in social systems, and therefore we must be especially critical of programs of social experimentation, even when we may remove from them their more Pavlovian overtones.

Of course, there has long been a continuing stream of thought opposed to the idea of isolated systems located in space and time. This stream has been largely outside the realm of science, but one might mention, as persons close to science, Berkeley and Leibniz in Newton's day through to Mach, White-head, Heisenberg, and Heitler in more modern times. Their protests went largely unheeded, and now the chickens have come home to roost! And not only in the social sciences and ecology, but within the archanalytical physical sciences, interestingly enough in the fields of the very small (quantum mechanics) and the very large (cosmology). I will merely mention the well-known fact that, in the quantum-mechanical description of atomic systems, isolated particles do not exist, but only probability functions which, in principle, permeate the whole Universe. I will take three examples from a field to which I am closer (note the geometrical metaphor), namely, cosmology.

Even within Newtonian mechanics there is an artificial element about the idea of an isolated body; the gravitational field of the Earth, for example, stretches to infinity—it is nowhere cut off. Consider then, a single body in an otherwise empty universe. How can we talk about its motion, for we have nothing against which to measure its motion? If we introduce a second body, we can say something about their relative motion along the line joining them, but nothing about rotation or revolution until we introduce a third body; and so on, and so on, until we have included the whole Universe! Newton's solution was to define an absolute space and time reference frame with respect to which motions could be measured—but there was always something peculiar about this absolute frame, since only accelerations relative to the frame could be measured, not velocities. Berkeley, and much later Mach, suggested that Newton's "absolute frame" was nothing but a surrogate for the rest of the universe, it being fortunate that in this matter of dynamics the local irregularities of the "nearby" universe are dominated by the sum effect of very distant parts of the universe, which have a high statistical degree of isotropy. Even today, many physicists who do not relish this direct challenge to the analytical method resist Mach's conclusion. To my mind, it is inescapable. A very simple (albeit far from rigorous) calculation can be made of the net gravitational force of the whole universe on a body accelerating relative to it. The calculation yields a net force opposing the acceleration. This is the *inertial* force, and it is due to the gravitational effects of the rest of the universe. Not only is the equality of gravitational and inertial mass explained (an otherwise contingent feature of Newtonian mechanics), but furthermore the constant of gravitation can be calculated from such purely cosmological quantities as the mean density of the universe and Hubble's constant. To an order of magnitude, which is all the observations will stand, the answer comes out right! Mach's principle was one factor that led Einstein to his General Theory of Relativity, the only theory to date that has had some success in coming to grips with a "holistic" universe.

My second example concerns Olbers's Paradox. Olbers showed that if the stars were stationary, similar, and uniformly distributed throughout infinite space, then the sky at night should be as bright as the surface of the Sun, because any line of sight would eventually hit a star. Olbers tried to explain the fact that the sky is dark at night by postulating sufficient interstellar matter to absorb the light of distant stars. His explanation will not work, because the absorbing matter would heat up until it reemitted just as much light as it absorbed. Of course, we know that the stars are not distributed at random through space, but form a system which we call our Galaxy; but Olbers's arguments can be applied *mutatis mutandis* to the system of galaxies. The point that I wish to emphasize is that, just as with Mach's Principle, it is the distant parts of the universe which, in sum, dominate the picture. The fact that the sky is dark at night tells us something about the remote parts of the universe, namely that they do *not* obey Olbers's postulates.

My third example is of a somewhat different character. Imagine that, within a box, we have a large number of molecules, thought of as perfectly-elastic spheres; the typical paradigm of the kinetic theory of gases. You well know that, if at some time all the molecules lie in one half of the box, after a lapse of time they will distribute themselves more or less uniformly throughout the box. The converse process, that of the molecules spontaneously collecting themselves into one half of the box, is not (often) observed. The statistical mechanical explanation goes something like this. If we do not allow ourselves the luxury of labeling every molecule and of knowing exactly where each molecule is situated and how it is moving at a given time, but allow ourselves to know only the statistical distribution of molecules within the box, we can say that a distribution with the molecules more or less uniformly distributed is more probable (in the sense of corresponding to more permutations of the different molecules among the different possible positions) than one with the molecules all in one half of the box. The uniform distribution is *less ordered* than the nonuniform distribution. Since we will more often, taking all possible observations for all time, observe the more probable distributions, we would expect (most often) to find the system changing from the less-probable to the more-probable distribution. We call this the Second Law of Thermodynamics, that a system evolves from a more-ordered to a less-ordered state; that its *entropy* increases.

This apparently gives a direction to time. Yet this is impossible, since the individual molecular interactions are, by our assumption, time reversible; it is logically impossible for a direction of time to be determined (in any absolute sense) by time-reversible processes. In fact, whether you go "forward" or "backward" in time, apart from statistical fluctuations, you will find the entropy of a system to increase. But notice that such considerations are valid only if one insists on being partially ignorant of the initial state of the system. According to the laws of dynamics which we are assuming, if we knew exactly the position and motion of every molecule at a given time, we could (in principle) calculate the state of the system at any time in the future (or the past). Then there is no question of the order of the system changing with the time, since each state prehends within it every other possible state of the system. Given a certain degree of ignorance of the details of the initial state, we can postdict the past of the system with exactly the same degree of uncertainty as we can predict its future. The so-called Law of Increase of Entropy is nothing more than the progressive unfolding of the consequences of our ignorance of the initial state. The box (which is an essential component of the argument) is a device for containing our ignorance within certain bounds.

The obvious significance of all this to cosmology is that we do find ourselves in a universe where matter is "organized" (i.e., in "improbable" states). But I remind you that "improbability" is a concept related to our ignorance of the full details of the system, and is of doubtful significance when applied to the universe as a whole. Because there can be no such thing as an isolated system, we will, perforce, be always in a state of greater or lesser ignorance, and there must always be an unfolding of the consequences of this ignorance in any system that we consider.

We have seen, then, examples of how, even within the physical sciences themselves, the limitations of the analytical method are becoming apparent. How much less likely is such a technique to succeed with the vastly more complex interrelationships of social systems. Is there an alternative? Some have supposed not. Thus Bertrand Russell wrote:

> I do not contend that the holistic world is logically impossible, but I do contend that it could not give rise to any empirical knowledge.[2]

Or again, Simone de Beauvoir:

> Monism ultimately leaves the sage no alternative but mysticism, fatalism inevitably leading to quietism.[3]

Now an apprehension of the unity of the universe does, maybe, require a certain quietude:

> When the mind is disturbed, the multiplicity of things is produced; but when the mind is quieted, the multiplicity of things disappears.[4]
> [There is a piquant irony in the fact that the state of mind which we vaunt as scientific wisdom is characterized by Buddhist philosophy as one of ignorance (*avidya*)!]

However, I believe that such pessimism regarding the viability of a holistic model of the universe is unwarranted—that it is in itself a symptom of the even deeper effects of the malaise that we are considering. Our logical processes of thought seem to be tied to an analytical framework. Do we not refer with approbation to a logical argument that it is "incisive"? Indeed, there is currently a specific debate as to whether or not the relationships between quantum mechanical variables can be accommodated within present analytical logic. If they cannot, so much the worse for present analytical logic! The analytical idealization has proved itself to be very useful in its proper place, as a tool. But usefulness is one thing—philosophical validity quite another. Yet a person's philosophy is of paramount importance as a filter of his or her experiences. So much information streams into our senses that we would quite literally go mad unless we unconsciously rejected the greater part of it. Our philosophy (recognized or unrecognized) is the filter. As Marshall McLuhan so graphically put it:

> The subconscious is the accumulating slag-heap of rejected awareness.[5]

So when you seek scientific "facts" to form a judgment or to bolster a prejudice, remember that there is almost no such thing as a "scientific fact," but only scientific "experiences" filtered by individual scientists. That these experiences have a certain degree of repeatability and cohesion is a necessary, but not sufficient, condition for science to be possible, in the sense of agreement among scientists. A further condition is the (implicit) agreement to a set of rules whereby a scientific statement is judged. Herein lies the supposed "objectivity" of science—an agreement as to rules of procedure. I do not make this statement pejoratively; these rules are a valuable discipline and such discipline should not be lightly foresworn. But insofar as the rules reflect the philosophies of scientists, they do not of themselves have any objective character. It is my belief that the most serious effect of the past successes of the analytical method is that the tool has usurped the functions of a philosophy, and that those who hold firmly to this philosophy are either incapable of recognizing within themselves any experiences which cannot be "explained" in analytical terms, or are at the least incapable of building them into a coherent world-picture. This leads directly to a separation of "people" from "the external world," and a lack of proper ecological concern. The terminology of science is too full of metaphors about "wresting Nature's secrets from her." It also leads directly to a separation of person from person, the elevation of a false conception of individuality which lies at the root of that sense of alienation so obvious in modern Western society. As Stephen Toulmin has pointed out in chapter 6, alienation is not peculiar to modern Western scientific communities, so science cannot carry all the blame. But the search for a single "cause" for a phenomenon is once more a reflection of our analytical predispositions. Certainly, we may say that the prevalence of an analytical scientific ideology can but *reinforce* such sense of alienation.

In short, as a tool the analytical method will have a part to play in the future of science. As a philosophy, it has had its fanfare here, and it is time that the trumpets sounded for it on the other side!

Where then to turn? Certainly we should not reject holus-bolus all the insights that analysis has given us. We might try to extend our analysis to consider systems, as in cybernetics or systems engineering. But this is not enough—it is still the building of a wristwatch out of molecules. No—what is needed is not a patch-up job of extending the reach of our previous awareness, but rather to tap for our world-picture new sources of awareness that lie in the Marshall McLuhan "slag-heaps." We must turn again to *intuition.* I almost said "emotion"—but "emotion" is such an emotional word.

I say "again" because science is no stranger to intuition—indeed, intuition is the driving force of science. Contrary to popular belief, science is not only nor even primarily a mining operation for "facts of nature." The reason is not far to seek. It is never possible to obtain by induction from experiment or observation a "valid" empirical law of nature for the simple reason that our observations are finite in number, and a finite set of observations can always be represented by an infinite number of different empirical "laws." Our choice of empirical law is based upon the aesthetic grounds of simplicity. At a later stage we have a physical model from which we deduce that certain results will follow from making certain experiments. Some of these experiments will have been already made, some yet to be made. Usually the predictions from a model are made in the form of mathematical equations, and it has been argued that the "model" is but the flesh and blood with which our imaginations clothe the essential mathematical relationships; only these relationships are tested. But there is more to it than this. The model is somehow *richer* in its potentialities for development than the bare equations themselves.

But how do we get from an "empirical law" to a "physical model"? Certainly not by explicit logic. No—by *intuition*. As if in a flash, the world is seen in a new perspective. It may take months or years to confirm and cross-check, but in a very real sense (and whatever the final verdict of posterity may be upon the value of the new perspective) one knows at the beginning that the answer will come out "right." As Thomas Kuhn has written:

> Scientists often speak of the "scales falling from their eyes" or of the "lightening flash" that "inundates" a previously obscure puzzle, enabling its components to be seen in a new way that for the first time permits its solution. On other occasions the relevant illumination comes in sleep. No ordinary sense of the term "interpretation" fits these flashes of intuition through which a new paradigm is born.[6]

Whence cometh this intuition? Why this consonance between the "external world" and the human psyche? I do not wish to examine this problem here, save to make one point. I remind you of Alan Watts's Parable of the Appleskin.[7] We might parody Mach's Principle by "This apple is as it is because the universe is as it is." But the appleskin is no more than a boundary between the apple and the universe, and such a boundary may be read either way. We might just as well formulate the Inverse Mach's Principle, viz., "The universe is as it is because this apple is as it is." Perhaps the universe and consciousness are but two sides of the same coin. We can see a striking confirmation of such a viewpoint in quantum mechanics. (The parallelism between Eastern philosophies and modern quantum mechanics has been argued cogently by Fritjof Capra.)[8] Quantum mechanics describes the world in terms of probability functions. Now *if* there were an underlying deterministic process (as Einstein believed), the probabilities could be defined with reference to this process. If, however, there is *no* underlying determinism (as most physicists believe), then the probabilities can be defined only in terms of predictions that we may make; these predictions depend not only upon our doing an experiment but also upon our *knowing the result*. Our cognition is an inseparable element of the real world, a point which Heitler (among others) has stressed.

Of course, there are problems with intuition; I quote Joseph Wood Krutch:

> The inner voice has whispered too many things to too many different men for me to have any conviction that it's always right or that it comes from anywhere except merely from within. But some voices one must listen to, and when this voice speaks to me its authority, however little it may be, is at least as great as the authority of the latest editorial in the latest weekly or monthly review. The hardest facts, as Havelock Ellis once remarked, are the facts of emotion. Joy and love, for example, cannot be doubted when one feels them. I know that they exist *in* me and *for* me when I heard the first peepers of spring and when I watched spring turn to summer. I cannot regret that I did so. I hope that whether the rest of the world is headed towards success or failure in its largest enterprises, I shall be permitted to watch with equal satisfaction at least one spring come again.[9]

However, I do not advocate the *undisciplined* use of intuition. Just as in past science, we must subject each intuition to the touchstone of whether it *works*—not merely in some general sense and not only for the "individual," although I would wish to see a greater willingness to consider different persons' intuitions not as competing with, but as complementary to each other. I do not advocate a break with the disciplined nature of past science—for rationality lies, I submit, in just such discipline—but only that we broaden the base of awareness upon which to build our understanding. It is but a historical accident that "science" has come to be associated with a spurious philosophy that has blinkered our perception of the unity of the world and ourselves. If we can but remove these blinkers, I believe that a truly rational discipline can be evolved that will *contain* past science but which will transcend it in being wider, more fruitful and more satisfying. If we can but do this, then we have many springs to which we may look forward, each richer than the one before!

NOTES

1. Arthur Erickson, "The Locusts that Eat Culture," *Vancouver Sun,* February 16, 1973, p. 5.
2. Bertrand Russell, *The Philosophy of John Dewey* (New York: Tudor Publishing Co., 1939).
3. Simone de Beauvoir, *The Long March,* trans. Austryn Wainhouse (Cleveland and New York: World Publishing Co., 1958).
4. Acvaghosa, *Discourse on the Awakening of Faith in the Mahayana,* trans. D. T. Suzuki (Chicago: The Open Court Publishing Co., 1960).
5. Marshall McLuhan, *The Gutenberg Galaxy* (Toronto: University of Toronto Press, 1962).
6. Thomas Kuhn, *The Structure of Scientific Revolutions* (Chicago: Chicago University Press, 1962).
7. Alan Watts, *The Book* (London: Jonathan Cape, 1969).
8. Fritjof Capra, "The Dance of Shiva," *Main Currents,* vol. 29, no. 1, (New York, 1972) p. 15.
9. Joseph Wood Krutch, *Baja California and the Geography of Hope* (New York: Sierra Club, 1968).

POSTSCRIPT

XII   Nicholas Copernicus:
The Gentle Revolutionary

FRANK H. T. RHODES

I AM HAPPY to take this opportunity of expressing the thanks of the whole University community to the sponsors of this exciting symposium. We thank the University of Michigan's Kopernik Five Hundredth Anniversary Celebration Committee, and especially Professors Steneck, Ehrenkreutz, Mohler, and Hiltner, and Dr. Ostafin, who have worked with such energy and dedication to bring about this symposium. We also thank the Michigan Academy of Science, Arts, and Letters, the Michigan Polish-American Congress, and the University Extension Service for their cooperation and support. And finally we thank the National Science Foundation and the American Council of Learned Societies, without whose financial sponsorship this symposium would not have been possible.

Our theme, "Science and Society: Past, Present, and Future," is sweepingly comprehensive. It is discussions such as this, spanning the traditional boundaries between the departments and colleges, that are so essential if we are to draw on the strengths inherent in a university, rather than succumbing to the fragmentation and isolation that arise from the quite proper but inevitably narrower preoccupations of our traditional disciplines and departments. Indeed, in the larger sense, it is debate on topics such as this which are essential if we are to bring what collective wisdom we have to bear on the problems that now confront us. Although no topic is of more concern as we face the problematic future of the human race than the relationship between science and society,

377

even that topic needs to be interpreted in the wider sense in which this symposium has addressed it. For we need not just technical expertise, but also wisdom. We must know science, even as science does not know itself, as one author has expressed it. You have discussed in depth the history of the Copernican revolution, the present implications of the relationship between science and society, and you have gazed into the problems that confront us in the future. At the end of three days of extensive lectures, what is there left to say?

As one looks at the symposium program it is interesting to see the change in tone that the topics represent. The symposium began on a note of celebration, which was reflected in the titles of the papers: "The World of the Young Copernicus," "The Impact of Copernicus on Man's Conception of His Place in the World," "The Assimilation of Science into Our Ways of Thinking and Living," and so on. That note of celebration gave rise to a note of boldness and opportunity as the symposium progressed. Thus we had lectures on "The Ecology of Knowledge," "Biological and Social Theories—A New Opportunity for a Union of Systems." Yet, what began as a celebration gives the impression of having rapidly become a wake. A symposium that began with an air of celebration now concludes on a note of pessimism and gloom. The final lectures are intriguing in the implications of their titles: "Science Education and the New Heretics," " 'Show Me a Scientist Who's Helped Poor Folks and I'll Kiss Her Hand,' " "The Twin Moralities of Science," "Has Science Any Future?"

Let me put this celebration and wake into a kind of perspective which I suspect most of us would accept. Copernicus has brought us from the closed world to the infinite Universe. The revolution which he initiated has changed our whole scale of thought, has given us such a new perspective that it represents a milestone in the long history of human understanding. It has liberated us from the first intuitive concepts of our senses; indeed, it has changed our whole notion of common sense, has vastly refined our perception, has revolutionized the way in which we see man in the relationship to the Universe, and has revealed pattern and order even in the depths of space. In a more general sense, science itself represents a particular form of reasoned, disciplined inquiry, which is an important component in the continuing identification and clarification of those perceptions and affirmations that constitute the basis of our understanding. Indeed many of the props of our present society represent the direct or indirect fruits of science and technology. Science, wisely and widely used, has provided freedom from toil and hunger, from need and fear, without which any discussion of the quality of life can be only a meaningless abstraction.

Yet, this optimism is scarcely one that we can justify without some qualification, in the face of the present condition of mankind. Science has brought not a brave new world, but a world divided. "I preach the Gospel of science," the editor of *Nature* wrote in 1959. "My father preached the Gospel of Socialism and my grandfather the Gospel of Christ." That statement, though comparatively recent in origin, has a pathetic and empty ring about it now. Science is now rarely regarded as benevolent. Impersonal, narrow, and mechanistic in its view, it is seen as much as a threat to man, as it is a means of salvation. Even the wise application of science and technology is seldom without its fringe hazards. The insecticide that wipes out malaria, also devastates bird populations. The dam that brings power to a vast area, also brings the scourge of schistosomiasis. It was John Lindsay who described us as the first generation to see on our television sets the landing of men on the moon, while in our cities we stand knee-deep in the garbage of our affluent living. The mistrust and the suspicion of science are part of the cynicism of our own age.

In one sense, the deepest problems we face are themselves the result of science and its success. We live in a society which is now threatened by the totality of its scientific dependence. The industrialized nations are wholly dependent on scientific and technological skills. Building materials, industrial processes, travel, communication, fabrics, fuels, fertilizers, personal health, and national strength now rest firmly on the foundation of science. Nor is the situation different in the developing countries. Dependent alike both on their own emerging technologies and technological assistance from the industrialized nations, the effects of science on the lives of their individual citizens are equally profound.

Society is also threatened by the health of its own members. Unlike every other species whose survival has depended on reproductive success, man is now the victim of his reproductive vigor. The burgeoning global population, still increasing at a rate of 2 percent per annum, poses a major threat to the survival of society as we presently understand it. If the present rate of population growth continues, it is clear that the planet cannot long support the numbers that will be produced. Given the present rate, at which world population now doubles every thirty-five years, widespread famine seems probable within a few generations.

Society is threatened by the damage and insult to the environment which our careless technology produces. We now face the pollution of the hydrosphere and the atmosphere on such a massive scale, that their recovery in some areas will only come about from drastic changes in our society and technology. Pollution is not just a passing slogan of the strident young. The environment is not just a prudent plank in the political platform. Pollution involves the air we breathe. The environment embraces the water we drink.

Society is further threatened by the extent to which it has plundered this planet. The environment has a capacity for self-renewal over long cycles, but our dwindling mineral resources, fuels, and constructional and industrial materials are nonreplenishable. The reserves of some commodities are now so depleted that most informed assessments give them an economic life expectancy of only thirty to fifty years. This is true of many of the most widely used metals, including lead and copper, and of our petroleum reserves. New energy sources, such as fast breeder reactors, are still far from developed. Synthetic materials, which may reduce the relentless consumption of metals, are often petroleum derivatives, and petroleum itself is in short supply. The responsibility to investigate this imminent material famine is unlikely to be assumed by private industry. It is scarcely even recognized by the major governments of the world.

Science-based society, threatened by the prospects of overpopulation, environmental destruction, and exhaustion of resources, provides science with a new context and a new challenge. The survival of science depends on the survival of society, though not on that alone. The survival of society depends on the survival of science, though not on that alone, but each depends on the other. But there is in this a paradox of global dimensions, for the revolution of Copernicus involved not only a revolution in man's understanding of the universe, but also in man's understanding and vision of himself. It was the work of Copernicus that displaced man from the center of the cosmos. Animalized by Darwin, mechanized and conditioned by Freud and Skinner, man is now a stranger to himself, an alien in a meaningless universe. This state of alienation is itself partly a product of the Copernican revolution. The paradox lies in the fact that, whereas the means of restoration of the life and health of society lie in the skills and remedies which only science can provide, the will and the wisdom to gain restoration by the application of those remedies, lie in the very virtues of compassion, wider vision, and more mature wisdom which science itself is said to threaten.

The discovery that man now faces an alien universe is not a new one, as several historians have reminded us. John Donne, writing in 1611 the first of two long elegies on the death of the fifteen-year-old Elizabeth, the only child of Sir Robert Drury of Hawstead in Suffolk, reflected the threatening impact of the increasingly general acceptance of the Copernican viewpoint.

> The new Philosophy calls all in doubt,
> The Element of fire is quite put out;
> The Sun is lost, and th'earth, and no man's wit
> Can well direct him where to look for it.
> And freely men confess that this world's spent,
> When in the Planets, and the Firmament
> They seek so many new; then see that this
> Is crumbled out again to his Atomies.
> 'Tis all in pieces, all coherence gone;
> All just supply, and all Relation:
> Prince, Subject, Father, Son, are things forgot,
> For every man alone thinks he hath got
> To be a Phoenix, and that then can be
> None of that kind, of which he is, but he.

Truly the Copernican revolution has cast a long shadow. In the despair and alienation of contemporary man, and in his loss of meaning and significance in a hostile universe, "tis all in pieces, all coherence gone."

So real is the protest against the reductionist spirit of science, and the conviction that it is chiefly "the myth of objective consciousness" that underlies our present ills and discontents that there are those who would reject the Copernican revolution. They tell us that science must now reform itself by repudiating those impersonal attitudes which are among its characteristics, because it represents an unbalanced and unhealthy development in human thought. This counterscientific attitude takes many forms, ranging from the anti-intellectualism of some to the genuinely humane and compassionate concerns of others. In its most typical form the movement represents a protest against scientism, about which I shall have more to say later, and the assertion that science must now undergo a new revolution, emerging, phoenix-like, as a "new science" in which social relevance and public utility will become the new criteria for scientific support and procedural priority. Science, it is argued, has ignored the needs and interests of man, and only a new revolution and a new science can correct this destructive tendency.

Much as I respect the sentiments of those who advocate this position, I believe that there are two fundamental weaknesses in its assumptions. The first is that it ignores the real basis for creativity in science, which is personal commitment and personal curiosity. Commitment takes many forms, and it is entirely probable and proper that there should be a substantial increase in the number of scientists whose research interests are guided and motivated by a sense of public concern. But to make this a fundamental requirement for all scientific activity is to ignore the essential loneliness of the scientific pioneer. No federal funding agencies, no professional society counsels, no multidisciplinary team projects should blind us to the fact that it is ultimately personal vision, individual hunches, and private insight which lie at the heart of scientific progress. If any testimony were needed to the truth of this, surely the life and work of Copernicus himself would provide it. Furthermore, "useless research," which has no immediate social or practical application, provides an essential reservoir of scientific knowledge and personal skills from which man will need to draw as he grapples with the new problems he must solve in the years ahead. Who could have foreseen, for example, that the chance discovery of radioactivity by Becquerel in 1896 could have had such far-reaching applications? From antibiotics to X-rays, the applications of scientific discoveries are seldom fully clear at the time the initial discovery is made. Indeed, the problems to which they later offer solutions may not even exist at the time of the original discovery. The implications and the applications of knowledge are not necessarily foreseeable. To destroy this bank of knowledge, this background of understanding, this established cadre of skilled practitioners, would have a crippling effect on the ultimate public usefulness of science.

There is another fundamental weakness in the position of those who argue that public utility should now become the chief criterion for science. This claim makes the assumption that social ends and public goals, if they are to provide an encompassing context and motive for scientific research and its public support, are themselves clearly evident. If, indeed, the goals of society are regarded as given and fixed, they are exempted from the kind of public scrutiny and clarification that free inquiry within democratic societies should involve and provide. One of the functions of the universities and their faculties is to act as agents of criticism of social practices and goals themselves; for only by standing apart from current social mores and priorities, can they provide the clarification, revision, and the refinement of social goals, which is the hope and birthright of each new generation.

We cannot reject and we should not repudiate the revolution that Copernicus initiated, however much some may wish to. The movement that he began now lies at the heart of science, and it is idle to pretend that either science or society would gain if we were to change the basis of the scientific enterprise. Copernicus did indeed lead us into a new world, a new universe, and we cannot deny our citizenship within it. We dare not bind the creative energies of man to needs presently conceived. Our surest safeguard against future adversity is a massive store of knowledge against problems as yet unrecognized and unknown. We must retain and respect the basis and methodology of contemporary science. The problem is not that the foundation of science is an unsatisfactory basis for science. It is rather that it is a wholly inadequate basis for life. Harry Woolf Ontio has written,

> However widely the ideal of strict detachment may help to guide science to ever new discoveries, we must not allow it to deprive our image of man and the universe of any rational foundation. All men, scientists included, must seek and hold on to a reasonable view of the universe and of man's place in it.

Our real problem is that the scientific attitude has degenerated into scientism, by carrying the attitudes of abstraction and reductionism over into fields where they are not logically appropriate. The limited view, the narrowed question, the restricted definition, the abstracted perception, all these are the secrets of scientific success. But they are disastrous as a basis for life or society. Henryk Skolimowski is right when he asserts that it is the superficial ideology of science and the inappropriate quantification of nature and experience which "atomize and mutilate the texture of our human and social life."

So we may not go back. We cannot undo the revolution of Copernicus, nor should we lament that we live in both its brightness and its shadow. For better or for worse we are the heirs of Copernicus, and also Brahe and Kepler, Galileo and Newton. We cannot return to the security of our medieval matrix. The Copernican view of the Universe is real. The reorientation which he provided was needed. The methods that ultimately formed the basis of modern science are powerful. We cannot and we should not deny the revolution. But we should not so idolize the revolution that we ignore the man behind it. How many books describing the Copernican revolution, for example, scarcely mention Copernicus himself. What we need is not a repudiation of the Copernican revolution but rather a return to the spirit of Copernicus the man. For science is a private venture, and if it is to be used for the benefit of man, it is the spirit of Copernicus which must underlie its practice. We know little of the private life of Copernicus. Almost all his correspondence has been lost, and we are forced to rely largely on glimpses of the man that we obtain through his writings, or the letters of his friends that refer to him, and on ecclesiastical records, and much later biographical fragments, which deal chiefly with his public life and clerical duties. I suggest that the spirit of Copernicus was represented by five characteristics of the man and his work. The first of these was a sense of fitness that arose from personal intuition and from a bold individualism. Copernicus complained of Ptolemy and the Ptolemaic scheme, not because it was unfitting to center the universe on the earth rather than the sun, but because the constructions by which Ptolemy had formulated his Universe violated Copernicus's sense of the "Principle of Regularity." Copernicus complained that these constructions were "inconsistent and unsystematic." Copernicus has been described as a man who attained the right answer partly for the wrong reasons. Indeed, it is ironic that Copernicus's sense of fitness involved his adoption of uniform circular motion, an explanation which has long since been abandoned.

It was this sense of fitness that produced the revolution and the Copernican revolution is itself important as a model. For science, despite all its public accreditation, is not self-renewing, or even cumulative in a strict sense. Science grows, as Thomas S. Kuhn has reminded us, not only by continuous accretion, but also by such revolutions as this—in which quantum transformations in knowledge result from the overthrow of established viewpoints, fashions, and paradigms. Minor progress comes from routine, production-line research: gaps in our knowledge are filled by orderly problem-solving, undertaken according to the rules of the game. In contrast, the great flashes of insight, which represent the turning points in science, come, often after periods of deep division and doubt, when the very rules and conventional orthodoxies of the scientific framework are challenged and a competitive and new world view is offered to those who will accept it in faith. This is why we must enlist the support and the energy of the iconoclastic young of each new generation. For the greatest progress in science has generally come from those who were either young enough or rich enough to be so uninvolved, so uncommitted, and so unindebted that they could dare to challenge the orthodoxy of the current scientific structure. And behind these transformations of such immense significance in mankind's understanding, there lies the sense of fitness of one solitary man.

The inbred schools of universities are sometimes poor places for the development of such an attitude. They tend increasingly to produce efficient but lifeless apprentices, rather than men of boldness and vision who will provide the continuing skepticism, insight, and disciplined testing on which the development of knowledge depends. Only by presenting science in our universities as an ongoing quest, rather than a completed edifice, a continuing encounter with nature, rather than a volume of established facts, can we hope to rectify the pedestrian, journal-filling plodding, which is so often accepted as an adequate alternative for genuine creativity. The Copernican sense of fitness, based on deeply personal intuition, is a reminder of our obligation to exemplify and to convey a sense of the flavor, fitness, and style of science, adequate to serve the needs of both scientists and nonscientists.

There is no simple formula for imparting this sense of style and fitness, though perhaps we can best hope to convey it by demonstrating the deeply personal, creative, and imaginative character of science. The average university science course presents its particular discipline as a neatly packaged catechism, all ends tied up, all ambiguities resolved. But this is not science. Science differs chiefly from other human insights in abstracting those aspects of experience which are common to all men. As such it is the extraction of wisdom from things: it deals with universals, rather than particulars. Yet science is not just "facts." There are no things labeled "facts" in nature. Science is rather an encounter with the created world, sometimes contrived, but often open-ended. As such, it demands as profound a groping for coherence in the depths of experience, as does any work of art: its difference lies only in its public verifiability, as opposed to the particular and private insights and assertions of the humanities.

When Copernicus sought to explain his own sense of fitness, he used a remarkable illustration, drawing on his experience as both artist and physician. He wrote:

> Nor have they been able thereby to discern or deduce the principal thing—namely the shape of the Universe and the unchangeable symmetry of its parts. With them it is as though an artist were to gather the hands, feet, head and other members for his images from divers models, each part excellently drawn, but not related to a single body, and since they in no way match each other, the result would be monster rather than man.

In a quite literal sense, Copernicus used man as an illustration of the measure of fitness. If, as I have argued, we must reject the demand to overthrow the present basis for science, we should at least accept the need to reintroduce man as a measure of scientific fitness. This has a particular urgency and concern for those of us who serve as members of universities. We have been slow to accept the fact that the measure of man provides an appropriate personal motive for a scientific vocation and for commitment to particular problems, just as the wider concept of gross human benefit furnishes a crude scale against which to measure the priority and appropriateness of applied science or technology. Providing the first choice is personal, and the application of this choice does not exclude the legitimacy of others, this will neither weaken nor destroy the basis of scientific method, and the freedom required to range widely in the scientific field. Science grew from the lonely gropings of isolated individuals who sought insight and meaning in the universe of the Creator they worshipped. Only later did the methods they thus developed become of more general application. Few great works of art, few masterpieces of literature or music were produced without an element of dedication and passion. The ultimate neutrality of the most creative men and women has yet to be proved. Far from endangering science, a new commitment to human dignity and survival may prove to be its salvation.

The second major characteristic of the Copernican spirit is that he sought and demanded unity. Copernicus was, as he himself wrote, "induced to think of a method of computing the motions of the spheres by nothing else than the knowledge that the Mathematicians are inconsistent in these investigations." Copernicus maintained that the Ptolemaic system of the universe was built on an incoherent series of constructions, depending for the solutions to particular problems on ad hoc constructions that possessed no overall harmony. His indignation about this derived not only from his sense of fitness, but from his recognition of a larger inconsistency that offended his sense of unity. For while Ptolemy had avowed his commitment to Aristotelian ideals of uniform circular motion, some of the constructions which he proposed departed from that principle. The principle itself is now recognized as invalid, but the search for consistency, so well exemplified by Copernicus, provides an important basis for modern science. This Ptolemaic inconsistency was sufficiently subtle to be either ignored or forgiven by other scholars, for Ptolemy's mathematical picture of the universe, although it did not conform in detail to Aristotle's physical picture, did not represent a major contradiction. It was rather, as Hall has so clearly shown, that there was something of a hiatus between the two. What was so impressive about the new Copernican universe was the wholesale interdependence of every part of the scheme that Copernicus proposed. He wrote, "Not only do the phenomena work out correctly, but the order and sizes of all the heavenly bodies and spheres, and that of the heavens themselves, become so interdependent. . . ."

Hall has shown that, measured by existing standards of knowledge and their logical application, Copernicus's arguments in favor of heliocentricity were illogical. Although he attempted to prove that the earth's movements did not conflict with Aristotelian principles, Copernicus did this only by interpreting those principles in such a new sense that he modified their basic meaning. His recognition of gravity, for example, as a universal property of spherical bodies, is a case in point. The real implication of the Copernican revolution was that the heliocentric system which he advocated could become acceptable only if the essential Aristotelian physics were to be rejected. With the acceptance of the Copernican universe, the whole fabric of Aristotelian thought was so rejected, the supremacy of ancient authority was undermined and the priority of personal experience was established. This was not an isolated event. It found its counterparts in the contemporary turmoil of the Reformation, which represented a closely parallel assertion of the primacy of personal faith over established authority. If today the validity of such a personal viewpoint seems self-evident, we might do well to ponder its origin in the patient assertion of Copernicus that, behind the apparently chaotic appearances of the world, man was entitled to search for a deeper order and unity.

Such a statement of unity, however praiseworthy in the abstract, would be of as much practical significance, and would find as little favor, with some contemporary experimental scientists, as it did with those who argued the validity of common sense at the time of Copernicus. E. A. Burtt, for example, has written "Contemporary empiricists, had they lived in the 16th century, would have been the first to scoff out of court the new philosophy of the Universe." In this sense, contemporary empiricism is a projection of the very common sense, whose defiance by Copernicus caused him so much anguish and concern. In his preface, for example, referring to his dedication of his book to Pope Paul III, he writes, "How I came to dare to conceive such motion of the Earth, contrary to the received opinion of the Mathematicians, and indeed contrary to the impression of the senses, is what your Holiness will rather expect to hear." And so, in one sense, Copernican unity was based upon something greater than the unity of the senses. It was almost a mystical insistence that compelled him to demand unity which harmonized in a more comprehensive system the fragmented schemes to which man's primary observations gave rise. Copernicus's new model of the Universe was, as Armitage and others have reminded us, based on choice, rather than the discovery of something new, or even the compelling influence of additional observations. It is in this choice and sense of fitness, rather than in his observational skills, that Copernicus is regarded as a pioneer of modern science. He produced relatively few new data—his great work involved the use of only twenty-seven of his own observations—and those observations he did make were not conspicuously more accurate than others that were available to him. Copernicus told Rheticus, for example, that he would be content to make celestial observations correct to ten minutes, even though it is possible to measure such angles to within one or two minutes without the use of a telescope. His instruments were no more sophisticated than those used by the Greeks a millennium before him, and were considerably less refined than those of some of his contemporaries, such as Regiomontanus and Walther of Nuremberg.

In defense of Copernicus, one should remember that his tower at Frombork, where most of his observations were made, romantic though it now appears, was scarcely a suitable site for an observatory. He himself was well aware of the handicaps under which he labored. "Fortune," he wrote, "did not give me as it gave Claudius Ptolemy, that beautiful opportunity of experience. For him the skies were more cheerful, while the Nile does not upbreed fogs, as does Vistula. Nature has denied that comfort to us, and that calm air. Therefore, due to the great density of the air, we see Mercury less frequently. . . . Much toil and effort was imposed upon me by this planet with its irregularities which I had to calculate." Despite repeated attempts, he was never able to observe the planet from Frombork.

The reverence that Copernicus showed for earlier observations scarcely characterizes the later scientific attitude. When in

1524 he published his open letter to Canon Bernhard Wapowski, attacking the work *De Motu Octavae Sphaerae,* published by Johannes Werner of Nuremburg in 1522, Copernicus wrote on the observations of Ptolemy, "We must closely follow their procedure and bow to their observations which have been handed down to us like a testament. And if anyone imagines that these observations cannot be fully trusted, at least in this respect the entrance to our science is closed to him." It was only in later years that Copernicus was to admit to Rheticus how his orthodox acceptance of ancient data had changed.

Yet even though Copernicus made relatively few original observations, and although his instruments were inferior to those of the Nuremburg astronomers, the motive and the nature of the observations made by Copernicus differed fundamentally from those of his contemporaries. The latter made more or less random observations, without any consistent overall plan, or even any general theoretical conclusions, whereas Copernicus related each of his observations to the more general theories with which he was concerned. In spite of the inadequacy of his own observations, what Copernicus did achieve was to take existing data and provide a blindingly new interpretation. In our present age, when science stands knee-deep in accumulated "facts," there is a most urgent need for more practitioners in the spirit of Copernicus, who are able to take the existing mass of data and provide new viewpoints and new perspectives, by reordering and reformulating their experiences and suggesting new questions and new syntheses. It was this quest for unity that Copernicus so triumphantly portrayed.

The third aspect of the character of the man behind the revolution was that Copernicus was an involved participant in the affairs of the world, and not an isolated observer. He was a man of great versatility rather than a narrow specialist, and in one sense this versatility was itself a factor in the scheme of the Universe that he developed. The contemporary of Magellan, Vasco da Gama, Columbus, Leonardo da Vinci, Michelangelo, Martin Luther, and Vesalius, Copernicus was a giant in an age of intellectual giants. His interests and activities represent an astonishingly full, varied, and successful life. In the sweeping range of his accomplishments, and in the balance and breadth of his interests, Copernicus was more typical of the many-sided universal man of the Renaissance (*"L'Uomo Universale"*) than of the competent scholar of the medieval world. Skilled not only in canon law, astronomy, mathematics, and economics, he also pursued the more practical arts of medicine and painting and even constructed many of his own pieces of astronomical apparatus.

Nicholas Copernicus was born on February 19, 1473, at Torun, an old Hanseatic League city on the right bank of the Vistula. Both the time and the place of his birth were to be of unusual importance in the development of his later career. Copernicus lived at a time of relative peace and affluence, of expanding trade and widening intellectual horizons. The development of printing, which had begun as early as 1473 in Cracow, gave new ideas rapid circulation. The voyages of discovery which marked this period brought a flood of new knowledge to the Old World. Bartholomew Diaz journeyed to the Cape of Good Hope in 1486, and six years later Christopher Columbus crossed the Atlantic. The fall of Constantinople to the Turks in 1453 led to the western exodus of Greek scholars, who brought with them their ancient manuscripts, including significant astronomical works. Their presence gave rise to a flowering of classical thought within the universities, especially those of Italy, which later produced a rediscovery of the great passion for knowledge, and a new emphasis on Humanism, which contributed to the Renaissance. Growing dissension in the Church ultimately gave rise to the Reformation and the Counter Reformation. The period also saw the gradual development of new nation states, which emerged from the endless conflicts between the armies of the smaller divisions of Medieval Europe.

The town of Torun, "Queen of the Vistula," now located in Poland, is still sometimes referred by its older German name of Thorn. At the time of Copernicus, the town stood on the boundary between the Polish and Prussian nations. Torun had been founded in the thirteenth century by the Teutonic Knights, a military order of German monks who fought constant "holy" battles with the pagan Prussians. Despised and feared by the Poles, this knightly order underwent a gradual decline in influence, so that by the time of Copernicus's birth, Torun had become a part of Poland. It was a prosperous trading city, having links not only with other parts of Poland, but also with Venice, Norway, Belgium, Germany, and England, although by the late fifteenth century, it had been eclipsed by Danzig.

Copernicus was the son of Mikolaj and Barbara Koppernigk. Both his father and his mother were of Silesian extraction, and his father was a leading merchant and citizen of the town of Torun, of which he served as a magistrate. Copernicus was the youngest member of a family of two boys and two girls. The elder sister became Abbess of Kulm, the younger married a Cracow merchant, and the elder brother, Andrew, was Copernicus's close companion during his studies in Italy. It was perhaps during his student years that Copernicus chose the Latinized version of his name, in preference to the alternative spelling.

In the year 1483, when Copernicus was only ten years old, his father died, and his mother's brother became the guardian of the family. This remarkable man, Lucas Watzenrode, a graduate of the Universities of Cracow, Cologne, and Bologna, was Canon of Wloclawck and Varmia, and was later to become Bishop of Ermland. Copernicus's debt to his maternal uncle was great, and was reflected by the dedication to him of Copernicus's translation of Theophylactus. At the age of seventeen, Copernicus was sent to Cracow, his father's old hometown and the Polish capital, where the university was one of the foremost in Europe. The University of Cracow was founded in 1364 by King Casimir the Great, who modeled it on the University of Prague. After Casimir's death the University declined, but in the year 1400 it was reestablished by Wladislaw Jagiello, Grand Duke of Lithuania and King of Poland, who modeled it on the older theological University of Paris, using funds provided in the will by his queen, Jadwiga.

The universities of Europe at the time of Copernicus had a rich and cosmopolitan life, drawing their students from many different countries. Cracow, as Rybka had shown, became something of a northern center for the humanist movement recently imported from Italy, with its skepticism of traditional sources of authority, and its emphasis on the validity of personal experience and rational principles. In the school of arts, which was a compulsory preparation for all other schools, the seven traditional liberal art subjects comprised the curriculum. These included the trivium—grammar, rhetoric, and logic—and the quadrivium, which included arithmetic, geometry, music, and astronomy or astrology.

We know little of Copernicus's studies at Cracow, although it seems probable that astronomy formed a part of them, since it was then the leading discipline of the faculty of liberal arts of the University. A list of astronomy lectures given at the University in the years 1491–95 shows how widely based was the curriculum in astronomy and mathematics. It was also probably at Cracow that Copernicus gained his first exposure to the Platonic philosophy, which differed so much from the prevailing Aristotelian orthodoxy of medieval universities. Although he was no longer teaching astronomy at Cracow at the time of Copernicus, the great astronomer and mathematician Albert Brudzewski (Albertus of Brudzew) may well have known Copernicus. Albertus himself remained an adherent of a geocentric theory of the universe, but he held critical views of some Ptolemaic constructions, and these were later quoted by Copernicus. There is no evidence that Copernicus completed his degree at Cracow. Indeed it was rare for most of those then admitted to the University to obtain a formal degree.

From Cracow, Copernicus traveled to the University of Bologna in the autumn of 1496. Copernicus's earliest observations were made while he was at Bologna, and it was there, in 1498 that his older brother, Andrew, joined him. Copernicus's formal intention in attending the University of Bologna was to study law, for which the University, founded as early as 1119, was the most renowned center in Europe. It was also a center for the new Humanistic tradition. The chapter of Varmia had allowed three years for Copernicus to complete his legal studies, but his interest in astronomy and the study of ancient writers and their views on the nature of the Universe delayed his progress.

Rheticus, in his *Narratio Prima* describes these years at Bologna as follows, "My teacher made observations with the utmost care at Bologna, where he was not so much the pupil as the assistant and witness of observations of the learned Dominicus Maria" (The Professor of Astronomy [Astrology]). It was at Bologna that Copernicus made one of his earliest and most important observations. Copernicus reasoned that, if the Ptolemaic theory were correct, the moon should be twice as near to the Earth during the quarters, as at full or new moon, in which case, the diameter of its disc would then have been twice as great. In a historic observation on March 9, 1497, when the moon eclipsed the star Aldebaran (α Tauri), Copernicus showed that this was not the case.

In the Jubilee year of 1500, when throngs of Christians from all over the world converged on Rome, we find Copernicus in the Holy City, where, Rheticus records, "he lectured on mathematics before a large audience of students and a throng of great men experienced in this branch of knowledge." Copernicus was elected Canon of Frauenburg Cathedral while he was at Bologna, and his brother Andrew was also elected to a similar post soon afterward. Both were formally installed in 1501, though they were immediately granted leave to continue their studies in Italy. Nicholas Copernicus then enrolled at the University of Padua, where, although he completed his legal studies, he did not graduate, choosing instead to take his doctorate from the University of Ferrara in 1503. After the completion of his legal studies, Copernicus returned to Padua to study medicine and it may be at this time that he acquired his knowledge of Greek. Copernicus was in his early thirties by the time he completed his studies, and I have discussed his long university career in some detail because it was clearly of great significance in his later life. What is perhaps less obvious is the encouragement it provides in our modern day insistence on the importance of nonvocational university studies. Copernicus's professional studies in law and medicine were no doubt important. But it was the "distribution requirements" of one fifteenth-century student at Cracow that changed our view of the Universe.

When Copernicus finally returned to Ermland, he was given a further leave of absence from his canonical duties to serve as secretary and private physician to his uncle. (It is worth noting in passing that, although Copernicus was appointed to a Canonry in 1497, his leaves of absence, which were periodically renewed, totaled fifteen years. This may provide some consolation to those university administrators who despair because sabbatical and other study leaves frequently do not produce immediate results.) Copernicus's new duties involved residence with the Bishop of Ermland that was to last some six years at Heilsberg Castle, the palatine residence, about forty miles southeast of Frauenburg, and later at Lidzbork. He left his uncle's service in 1510, and it was at some time during this period, perhaps about 1512, that Copernicus wrote his *Commentariolus,* the first statement of his heliocentric theory.

Copernicus was forty years old when he finally took up his clerical duties in Frauenburg. As a Canon of the cathedral there, he served the church as a statesman and diplomat in peace and war. He was an ecclesiastical administrator and legal expert of considerable skill, and occupied a number of diplomatic and administrative offices. In 1516, he became treasurer of the Chapter at Frauenburg, and was appointed to manage the affairs of two outlying Diocesan estates, Mehlsack and Allenstein. Details of land transactions, recorded in his own hand, still survive from this time.

The various battles that were fought around the Frauenburg area also involved Copernicus from time to time. In the autumn of 1520, for example, he returned to Allenstein as Commissioner of Ermland, and after unsuccessful negotiations, in which he participated, the Teutonic Knights laid siege to the castle. Only one of the other Canons remained with Copernicus in Allenstein Castle. The siege continued through the long winter—and correspondence dating from that period contains offers to send Copernicus further provisions and firearms, with the promise of powder and lead to follow. Peace finally came in the spring of 1521, and Copernicus began the task of collecting details of the devastation which had been caused in the area, and of drawing up a memorandum as a basis for peace.

It was in these offices that Copernicus first became acquainted with King Sigismond of Poland, urging him to restore the independence within the Episcopal territory which had been temporarily annexed to Poland. In spite of his lack of any formal training in what we should now call economics, Copernicus was a capable financial administrator and theorist. He wrestled in the sixteenth century with the problem of inflation, which was as real then as it is today. The Prussian currency, minted by various local authorities, was becoming increasingly debased because of the decline of the content of gold and silver used in the coins. Copernicus condemned this practice, arguing that the government which sought such short-term gain was like the farmer who hoped to profit by sowing inferior seed, because it was cheaper than good seed. It is for this reason that some have suggested that the law that bad currency drives out good should be named after Copernicus rather than Sir Thomas Gresham, its subsequent "discoverer."

Copernicus urged that common action was needed to restore stability of the currency, and argued strongly for a monetary union of the various states involved. He probably began writing on the topic as early as 1519, and presented reports in 1522 and 1528. In 1522, Copernicus attended a conference at Graudanz, at which he offered his proposal for the currency unification of East and West Prussia. Although the proposal was rejected at this conference, a modified and more comprehensive scheme which he proposed in 1528 was subsequently adopted.

One quotation, translated by Kesten from the 1527 Treatise of Copernicus on Currency Reform, shows something of the humanity of Copernicus in his concern for the needs of the poor and the quality of his country's life. He wrote: "Moreover, trade and communication, arts and crafts flourish where money is good. With bad money, however, the people grow slack and inert, they neglect cultivating the spirit. We can still remember how inexpensive everything was in Prussia when good money was circulating. But now the value of all means of subsistence has risen. It is quite clear that light money furthers laziness and in no wise relieves misery. . . ."

It is easy, on the one hand, to dramatize the significance of Copernicus's administrative work, and, on the other, to decry the trivialization of his life that it produced. Kesten parodies the second viewpoint in the following remarks.

> There we see a man great enough for ten whole centuries sent into a stupid province, occupied with trifles, frittering his life away in keeping here the accounts of a few bonded peasants or collecting and delivering a few marks, treating there a poor man's mange (with or without success), going to the same church time and again, saying the same prayers, conversing with the same dozen colleagues; now he has to make out a complaint against a powerful neighbor; now he drafts an essay about the soundness of money; at another time there is a lunar eclipse, at last; then he has to protect a ridiculously small town; a king lists him among his candidates for a bishopric—it is only a derisive gesture, the other one becomes bishop; robbers and soldiers pillage. All this time Copernicus has to be active; constantly he

has to spend his time, his strength, his knowledge—and so a great life is frittered away.

Is it really frittered away? Does not a river draw its strength from a hundred sources? Does not the wise man know how to use everything?

One recalls the remark of John Milton,

I call, therefore, a complete and generous Education that which fits a man to perform justly, skillfully, and magnanimously all the offices both private and public of peace and war.

As a physician Copernicus appears to have been careful, conscientious, and successful, though scarcely innovative in his treatments. Some of his medical books are still preserved, with his own annotations made in the margins. Copernicus's earliest patient was his uncle, but he later served many other members of the chapter, including successive bishops of Ermland. Copernicus's medical skills were, however, more widely known. Only two years before his death, for example, Copernicus traveled to Konigsberg, at the request of Duke Albert of Prussia, to treat one of his courtiers.

Behind these several activities there is the character of Copernicus himself, revealed repeatedly as wide-ranging in his interests, liberal in his views, and humane and compassionate in his sympathies and commitments. In both his writing and his friendships, these attitudes are revealed.

Not the least important indication of Copernicus's sympathy for the new Humanist tradition was his publication of the Greek poet Theophylactus Simocatta, a collection of eighty-five brief poems, written about 630 A.D., which Copernicus translated into Latin. It is typical of Copernicus that, as he undertook the journey to deliver his manuscript translation to the publisher, Johann Haber in Cracow, on June 2, 1509, he also took the opportunity it provided to observe a lunar eclipse on the way. Some of Copernicus's contemporary scholastic colleagues feared that the revival of classical learning might also lead to the resurgence of ancient paganism, but Copernicus apparently felt strongly that the venture was worth the risk.

This liberality of spirit and tolerance of judgment were also shown in the attitude of both Copernicus and his close friend and fellow Canon, Tiedmann Giese, toward the Reformation. Luther made his stand in 1521, and the Protestant faith made rapid gains throughout Western Europe. Although Copernicus himself wrote nothing on this subject, it is clear from Giese's book of 1525 that Copernicus shared his liberal outlook. Both sought to preserve the ultimate unity of the church, and deplored the excesses of both Protestants and Catholics. Fabian von Lossainer, who succeeded Lucas Waczenrode, was a moderate man, but in 1523 he was succeeded as Bishop of Ermland by Mauritius Ferber, a strong Protestant antagonist, as was his successor, Danticus. From this time, it is probable that Copernicus's liberal views would not have been welcome in Frauenburg. He lived a rather solitary life, and he may also have suffered to some extent from the additional latent suspicion of his revolutionary astronomical views, and his befriending of the Protestant scholar Rheticus.

George Joachim von Laucher (Rheticus), Professor of Mathematics at the recently established Protestant University of Wittenberg, was forty-one years the junior of Copernicus. It was his visit in the spring of 1539 and his account of the Copernican theory, published in 1540 with permission of Copernicus, that seem finally to have persuaded Copernicus to publish *De Revolutionibus.* Copernicus also refers in the preface of his book to the influence of Giese, and of Cardinal von Schonberg, who wrote to him from Rome urging him to publish details of his new astronomic scheme. Rheticus had a hand in the final publication of Copernicus's masterpiece, and chose a friend, John Petrejus of Nuremberg, to act as publisher. It was also the unexpected appointment of Rheticus to a new post at Leipzig University that forced him to entrust the task of seeing the book through the press to a friend, Andrew Osiander, a Lutheran clergyman, who wrote the anonymous and notorious preface to the book.

During the latter part of 1542 and in 1543, Copernicus was repeatedly stricken with hemorrhage and apoplexy, which left him partly paralyzed. He lingered on until May 24, 1543. Giese described the close of Copernicus's life in a letter to Rheticus: "he had lost his memory and mental vigor many days before; and he saw his completed work [*De Revolutionibus*] only at his last breath upon the day that he died." Copernicus lived his last days in solitude and perhaps even neglect. On December 8, 1542, Giese wrote to Copernicus's friend and fellow Canon George Donner,

> Since Copernicus even in his days of good health liked retirement, only a few friends would probably stand sympathetically at the side of the gravely ill man, yet we are all his debtors on account of his pure soul, his integrity and his extensive learning. I know that he always counted you among the most faithful of his associates. Therefore I beg you to stand by him protectingly if his fate requires this, and to assume the care of the man whom together with me you have always loved, in order that he may not lack brotherly help in his affliction, and that we do not appear ungrateful toward a friend who has abundantly earned our love and gratitude. (Translated by Kesten)

Such was the man: Canon of the Church, administrator, negotiator, military consultant, physician, economist, poet, linguist, artist, and humanist. Yet Copernicus is remembered for none of these things. Almost as a hobby, in the midst of this astonishingly busy, successful, and varied, though rather solitary life, he also found time to reorder the Universe. Though slow in its coming, for it was thirty years in the making, and slower still in the acceptance of its theory, the publication of *De Revolutionibus Orbium Coelestium* changed forever our view of the world in which we live, and of our place within it. By the boldness of the theory and the vigor and harmony of the arguments he used, Copernicus established a new style of knowledge—that which we now call science. The life of Copernicus, in all its richness and diversity, is a standing reproach toward contemporary men of science who so often withdraw themselves from the affairs of the world. May it not be too high a price to pay for our present scientific success that we have tended to accept as inevitable the insulation of scientists from public affairs. Perhaps at no other time has there been a greater need for the involvement of the scientist in national and international affairs, not because he has some greater wisdom or more ingenious solution to offer than other men, but because most problems of national and international policy now involve scientific considerations, in which the expert knowledge of the scientist is of critical importance. Indeed, as C. P. Snow has argued, the scientist also needs to contribute foresight, disciplined skepticism, and pragmatism to the short-term balanced compromise solutions which are so often the typical products of administrators and governments.

The fourth aspect of the life of Copernicus that I find significant is that his achievement was at least in part a personal response to the frustrations that he felt at the imperfections and the limitations of the practical application of science. Certainly this was not his only or perhaps his major motive in seeking a larger synthesis, but there is some evidence that it was an important consideration in his determination to resolve the inconsistencies of the Heavens.

He wrote in the preface of his book, "nothing urged me to think out some other way of calculating the motions of the spheres of the world but the fact that, as I knew, mathematicians did not agree among themselves in these researches. For in the first place they are so uncertain of the motion of the Sun and Moon, that they cannot observe and demonstrate the constant magnitude of the tropical year."

The practical consequences of these limitations were severe. Astrological predictions, for example, were unreliable. Lunar and tidal tables for seamen proved to be inaccurate, and calculations of longitude and latitude were imprecise. Worldwide voyagers could no longer rely on coastal navigation based on familiar landmarks. The calendar was hopelessly out of joint, and the need for its reform plagued the medieval world. The calendar which was then in use had originally been introduced by Julius Caesar in the first century B.C. He had established the length of the normal year as 365 days, adding an extra day to February in every fourth (leap) year, and thus making the length of each year an average of 365 1/4 days. In fact, this is about eleven minutes longer than the true length of the year, and although the differences are insignificant over one or two years, the cumulative error became substantial over several centuries. The church began to experience grave difficulties in this, including problems in establishing the date of Easter. Copernicus was invited to visit Rome in 1514 to help resolve the problem, but he declined the invitation because he claimed that the difficulties of the calendar could not be resolved until the laws of celestial motion had been remedied, and it was to this task that he turned his hand. It is interesting that Reinhold's "Prussian Tables," which were based on the work of Copernicus, were ultimately used to establish the new Julian Calendar approved by Gregory XIII in 1582.

In all this, one sees something of the need for balance. I have argued that social applications should not and cannot be the only justification and motive for scientific research, yet the scientist as a person surely cannot ignore their significance. All science, however remote from present problems, has an ultimate potentiality for practical application, and one factor in leading the scientist to select the particular problems on which he works has often been that of practical application. It is in this particular choice that the practical value of science must be a component of personal decision concerning priorities in scientific research, though not necessarily in terms of their public support.

The fifth aspect of the life of Copernicus that I find revealing is that he was a man with a finely developed sense of reverence and awe in the created Universe, a man who had found meaning and significance in life, and in whom the ultimate basis for scientific harmony was a world view which was deeply rooted in a religious sense of perspective. It is fashionable today to decry this, or divorce it from Copernicus's scientific contributions by regarding it as an accidental aberration or blemish, having no real significance for his scientific work. Indeed the whole religious viewpoint which he represented is often regarded as inconsistent with the mechanistic Universe which he revealed. I believe we need to hasten slowly in this, for it is at least arguable that it was this sense of the orderliness and rationality of God that was the ultimate source of Copernicus's insistence on the orderliness and rationality of the Universe. Copernicus himself suggests that his sense of fitness and unity in science had a root in this deeper religious commitment. Let me quote again from Copernicus's preface.

> I pondered long upon this uncertainty of mathematical tradition in establishing the motions of the system of the spheres. At last I began to chafe that philosophers could by no means agree on any one certain theory of the mechanism of the Universe, wrought for us by a supremely good and orderly Creator, though in other respects they

investigated with meticulous care the minutest points relating to its
orbits. I therefore took pains to read again the works of all the
philosophers on whom I could lay hand to seek out whether any of
them had ever supposed that the motions of the spheres were other
than those demanded by the mathematical schools.

The influence of religion in the form of Judaism and Christendom,
upon the overall development of science has been recognized by
many philosophers and historians. Herbert Butterfield, M. B.
Foster, J. MacMurray, R. G. Collingwood, and Alfred North
Whitehead, all testify to this effect. Alfred North Whitehead, for
example, inquiring into the convictions which underlie the formation of the modern scientific movement wrote,

> When we compare this train of thought in Europe and the attitude
> of other civilizations when left to themselves, there seems but one
> source of origin. It must come from the Medieval insistence on the
> rationality of God, conceived as with the personal energy of Jehovah
> and with the rationality of a Greek philosopher. Every detail was
> supervised and ordered; the search into nature could only result in
> the vindication of the faith and rationality. Remember that I am not
> talking of the expressed beliefs of few individuals. What I mean is
> the impress on the European mind of the rising of the unquestioned
> faith of centuries. By this I mean the instinctive tone of thought and
> not a mere creed of words. . . .

Clearly this does not imply that only those who subscribe to some
doctrinal basis of faith can succeed in the world of science, but it
does suggest that a larger world view is a prerequisite of those who
would make the quantum leap forward characterized by the
revolution of Copernicus. We have too easily supposed that the
Copernican Universe implied the end of the significance for the
meaning and life of man. For Copernicus, it meant the discovery
and reinforcement of new meaning and new significance.

Copernicus's introduction to Book One of *De Revolu-*
tionibus makes clear the way in which he saw the relationship of
his astronomical studies to the wider human experience.

> This art [Astronomy] which is as it were the head of all the liberal
> arts and the one most worthy of a free man leans upon nearly all the
> other branches of mathematics. Arithmetic, geometry, optics,
> geodesy, mechanics, and whatever others, all offer themselves in its

service. And since a property of all good arts is to draw the mind of man away from the vices and direct it to better things, these arts can do that more plentifully, over and above the unbelievable pleasure of mind [which they furnish]. For who, after applying himself to things which he sees established in the best order and directed by divine ruling, would not through diligent contemplation of them and through a certain habituation be awakened to that which is best and would not wonder at the Artificer of all things, in Whom is all happiness and every good? For the divine Psalmist surely did not say gratuitously that he took pleasure in the workings of God and rejoiced in the works of His hands, unless by means of these things as by some sort of vehicle we are transported to the contemplation of the highest Good. (Translated by Charles Glenn Wallis)

Religion in this wider sense involves the overall integration of all experiences, including the experience of science itself. Indeed it is not science as science that our contemporaries have rejected, but the narrow projection of science into a world view, to which the name scientism has been given. Presumably one of our tasks as those who are both scientists and members of universities is to demonstrate the wider implications and relationships of what we teach. For if the survival of society depends on science, we must learn not only how to use it wisely, but also how to relate it to every other aspect of human experience, and how to recognize it as an account of human experience which is complementary to every other.

This will not be easy for us. Science has achieved much of its success by division and abstraction. Isolation and abstraction are still the essential methods by which scientific knowledge is extended, but, even in science, they have their limitations. "When we try to pick out anything by itself, we find it hitched to everything else in the universe." As I have already argued, taken as a life-style rather than a scientific methodology, isolation and abstraction are disastrous, for they lead to a fragmented vision of man and a diminished sense of the coherence of life. It is often for this reason that the thoughtful student rejects science. He sees the scientist as one so preoccupied with plumbing the depths of his own limited abstractions, that he either ignores the rest of life, or treats it with such superficiality that his values and judgments become suspect. What little integration and coordination there is, is left increasingly to those who lack the will, the character, or the skills to succeed in science.

So the Copernican revolution continues. It is idle to pretend that we can push the clock back, or deny the ancestry of the revolution in whose shadow we live. We are its heirs, but there is danger that in our enthusiasm to reassert the significance of the revolution, we may overlook the humane and compassionate character of the revolutionary. We shall best serve the needs of man neither by denying the reality of the Copernican revolution, nor by redesigning a new man-centered science, nor by repudiating science as such. We shall best serve the needs of man by retaining the power of the Copernican revolution but joining it to the humanity of Copernicus the man.

This does not mean the abandonment of science. The appeal of the social sciences to the contemporary student generation lies in their humane orientation, their wide-ranging boundary contacts with other disciplines, and their unashamed striving for a general synthesis. If science is presented creatively, it can offer all these, and more, for it has all the intellectual excitement, commitment, and elegance of these other disciplines. But we shall best enhance science, neither by abandoning it for technology or sociology, nor by loosening or reducing the rigor, abstraction, and empiricism of its approach, but rather by preserving all these as precious insights, and then pursuing the scientific enterprise in the context of human hopes and needs. In doing this, we shall serve the cause of science itself, and ultimately the cause of man.

In all this, we seek neither a Brave New World run by scientists, nor a 1984 world of George Orwell. We seek, rather, an enlightened democracy where the collective vision of man will give new impetus and vigor both to science and to society, and where the growing power of science, the beauty of its concepts, the austerity of its discipline, and the wise and compassionate applications of its achievements give hope and help to mankind as it faces the challenge of the future. For, now, as never before in the history of our race, mankind faces a choice. We may, on the one hand, continue to blunder on, ignoring the danger signs until we stumble into a polluted, plundered, over-crowded world, the home of a dying race which once knew greatness; or, we may recognize the peril and harness all the creativity, power, vision, skill, and goodwill of man into a national and international program of scientific reclamation. It is unlikely that this will come about easily, for wisdom, like truth, is seldom decided by majority vote. Yet surely the challenge to science is now greater than ever before. Without its help, mankind is doomed. With its help, we may hope to design and establish for all mankind a chain of interacting dynamic communities, living in internal harmony, in harmony with one another, and in harmony with nature; communities in which each man may identify and embrace that excellence of spirit, which constitutes for him the good life.

The Copernican revolution removed man from the center of the Universe, but the Copernican spirit retains man in the central position as the motive and measure of the scientific search. We have beatified the revolution and forgotten the spirit of the man who produced it. Our best hope for the future is the reassertion of the humane basis on which Copernicus established his new theory, and revealed the infinite universe in which we dwell.

Notes on Contributors

J. M. BOCHENSKI is professor at the University of Fribourg in Switzerland. Among his latest publications are *Marxismus-Leninismus* (Munich, 1973), and *Guide to Marxist Philosophy* (Chicago, 1972). Professor Bochenski is also the editor of *Studies in Soviet Thought* (quarterly), and *Sovietica* (book-series).

CARL COHEN is professor of philosophy, Residential College, The University of Michigan. He has published works on *Democracy* (New York, 1973) and *Civil Disobedience* (New York, 1971), and is presently studying the philosophical issues that arise from efforts to maximize democratic government in contemporary communities of different kinds.

THOMAS J. COTTLE is a research sociologist and practicing psychotherapist affiliated with the Children's Defense Fund of the Washington Research Project. His interest in the lives of families in poor urban and rural areas relates to his publications, *Time's Children: Impressions of Youth* (Boston, 1971); *The Abandoners: Portraits of Loss, Separation and Neglect* (Boston, 1973); *The Voices of School: Educational Issues Through Personal Accounts* (Boston, 1973); *Black Children, White Dreams* (Boston, 1974); *The Present of Things Future: Explorations of Time in Human Experience,* with Stephen L. Klineberg (New York, 1974); and the forthcoming, *A Family Album: Portraits of Intimacy and Kinship* (New York, 1974).

JOSEPH CROPSEY is professor of political science, The University of Chicago. He is the author of *Polity and Economy* and *Hobbes' Dialogue Between a Philosopher and a Student of the Common Laws of*

England (Chicago, 1971), and *Ancients and Moderns; Essays on the Tradition of Political Philosophy in Honor of Leo Strauss* (New York, 1964) and is currently working with the history of classic and modern political philosophy.

A. HUNTER DUPREE is George L. Littlefield Professor of History, Brown University. His current research interests extend to the history of science and technology in America, the history of measurement, and social and biological theory. His published works relating to these interests include "The Role of Technology in Society and the Need for Historical Perspective," *Technology and Culture*, 10 (1969), 528–34; and "The Pace of Measurement from Rome to America," *Smithsonian Journal of History*, 3 (1968), 19–40.

JOHN M. FOWLER is visiting professor in the Department of Physics and Astronomy, University of Maryland. Recently he has been involved with evaluations of instruction in science and the energy crisis and has published articles on "What Our Left Hand Has Been Doing," *Physics Today*, 23 (1970); "Content and Process in Physics Teaching," *American Journal of Physics*, 37 (1969); "The Interdisciplinary Curriculum," *Physics Today*, 21 (1968), and "Energy and the Environment," *The Science Teacher*, 39 (1972). He is presently completing a book, "Energy and the Environment" for fall 1974 publication by McGraw-Hill.

C. A. HOOKER is professor of philosophy and environmental science, University of Western Ontario. His particular interest in systems analysis of environmental and social systems and social management theory derives from broader interests in physics, the philosophy of science and epistemology. His publications along these lines include "The Systematic Significance of Environmental Problems," *Proceedings of the Law Society of Upper Canada*, Continuing Legal Education Conference (May, 1972), and "Systematic Realism," to appear in *Synthese*.

WILHELMINA IWANOWSKA is professor of astrophysics and director of the Institute of Astronomy, Nicolaus Copernicus University, Toruń, Poland. Her current research interests concern the chemical evolution of the Galaxy. Her most recent published works include "Diffusion of Elements in the Galaxy," with E. Basińska, *Proceedings of the First European Regional Meeting in Astronomy* (1974) and

"Statistical Population Indices of White Dwarfs," *Bulletin of the Astronomical Observatory in Toruń*, no. 50 (1973).

NOLAN PLINY JACOBSON is chairman and professor, Department of Philosophy and Religion, Winthrop College, Rock Hill, South Carolina. His research on the relationship between self-corrective life-styles and science has resulted in a wide range of publications, including *Nippondō (The Japan Way)* (Tokyo, 1974); "Buddhism, Modernization, and Science," *Philosophy East and West,* 20 (1970), 1–23; and "The Cultural Meaning of Science," *Educational Theory* (1968), 23–31.

PAUL W. KNOLL is chairman and associate professor, Department of History, University of Southern California. His research on the University of Cracow stems from broader interests in Polish history, as evidenced by his publications *The Rise of the Polish Monarchy: Piast Poland in East Central Europe, 1320-1370* (Chicago, 1972); "Casimir the Great and the University of Cracow," *Jahrbuecher fuer Geschichte Osteuropas,* 16 (1968), 232–49; and "Poland as 'Antemurale Christianitatis' in the Late Middle Ages," *The Catholic Historical Review* (October, 1974).

DAVID C. LINDBERG is professor of the history of science, University of Wisconsin. His published works include *John Pecham and the Science of Optics* (Madison, Wis., 1970); "Alkindi's Critique of Euclid's Theory of Vision," *Isis,* 62 (1971), 469–89; and *A Catalogue of Medieval and Renaissance Optical Manuscripts* (forthcoming). These works stem from a broad interest in medieval science and the Scientific Revolution and a more particular interest in early theories of vision.

ALASDAIR MACINTYRE is University Professor of Philosophy and Political Science, Boston University. He is interested in the methodology of the social sciences and the history of moral concepts and beliefs and has published *Against the Self-Images of the Age* (New York, 1971).

ORREN C. MOHLER is professor of astronomy, The University of Michigan. His interests within astronomy center around solar observation and the history of solar observation and have resulted in publications such as his most recent article on "A Comparative Study of Solar Cycles 18, 19, and 20," with H. W. D. Prince and R. E. Hedeman, *Publications of the American Geophysical Union* (in press).

MICHAEL W. OVENDEN is professor of astronomy, Department of Geophysics and Astronomy, University of British Columbia. His primary field of research is celestial mechanics.

DENNIS C. PIRAGES is associate director of the Program in Science, Technology and Public Affairs, University of California, San Diego. His published works include *Ark II: Social Response to Environmental Imperatives,* with Paul Ehrlich (New York, 1974), *Modernization and Political Tension-Management* (New York, 1972), and *Seeing Beyond: Personal, Social, and Political Alternatives* (Reading, Mass., 1971). These works stem from broader interests in science and international systems, the social impact of technology, and decision-making in industrial societies.

GERALD M. PLATT is associate professor, University of Massachusetts, Amherst. He has published works on *The American University,* with Talcott Parsons (Cambridge, Mass., 1973), and *Psychoanalytic Sociology: An Essay on the Interpretation of Historical Data and the Phenomena of Collective Behavior* (Baltimore, Md., 1973).

FRANK H. T. RHODES is dean of the College of Literature, Science, and the Arts and professor of geology and mineralogy, The University of Michigan. He publishes extensively in scientific journals. Recent books of general interest include *Conodont Paleozoology* (Boulder, Colo., 1973), *Geology* (Racine, Wis., 1973), and *The Evolution of Life,* rev. ed. (Baltimore, Md., 1974). His interest in the strategy of science and educational policy is reflected in the Postscript he contributed to this work.

EDWARD ROSEN is professor of the history of science, City University of New York. He has edited a number of works dealing with the history of astronomy, including *Three Copernican Treatises,* 3d ed. (New York, 1971), *Kepler's Somnium* (Madison, Wis., 1967), and *Kepler's Conversation with Galileo's Sidereal Messenger* (New York, 1965), and is presently the editor of *Nicholas Copernicus, Complete Works,* Vol. I (London, 1972), and Vol. II (in press).

HENRYK SKOLIMOWSKI is professor of philosophy, The University of Michigan. His interests in the history and philosophy of science focus on the philosophy of technology, and science and society. He has published articles dealing with "The Scientific World View and the Illusions of Progress," *Social Research* (April, 1974); "Technology versus Nature," *The Ecologist* (February, 1973); "Science and the Modern Predicament," *New Scientist* (February, 1972); and "Science in Crisis," *Cambridge Review* (January, 1972).

NICHOLAS H. STENECK is assistant professor of history, The University of Michigan. His work and interest in late medieval science has led to publications on "Albert the Great on the Classification and Localization of the Internal Senses," *Isis* (1974); "The Sense of Vision and the Origins of Modern Science," with David C. Lindberg, *Science, Medicine and Society in the Renaissance* (New York, 1972); and "A Late Medieval Debate concerning the Primary Organ of Perception," Proceedings of the XIIIth International Congress of the History of Science, Moscow, 1971.

STEPHEN TOULMIN is professor of social thought and philosophy, The University of Chicago. His most recent books are *Wittgenstein's Vienna,* with Allan Janik (New York, 1973), and *Human Understanding,* Vol. I (Princeton, 1972), and he is currently pursuing research dealing with the intersections of epistemology and science.

ROBERT S. WESTMAN is assistant professor, Department of History, University of California, Los Angeles. He has written a number of articles on post-Copernican astronomy, including "Continuities in Kepler Scholarship: The European Kepler Symposia, 1971, in Historiographical Perspective," *Vistas in Astronomy,* 18 (London, 1974); "Kepler's Theory of Hypothesis and the 'Realist Dilemma'," *Internationales Kepler Symposium 1971* (Hildesheim, 1973), 29–54; "The Comet and the Cosmos: Kepler, Maestlin and the Copernican Hypothesis," *Studia Copernicana,* 5 (Warsaw, 1972), 7–30; and is editor of a volume, *The Copernican Achievement* (Berkeley and Los Angeles, forthcoming).

L. PEARCE WILLIAMS is professor of the history of science, Cornell University. He has written books on *Michael Faraday, A Biography* (New York, 1965), *The Origins of Field Theory* (New York, 1966), and is presently working on a biography of André-Marie Ampere and scientific education in the nineteenth century.

JERZY A. WOJCIECHOWSKI is professor of philosophy, University of Ottawa, Canada. His wide-ranging interest in the problems of knowledge (science) in relation to culture have led to such publications as

Proceedings of the Ottawa Conference on the Conceptual Basis of the Classification of Knowledge (Munich, 1974); "Philosophie de la culture. Remarques critiques," *Revue de l'Université d'Ottawa*, 43 (1973); and *Survey of the Status of Philosophy in Canada* (Ottawa, 1970).

St. Scholastica Library
Duluth, Minnesota 55811